The Ecology and Evolution of Hominin Geographic Ranges

Setting a context for archaeological interpretation using comparative analysis

Katharine MacDonald

BAR International Series 1550
2006

Published in 2016 by
BAR Publishing, Oxford

BAR International Series 1550

The Ecology and Evolution of Hominin Geographic Ranges

ISBN 978 1 8417 1979 5

BAR Publishing is the trading name of British Archaeological Reports (Oxford) Ltd. British Archaeological Reports was first incorporated in 1974 to publish the BAR Series, International and British. In 1992 Hadrian Books Ltd became part of the BAR group. This volume was originally published by Archaeopress in conjunction with British Archaeological Reports (Oxford) Ltd / Hadrian Books Ltd, the Series principal publisher, in 2006. This present volume is published by BAR Publishing, 2016.

Printed in England

BAR
PUBLISHING

BAR titles are available from:

BAR Publishing
122 Banbury Rd, Oxford, OX2 7BP, UK
EMAIL info@barpublishing.com
PHONE +44 (0)1865 310431
FAX +44 (0)1865 316916
www.barpublishing.com

Table of Contents

List of Illustrations

List of Tables

Acknowledgements

I owe thanks to many people. I received a NERC studentship for my research. My supervisor, James Steele, has been a constant source of ideas and enthusiasm, and has given crucial support and criticism. Simon Reader provided a copy of his primate innovation, social learning and tool use frequency database. I am grateful to Simon and to Kevin Laland for stimulating discussion both of my research plan and interpretation of the primate analyses. Yvonne Marshall and David Wheatley made very constructive comments on my upgrade paper, and Clive Gamble, Carina Buckley and Sonia Zakrzewski gave helpful suggestions for a number of thesis chapters.

My family, friends and fellow research students have provided emotional support, good company, good food and other such essentials. Not least, their varied perspectives have given me new insights into my own work. Finally, Alec has put up with me while writing his own thesis, and produced American-style pancakes every Sunday, for which I am deeply grateful.

Preface

Since I completed this thesis in 2003, a number of paleoanthropological and archaeological discoveries have contrived in various ways to shake our convictions about all aspects of hominin range expansion. The discovery of fossils of a small-bodied, small-brained hominin on the island of Flores in Java indicated that one more species of hominin existed outside Africa before humans, and perhaps more importantly challenged our ideas about the importance, and buffering effects, of intelligence and technology in hominin occupation of diverse environments. New discoveries, and reanalysis of the documentation of old discoveries, in Java and Northern China, has rendered an early chronology for hominin presence outside Africa extremely convincing if patchily documented, where it earlier seemed tenuous. On a smaller scale the discovery of stone tools in the Cromer Forest beds in south-west Britain, 200 years before the first evidence north of the Alps, has refined our view of the first occupation of Europe.

As I comment in my introduction, our understanding of hominin distribution is highly reactive, shaped by factors of preservation and discovery. These recent discoveries have changed our views of the species involved, timing and areas occupied in past hominin dispersals. Perhaps more importantly, they have led some researchers to reevaluate the state of the evidence for the classic case of hominin range expansion, 'Out of Africa 1', the expansion of *H.erectus* from Africa into previously unoccupied areas of Eurasia. In their recent Nature review, taking an Asian perspective on hominin dispersals, Dennell and Roebroeks (2005) highlighted the fact that our current knowledge of the hominin fossil record is strongly influenced by historical factors, particularly colonial history. As a result, we have limited archaeological and fossil evidence for the early presence of hominins in Asia, and even less prior evidence of absence, particularly considering the distances involved. These authors argue that based on the current evidence, it is quite possible that there should have been an earlier hominin presence in Asia, that different, perhaps multiple hominin species should have been involved, and even that *H.ergaster* could have migrated into Africa – we simply do not know. While the authors focus on expansion out of Africa, similar caveats apply to other distribution patterns in difference areas and periods. Based on this discussion it seems that from current evidence the pattern of hominin range expansion is barely known.

So where does a thesis that set out to explain patterns in the distribution of early hominin species stand now? It could be argued that it is too early for interpretation. Certainly many models of hominin range expansion closely based on first and last appearance dates for *H.erectus* in Africa and Asia seem premature. However, for a long time our interpretation of the fossil record has been based on a combination of assumptions about what is important in range expansion (with particular reference to human colonization) and reaction to certain key, well-documented and well-publicized discoveries. It could therefore also be argued that exploratory work, based on studies of comparative data, could have an important role in establishing expectations and setting constraints on our assumptions about early hominin geographic ranges. This will critically include research on the conditions for preservation and discovery of hominin fossils in little explored areas, as Dennell and Roebroeks suggest. It can also include comparative study of the factors shaping range size, expansion and contraction, and the survival of populations in other species.

Chapter 1

INTRODUCTION

Early hominin geographic ranges

One of the most striking characteristics of modern humans is our near-global dispersal. This remarkably wide distribution, and occupation of all sorts of extremely different habitats, reflects our human ability to manipulate our own environment and to moderate the effects of environmental change. If we can describe early signs of this trend in human prehistory or even in early hominin species, and how these are connected to changes in behaviour, phenotype or environment, we may be able to isolate the unique factors of human distribution and behaviour today.

At the same time, geographical distribution provides a valuable source of information on the evolution and behaviour of extinct hominin species. Spatial data is an important part of any archaeological investigation, at a range of scales from local to global. Such data is particularly valuable for the Palaeolithic period, for which the forms of archaeological information available are relatively specialised and limited. In addition, interpreting the behaviour of extinct species presents particular challenges.

The geographic range (the extent of distribution) of a species is highly distinctive, and is influenced by that species' behavioural and physical characteristics and the external environment. Palaeolithic archaeology is often concerned with the analysis of hominin geographic ranges as they change over time. Interpretation focuses on the cultural evolution of social and technological adaptations that permitted hominins to extend those ranges, and the environmental factors that ultimately limited them. Understanding this interaction can inform us about early hominin behaviour and their relationship with their environment, and help us to understand the evolutionary processes underlying hominin adaptations.

There have been numerous changes in hominin geographic ranges in the course of human evolution, indicated by the distribution of fossil and archaeological sites. Recent discoveries in Chad indicate that even the earliest hominins were widely distributed in sub-Saharan Africa. Early expansions (such as the first appearance of australopithecines in southern Africa) were accompanied by speciation. Early *Homo* was probably the first hominin species to expand its range to include both East and Southern Africa. The emergence of *H.ergaster* was followed by range expansion into more arid and high altitude habitats within Africa. Hominin populations may have moved out of Africa as early as 1.8 my ago, and over the following million years western and eastern Asia became part of the hominin geographic range.

Evolutionary processes such as adaptation, speciation and extinction are closely connected to geographical processes. Evidence for geographical distribution and dispersal prior to 1.8 my ago has primarily been discussed as a source of information on patterns of evolution within the hominin group (Turner and Wood 1993; Bromage and Schrenk 1995; Strait and Wood 1999). The large volume of research on the first range expansion into Eurasia focuses in more detail on the factors that interact to shape a species geographic range. A number of authors examine the effect of ecological characteristics on the distribution of hominins and other species (Jablonski, Whitfort *et al.* 2000; Anton, Aziz *et al.* 2001). Others highlight the social adaptations (Gamble 1993; O'Connell, Hawkes *et al.* 1999) and technological advances (Klein 1999) that may have permitted hominins to expand their ranges. Another body of research investigates the environmental factors that may have ultimately limited hominin geographic ranges: for instance, community structure (Turner 1984; Turner 1992; Arribas and Palmqvist 1999), or disease (Bar-Yosef and Belfer-Cohen 2001). Hominin range expansion inspires research interest as a potential source of information about the behaviour and ecology of extinct hominin species.

Interpretation of the distribution of fossil and archaeological sites in terms of hominin adaptations is difficult. For a start, information about distribution is largely driven by factors of preservation and discovery. For instance, recent finds of early hominin fossils in central Africa have provided the first and only indication that these species ranged outside south and east Africa (Brunet, Beauvilain *et al.* 1997; Brunet, Guy *et al.* 2002). The relative scarcity of early fossil and archaeological sites, particularly in Asia (Dennell and Roebroeks 2005) means that there are large gaps in the record of early hominin distribution. For instance, our understanding of early Pleistocene dispersal out of Africa depends on a few possibly very early fossil and archaeological sites in Georgia, Pakistan, China and Java (Dennell, Rendell *et al.* 1988; Swisher, Curtis *et al.* 1994; Huang, Ciochon *et al.* 1995; Gabunia, Vekua *et al.* 2000; Vekua, Lordkipanidze *et al.* 2002). Many early Palaeolithic sites are characterised by heavily time-averaged assemblages. Dating is difficult for early Palaeolithic sites and the quality of dating varies geographically, with absolute dates rare particularly outside the east African rift valley (Singhvi, Wagner *et al.* 1998).

Secondly, translating archaeological data into testable inferences of hominin behaviour is difficult. The types of archaeological information available for the Palaeolithic period are particularly limited. An important approach to reconstructing hominin behaviour is taphonomic or middle-range research, which documents the linkage between a modern behavioural process and the traces it would leave in the archaeological, geological or paleontological records (Blumenschine, Cavallo *et al.* 1994, p.200). Middle-range research has been fruitfully applied to elucidating the record of hominin behaviour in the fields of stone tool technology (Schick 1987; Toth 1987), diet (Kay and Grine 1988; Sillen 1992) and locomotor strategy (Susman and Stern 1982). However some aspects of behaviour such as flexibility, social learning or social organization are more difficult to arrive at from the sort of archaeological record characteristic of the Palaeolithic period. Archaeological interpretation is complicated by the fact that the cognitive make-up of early hominin species was different from that of extant humans and primates.

How have other workers attempted to test hypotheses about hominin range expansion? The problems of chronological resolution and partial distribution patterns make it very difficult to use the archaeological data to distinguish between detailed models of range expansion. In terms of archaeological evidence, the focus of research has been on refining the dating of particular sites in order to gain some idea of the sequence of arrival in different areas. There is increasing evidence pointing towards the possibility of a Late Pliocene to Early Pleistocene first expansion 'Out of Africa' and this has generated intense debate about the dating of a few important sites. For instance, the dating of the fossil hominins from Sangiran and Mojokerto in Java has been subjected to reanalysis and reinterpretation (DeVos and Sondaar 1994; Swisher, Curtis *et al.* 1994; Anton 1997; Anton and Franzen 1997; Langbroek and Roebroeks 2000). Analysis of stone tool technology provides some evidence for behavioural parameters. In addition, stone tool assemblages are discussed in relation to range expansion, as evidence for the cultural (Aguirre and Carbonell 2001; Saragusti and Goren-Inbar 2001) and biological (Foley 1987; Foley and Lahr 1997) connections between populations.

Reconstruction of local and regional environments provides important evidence for processes of range expansion. Such evidence is critical for establishing dispersal routes followed by hominins in range expansion (Rolland 1998). Analyses of the changing regional environment of the time can give us information about potential environmental limits to distribution. Examples include studies of changing community structure (Turner 1984; Turner 1992; Arribas and Palmqvist 1999), and the expansion of grassland environments (Dennell 1998; Holmes 2002).

Some researchers have turned to other disciplines for insight into hominin geographic ranges. Archaeologists studying migration and colonization in modern humans use models derived from biology and sociology. The potential of computer simulation for evaluating ecological and environmental factors in hominin dispersal has been investigated by Mithen and Reed (2002). A number of workers have explored the potential of data on the distribution of extant species for developing and testing models of processes of hominin range expansion: analysing primate distribution (Eeley 1994; Eeley and Foley 1999; Eeley and Lawes 1999), or developing dispersal equations (Anton, Aziz *et al.* 2001).

Limitations to the archaeological data and interpretation provide a strong incentive for using lines of evidence from other, convergent, disciplines to contextualize the archaeological data. Independent studies have the potential to set limits to archaeological speculation (Steele and Shennan 1996; Roebroeks 2001). A wide range of possible distribution processes can be imagined for early hominin species. Comparative evidence of living non-human primates and other mammals can be used as a basis for specifying which of these processes are plausible given the behavioural and biological characteristics of early hominin species. Such studies can define the 'degrees of freedom' in reconstructing hominin behaviour and distribution processes. However fossils and archaeological remains are central elements in any reconstruction of hominin behaviour. Ideally, by combining data and ideas from a number of sources, it may be possible to make some predictions on the character of the archaeological record, which can be tested in the future.

Theory in studies of early human evolution

The unique features of modern humans, such as language and complex culture, prompt questions concerning the applicability of ecological and evolutionary theory to the hominin group. However the existence of an amazing ability unique to a particular living species is far from unusual in the animal kingdom (Foley 1987). Evolutionary theory remains the only theoretical framework capable of explaining complex adaptations, such as human language (Pinker 1995).

The modern synthesis, which combines the ideas of Darwin and Mendel, describes descent in terms of genetic inheritance and modification in terms of natural selection. While there have been challenges to standard Darwinian evolution (Eldredge and Gould 1972), the effect of such challenges has been to clarify elements of the modern synthesis rather than to overturn it (see Dennett (1996, Chapter 10)). There is widespread professional resistance to the idea that humans are subject to the algorithmic processes of evolution (Dennett 1996), and this may be influenced by concerns about its political misuse or religious implications (Dennett 1996; Szalay 1999). According to Foley (1999), theory development in early hominin evolution has been primarily non-Darwinian. This bias is likely to have limited the explanatory power of archaeological and anthropological interpretations of early human origins.

In addition, as Foley (1999) has pointed out, the macroevolutionary approach has been particularly prevalent in studies of early human evolution. This approach focusses on major evolutionary change, usually over a long period, and the evolution of genera or higher taxa. However, such large-scale patterns are mediated through local events, and specific contexts are important (Foley 1999). The ideal theoretical framework will make links between the local events and specific contexts, and the large-scale, long-term processes of evolution and range expansion.

Ecologists study the relationships between organisms and their environments, in order to elucidate the principle governing those relationships (Foley 1987). The principles of ecology are strongly connected to the theory of evolution, because it is natural selection that in the long term shapes ecological relationships through differential reproductive success. Thus ecological and evolutionary theory can provide the necessary connection between different scales of analysis. Behavioural ecology has produced several models that relate the variability in a behavioural or ecological parameter to an environmental or social condition (Foley 1992). By examining such patterns of association, it is possible to make predictions of the behaviour or adaptations of the hominins that can be tested in the archaeological and fossil record (Foley 1987; Oliver, Sikes *et al.* 1994).

Biogeography is the science of the geographical distribution of living things. The study of the structure and dynamics of geographic ranges is an important part of biogeographical research (Brown, Stevens *et al.* 1996). The spatial and temporal distribution of organisms is studied in relation to behavioural and environmental variables in order to answer questions about the underlying processes (ibid.). Studies in biogeography could potentially be very useful in interpreting early hominin geographic ranges, by identifying plausible underlying processes and patterns of association, and making predictions that can be tested in the archaeological record.

A number of authors have used biogeographical theory and method to interpret early hominin geographic ranges (Bromage and Schrenk 1995; Foley 1999). Two recent studies have suggested that biogeographic trends are present in the cultural diversity of modern humans (Collard and Foley 2002; Moore, Manne *et al.* 2002). Most interestingly, Harriet Eeley (1994) carried out a quantitative analysis of primate geographic range area as part of her doctoral research, and applied these results to provide a context for interpretation of early hominin distribution. I wish to extend this approach in my own thesis.

Evolutionary theory, and its subdisciplines such as behavioural ecology, is the best framework for interpreting human origins, and therefore should underpin the study of early hominin ranges. The combination of ecological and evolutionary theory provides a link between different scales of analysis. In particular, biogeographical studies can potentially illuminate the patterns and processes of early hominin distribution.

The research project

Theoretical summary

Based on the arguments above, in this thesis I use an evolutionary and ecological approach to interpreting early hominin geographic ranges. Specifically, I integrate theory from biogeography and studies of human evolution. Where possible, I back up the integrated models with reference to examples of trends in the fossil record.

I argued above that gaps in the archaeological data on hominin distribution and behaviour provide a strong incentive for the use of comparative data and theory from convergent disciplines. In this thesis, I carry out comparative analyses of trends in the geographical distribution of modern primate species, and of African mammal taxa. This approach will establish whether theoretical models of hominin range expansion are plausible based on evidence from extant species, and test whether early hominin species fit the primate or carnivore trend. If they do, plausible scenarios for range expansion can be specified for hominin species based on their ecological or behavioural characteristics. Alternatively, knowing when and in which species hominins diverged from the primate or carnivore trend will help us to understand why. For instance, a novel adaptation, innovation, or exponential change in some character may have removed the constraints acting on other species.

Modern primates are often used as a source of analogy for early hominin behaviour. They are qualified as analogues by their close phylogenetic relationship. In addition, they share characteristics such as a large brain size, and complex social interactions, which are seen as particularly important in human evolution. Substantial datasets from field studies now exist for behaviour such as hunting (Boesch and Boesch 1989; Fedigan 1990; Wrangham and Bergmann-Riss 1990; Stanford, Wallis *et al.* 1994; Rose 1997; Uehara 1997) and tool use (reviewed in Schaik (1998)). Geographic variation in chimpanzee and orangutan behaviour has been interpreted as evidence for culture (Whiten, Goodall *et al.* 1999; Schaik, Ancrenaz *et al.* 2003). In addition, a new database of instances of primate innovation, tool use and social learning provides a comparative measure of behavioural flexibility (Reader and Laland 1999; Reader 2000; Reader and Laland 2002). These additions to the data on primate behaviour emphasise the importance of these species to understanding early hominin behaviour.

The comparison of carnivore and *Homo* ecology has been made by other workers, notably Shipman and Walker (1989). Analogy with carnivores, based on similarity in adaptation rather than family relationships, is appropriate for a discussion of aspects of the hominin ecological niche, such as diet. While some primates do eat meat, it rarely makes up an important part of their diet. Carnivore ecology provides more useful information about the implications of a diet focused on meat.

These comparative analyses will also add to our understanding of the distribution of contemporary species. The biogeography of primate species is of considerable current interest (Eeley 1994; Ruggiero 1994; Cowlishaw and Hacker 1997; Eeley and Foley 1999; Eeley and Lawes 1999; Harcourt 2000). The primates provide data on the distribution of a tropical order, and therefore may provide answers to biogeographical questions. In addition, many primate species are under threat of extinction, and understanding the factors shaping a species' geographical distribution is important for conservation (Cowlishaw and Dunbar 2000). Previous analyses have focussed on variables to do with the environmental and ecological niche (Eeley 1994; Ruggiero 1994; Cowlishaw and Hacker 1997; Eeley and Foley 1999; Eeley and Lawes 1999; Harcourt 2000). I intend to increase our understanding of the processes shaping primate geographic ranges by looking at the effects of previously unexamined variables related to cognition, social behaviour and life history.

I also intend to clarify the effects of dietary niche, particularly meat eating, on geographical distribution. In order to do this, I compare the distribution of the carnivores with that of species with a different dietary niche. Taxonomic and functional groups of mammals tend to differ in the scale of geographic range sizes and in adherence to ecological rules. Studies of variation in higher taxa, and the processes behind it, are rare. I contribute to this issue in biogeography, by documenting such variation in African carnivores, primates and ungulates, and exploring one plausible explanation for this variation.

The only direct evidence of changes in hominin geographic ranges over time, and of hominin behaviour, ecological niche and environmental limitations, comes from the archaeological record. I argued above that comparative models are most useful when they produce predictions that can be tested using the archaeological record. In order to assess whether it is possible to test the comparative models developed in this thesis, I survey the data on hominin distribution, niche and behaviour for a particular time period. My aim is not to provide a definitive interpretation of hominin geographical distribution: however I do intend to make some conclusions about the evolution and ecology of hominin geographic ranges during a limited period.

I focus on the period 1.8-0.6 my ago in Africa, which is of particular interest for a number of reasons. The earliest fossil sites outside Africa date to this period, suggesting a major range expansion, although we need earlier evidence of absence to test this hypothesis (Dennell and Roebroeks 2005). The appearance of early African *H.erectus* marked a major shift in hominin adaptive strategy, involving morphological and behavioural innovations. More than one hominin species was present in Africa during this period, and a comparison of the data for *H.erectus* and the robust australopithecines is useful for characterising the niches of these species.

Research methodology

Comparisons among species are frequently used to test hypotheses of how organisms are adapted to their environment. Comparative studies identify evolutionary trends by comparing the values of some variable or variables across a range of data. This method has been applied to good effect in studies of human origins, for instance in predicting hominin group sizes (Aiello and Dunbar 1993; Steele 1996). The nature of comparative data presents problems for statistical analysis. However comparative biology has been revolutionised by the development of statistical techniques to account for phylogenetic relationships (Purvis, Gittleman *et al.* 1994).

The use of Geographic Information Systems (GIS) methods provides a number of advantages in the analysis of geographic ranges. By taking the effects of projection into account, I can derive a more accurate database of primate geographic range size than can be found in the literature or has been used in related investigations (Eeley 1994; Eeley and Foley 1999; Harcourt 2000). Using GIS capabilities it is possible to analyse primate distribution in relation to other factors such as vegetation, rainfall or climatic variability. For instance, the GIS software can be used to calculate the mean rainfall experienced by each primate species, by overlaying primate range and climate maps. Maps of the variation in factors such as climate, species distribution and ecological characteristics aid interpretation. A range of useful GIS data is now distributed via the Internet, including maps of the geographic ranges of African mammals (IEA 1998).

A key part of my research is a comparative study of primate distribution in relation to physical and behavioural characteristics and environmental factors using GIS and statistical techniques. The comparative method makes it possible to test hypotheses of adaptation and correlated evolution of environmental tolerance, geographical distribution, and certain key behavioural characteristics. The breadth of data available from studies of extant primates makes it possible to analyse the role of a range of behavioural and life history characteristics in the evolution of primate geographic ranges. For example, a database of primate innovation, tool use and social learning frequencies has been provided by Simon Reader (2000).

It could be argued that a broad scale, cross-species analysis of patterns in behaviour would be best followed by more detailed case studies to flesh out the data. An investigation of particular wide-ranging primates would improve our understanding of modern primate distribution and could also give insights into early hominin geographic ranges. In this thesis I aim to give an overview of the patterns and processes of distribution relevant to early hominins, so I concentrate on a larger scale.

I investigate large-scale patterns in the distribution and ecology of modern African mammals in order to illuminate the role of dietary niche, and particularly meat eating, in distribution and range expansion. Maps of the spatial distribution of ecological variables can be used to show spatial trends (Ruggiero, Lawton *et al.* 1998; Cowlishaw 1999; Eeley and Foley 1999; Eeley and Lawes 1999). Comparison between maps of the carnivores, primates and ungulates can indicate whether geographical trends are broadly related to dietary niche. This method provides an efficient way of dealing with a large-scale comparison, although it is less useful in terms of producing predictive models.

Models based on a theoretical framework and analysis of modern species are only useful if we are able to get comparable information from the archaeological and palaeoanthropological dataset. I examine the data available for hominin niche breadth, biomass, diversity, distribution, and other relevant variables during this period. I use a broad range of information, including site distribution, landscape use studies, fossil density and abundance, and raw material transfer data. While the data on the shape and extent of the geographic range is limited, the archaeological and fossil data provides information about other aspects of distribution, such as habitat preference and tolerance, which can be used to test predictions about the evolution and ecology of the early hominin geographic range.

Overview

In Chapter 2, I survey the literature on the modern distribution of species, to determine which factors have been identified as important and how they interact. I focus on spatial patterns in the geographic range size of species, which are relevant to questions of change over time. I identify a number of areas in which our understanding of these processes can be expanded, and will address these in Chapters 4 and 5.

In Chapter 3, I outline changes in early hominin geographic ranges over time. I then present three models of the evolution of hominin geographic ranges. These models are based on theories about human evolution, and the spatial processes described in Chapter 2, and are backed up where possible with examples of trends in the fossil record. The models will be refined and evaluated using relevant comparative data (Chapters 4 and 5).

In Chapter 4, I describe a comparative analysis of primate distribution in relation to physical and behavioural characteristics and environmental factors, conducted using GIS and statistical techniques. I am particularly interested in the interaction of the following factors: behavioural flexibility and cultural transmission, geographical distribution and environmental variability, and life history parameters. This analysis will allow me to test the predictions of the first two models presented in Chapter 3.

In Chapter 5, I describe an investigation of large-scale patterns in the distribution and ecology of modern African mammals. The aim is to assess the role of dietary niches, particularly meat eating, in species distribution. This will allow me to test the predictions of the third model presented in Chapter 3.

The aim in Chapter 6 is to assess the relevance of the models developed through theoretical discussion and comparative analysis, by examining the data on early hominin distribution and ecological niches for a particular case study. A further aim of this Chapter is to assess hypotheses of hominin range expansion during the period 1.8-0.6 my ago in Africa, by comparing palaeoanthropological and archaeological data with the predictions of the models.

Chapter 2

SPECIES GEOGRAPHIC RANGES

Introduction

My aim in this thesis is to determine the factors that shaped the geographic ranges of early hominin species. The geographic range of a species is defined as the total extent of its distribution. There is interspecific variation in the size, shape, boundaries and internal structure of a species' geographic range (Brown *et al.*, 1996). In this review, I will focus on patterns in area, on the grounds that the factors that influence spatial variation in geographic range size are likely to also be useful in explaining changes over time such as range expansion. The area over which a species may expand its geographic range is limited by environmental variables such as climate, competition, and the distribution of biotic and topographical barriers. However the effect of such environmental variables is likely to be modified by species level biological, ecological and behavioural characteristics. These factors are dynamic and interact over time. The aim of this Chapter is to summarise current understanding of these complex processes, through a survey of the literature on biogeography.

It is clear from literature survey that the Primates provide an interesting case study for biogeography, giving much needed data on the distribution of a primarily tropical mammal, and also being very well studied. A number of the general patterns established in the field of biogeography have been demonstrated for primates (Cowlishaw and Hacker, 1997, Eeley and Foley, 1999, Harcourt, 2000, Ruggiero, 1994). Investigations have identified relationships between range size, species richness, niche breadth and latitude, particularly in the African primates (Cowlishaw and Hacker, 1997, Eeley, 1994, Harcourt, 2000). I plan to improve the data available on primate geographic range area worldwide, and generate new data on the climatic variability experienced by primates at a range of spatial and temporal scales.

In addition, I plan to investigate the effects of a wider range of behavioural factors on primate geographic range size and tolerance. This survey of the literature also identifies a tendency for taxonomic and functional groups of mammals to differ in scale of geographic range sizes and in adherence to ecological rules. Studies of this higher taxa variation, and the processes behind it, are rare. I propose an analysis of African mammal distribution that will clarify the effects of ecological niche on geographical distribution.

Patterns in species geographic range size

Frequency distribution

There is enormous variation in the geographic range size of individual species, as much as 12 orders of magnitude (Brown *et al.*, 1996). For example, among the primates, the tufted capuchin (*Cebus apella*) occupies northern and central South America, an area of approximately 11 million km^2, while the lar gibbon (*Hylobates klossi*) is restricted to the Mentawai Islands in Indonesia, an area of just over 6000 km^2.

The distribution of geographic range size data across orders of mammals tends to be strongly right skewed (Brown *et al.*, 1996). A growing number of studies suggest that species range size distributions are, in general, approximately lognormal (Gaston, 1996). That is, relatively few species have very large or very small ranges. Small ranges tend to be associated with islands (where coastlines act as barriers to expansion) and with lower latitudes, as discussed below. This applies to a wide range of taxa, from plants to mammals. As can be seen in Figure 2.1, this is broadly true of the primates, although on further examination, it appears that the logarithmic transformation is significantly left-skewed (Eeley and Foley, 1999). This pattern has also been observed among New World bird species (Blackburn and Gaston, 1996). It might be expected that the frequency distribution would be truncated at larger sizes due to the limits of the continental area (Pagel *et al.*, 1991). However Blackburn and Gaston (1996) argue that this cannot be the case for New World birds as none occupy all of the landmass, and this is also true of primates in Africa, South America and Asia. What does limit maximum range size is thus unclear.

The latitudinal gradient

The geographic extent of species tends to decrease with decreasing latitude and decreasing elevation in terrestrial

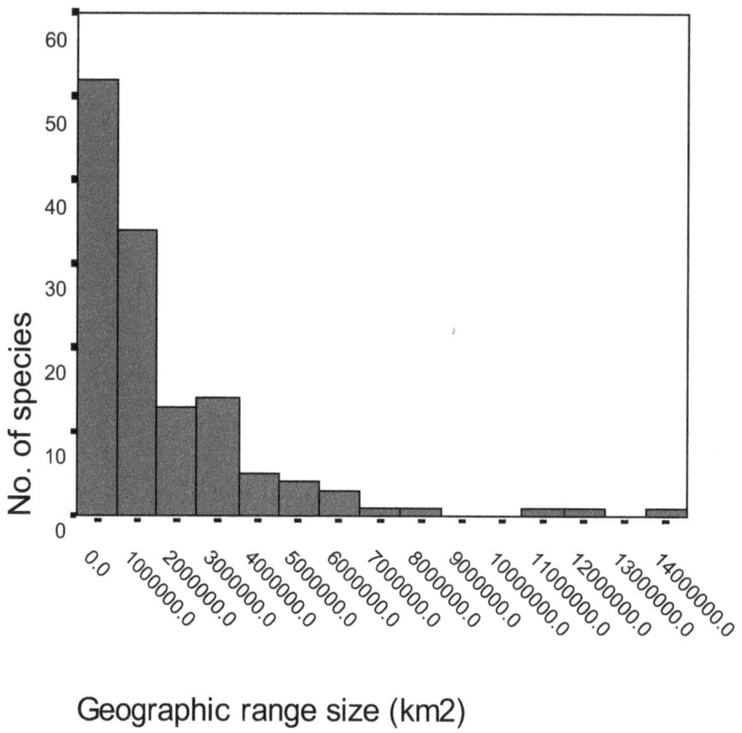

Figure 2.1

Frequency distribution of primate geographic range size in km^2 (data from Wolfheim 1983)

environments (Brown *et al.*, 1996). The latitudinal pattern was documented for subspecies by Rapoport (1982) and subsequently shown to hold for species within higher taxonomic groups (Stevens, 1989). Stevens (1989, 1992) has explored the generality of the relationship and its occurrence in elevational and depth gradients. He suggested that the correlation between geographic range and latitude be called 'Rapoport's Rule'.

However, there is some argument over the generality of Rapoport's rule. Gaston *et al.* (1998a) collated results from published studies of the relationship between range size and latitude. They conclude that overall the evidence for a general pattern is equivocal. According to Rohde (1996), it is a local phenomenon limited in expression to the Palaearctic and Nearctic above latitudes of 40-50° North. Gaston *et al.* (1998a) point out that Rapoport's Rule may be viewed as a local phenomenon partly because few studies extend to lower latitudes, or focus on a taxon that is predominantly tropical.

Given the paucity of studies of range size and latitude in lower latitudes, studies of primates are of particular interest. Cowlishaw and Hacker (1997) have examined latitudinal gradients in the geographic range and diversity of the African primates. Regression of the species geographic extent on its latitudinal midpoint indicated that African primates do not adhere to Rapoport's rule (ibid.). However reanalysis of the data using northern and southern subsets showed that the Rapoport effect is present in those species whose range centres are to the south of the equator (Cowlishaw and Hacker, 1997). It is likely that, as Eeley and Lawes (1999) point out, the existence of a major biotic barrier in the form of the Sahara desert limits range expansion in northern latitudes. Harcourt (2000) concludes that the Rapoport effect is present in African primate genera, taking phylogenetic relationships into account.

Ruggiero (1994) analysed range size and latitude in a number of orders of South American mammals. There was a general tendency for the average latitudinal range of species to decrease with decreasing latitude, but the Rapoport effect was consistently supported only by *Carnivora* and *Primates* (Ruggiero, 1994). This study is interesting in suggesting that South American primates adhere to Rapoport's rule to an unusual extent (Eeley and Lawes, 1999). However, according to Harcourt (2000) the Rapoport effect is not present in South American primate genera. In addition, further analysis suggests that the Rapoport effect is present worldwide (excluding Madagascar) (Harcourt, 2000).

The studies described above suggest that Rapoport's rule is at least partially relevant to primate distribution. African primates seem to provide a particularly good example of this effect. The differences in results may be largely a result of differing methodology: for instance, in measuring geographic ranges, and in statistical method. The fact that Harcourt found a Rapoport effect in Africa north of the equator, while Cowlishaw and Hacker (1997) did not, is explained by the omission of *Macaca* and *Theropithecus* as outliers in the former analysis (Harcourt, 2000). These genera are high latitude, northern, and monospecific, with small geographic ranges (ibid.). In addition, Harcourt (2000) makes comparisons at the genus rather than the species level. The differences between the two studies of South American primates may be caused by the fact that Ruggiero (1994) used a different phylogeny, and excluded all Central American species (Harcourt, 2000). In addition, Ruggiero used the average latitudinal extent of species found within different latitudinal bands: the strength of relationships in this study could be biased by spatial autocorrelation (Cowlishaw and Hacker, 1997). However, the general impression given by these studies is that the Rapoport effect is variable where primates are concerned, and this is consistent with conclusions from other taxa (Gaston *et al.*, 1998a).

Species richness

A further pattern in species distribution is the tendency for species richness to increase with decreasing latitude. There is a significant relationship between latitudinal extent of range and species richness in South American bats, primates and carnivores (Ruggiero, 1994). There is a latitudinal gradient in species richness both north and south of the equator for African primates (Cowlishaw and Hacker, 1997, Eeley and Foley, 1999).

Furthermore, a latitudinal pattern and environmental determinants have been identified in human cultural diversity (Collard and Foley, 2002). While the authors do not view human cultures as like species, they argue that this congruence may reflect a similar mechanism, perhaps the need and potential for niche separation (Collard and Foley, 2002). This also suggests that such geographical patterns are likely to have been present in early human ancestors, as the result of a variety of mechanisms.

Variation in geographic range size above the species level

> 'Among organisms as a whole, range size varies by more than 12 orders of magnitude. Within genera, families, orders, and classes of plants and animals, range size often varies by several orders of magnitude, and this variation is associated with variation in body size, population density, dispersal mode, latitude, elevation, and depth.'
>
> (Brown *et al.*, 1996, p.597)

I have discussed variation in geographic range size between individual species. Brown (1996) notes that the range of variation in geographical extent is specific to particular taxonomic or functional groups, and this is apparent at different levels from closely related species to different orders.

Different taxa within a region may vary considerably with respect to how they conform to Rapoport's rule (France, 1992, Rhode *et al.*, 1993, Ruggiero, 1994). In a taxonomic comparison of South American mammals, only *Carnivora*, *Primates* and *Chiroptera* support the hypothesis that Rapoport's rule is a consequence of selection for broader climatic tolerances of species at higher latitudes (Ruggiero, 1994). Ruggiero (1994, 1998) suggests that a taxon's evolutionary history and ecological characteristics are a significant determinant of how well it conforms to Rapoport's rule.

The shape of the frequency distributions of geographic range size may also vary between taxa. In South American mammals, frequency distributions of geographic range size tend to be lognormal: however for the *Carnivora* and *Chiroptera*, widespread species are more common. These orders may contain a greater proportion of climatic and habitat generalist species making up their species assemblage in South America (Ruggiero, 1994).

Thus taxa with consistently differing characteristics are likely to have a different scale of geographic range size variation, and may also differ in their adherence to biogeographical rules and have a different frequency distribution. There has been relatively little research into the processes underlying such differences. One characteristic that might be expected to have such an effect is dietary niche or trophic level. Trophic level affects a number of other ecological variables, including body mass, home range size and population density. Both small geographic range size and trophic level are correlated with extinction risk in contemporary species (Purvis *et al.*, 2000). This suggests that there may be constraints on species at higher trophic levels: successful species may have to have larger geographic ranges. Given the varied implications of this characteristic, I suggest that dietary niche may affect the scale of variation in geographic range size and adherence to ecological rules of particular taxonomic or functional groups.

The role of species' characteristics

Environmental variability and niche breadth

Environmental variability

The cases discussed above suggest that the Rapoport pattern is more complex than a simple relationship with latitude. Why does the Rapoport effect exist in some but not all cases? Stevens (1989) suggested taxa that did not experience the full range of environmental variability should not exhibit the pattern: for instance, species that hibernate during the winter. This explanation is based on the idea that it is variability that produces the Rapoport effect.

According to Stevens' (1989) environmental variability hypothesis, individuals at higher latitudes experience more variable, seasonal environments (see Figure 2.2). Therefore natural selection in those areas favours those individuals with a wider niche breadth and more generalist characteristics, leading to the spread of these characteristics through high latitude species. Such wide tolerance of environmental variability would allow species to occupy a variety of habitats and thus expand their range. Measurements of climatic variability have been used to explore this hypothesis. Letcher and Harvey (1994) found that for Palaearctic mammals annual temperature range at the centre of the geographic range of a species is positively correlated with range size, but that latitude was a better predictor. Based on a survey of such studies, Gaston *et al* (1998a) argue that Stevens' theory does not explain the pattern of exceptions to the rule. For instance, some migrant bird species show Rapoport's rule, contrary to predictions (Gaston *et al.*, 1998a).

Again, there are a number of studies that investigate whether primates follow this pattern. In their analysis of African primates, Cowlishaw and Hacker (1997) used six climatic variables, in addition to the latitudinal midpoint itself, as potential determinants of geographic range size. These included altitude, mean daily temperature, daily temperature range, annual temperature range, annual rainfall, and the proportion of annual rainfall that falls in the wettest quarter. They found that the geographic range of African primate species increases with the seasonality of rainfall in both directions from the equator. The effect of climatic variability statistically subsumed that of latitude (Cowlishaw and Hacker, 1997). These results are backed up by the results of Harcourt's analysis (2000). This supports the hypothesis that geographic range size is determined by adaptation to climatic variability in African primates.

Ruggiero (1994) analysed range size and latitude in a number of orders of South American mammals. She found that bats, primates and carnivores showed a strong positive association between the size of the geographic range and both species richness and the range of temperature extremes recorded in each latitudinal band. However a similar analysis by Harcourt (2000) suggests that the relationship between climatic variation and geographic range size is influenced by phylogeny in South American primates. These contradictory results are similar to those described above for tests of the Rapoport effect.

Different trends in African and South American primates may reflect the different distribution of habitats within these continents. All South American primates are forest dwelling species, and tropical forest is less variable climatically than savannah (Harcourt, 2000). It should also be noted that these studies use a number of different measurements of climatic variability, based on monthly and seasonal variation. Different scales of variation, as well as different aspects of the environment, may have different effects on primates.

Niche breadth

An alternative approach to this question is to look at patterns in species' niche breadth and distribution. Based on Stevens' (1989) environmental variability hypothesis, we would expect to find that niche breadth would increase with latitude, and that geographic range size would increase

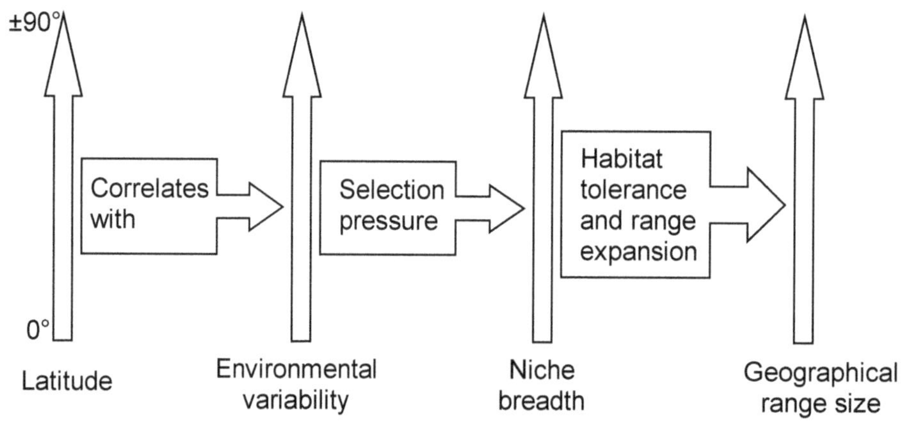

FIGURE 2.2.
THE GENERAL THEORETICAL RELATIONSHIP BETWEEN LATITUDE, NICHE BREADTH AND GEOGRAPHIC RANGE SIZE

with niche breadth. Eeley and Foley (1999) have identified a general latitudinal increase in both mean habitat and dietary niche breadth in African catarrhines. Niche breadth in turn is strongly correlated with range size both in a species averaging, cell based analysis and in a cross species analysis (including when phylogeny is taken into account) (Eeley and Foley, 1999). This relationship would suggest that the broader tolerance of primate species at higher latitudes allows them to expand their range. These results are backed up by the independent results of Harcourt's (2000) analysis. He found that primate taxa with a broad latitudinal extent often showed greater adaptability, when phylogeny was controlled for (as measured by number of dietary types, number of habitat types and number of species per genus).

Species richness

According to several studies, high species diversity in equatorial regions may be a consequence of environmental factors favouring specialisation. Both Stevens (1989) and Brown (1995) have argued that latitudinal gradients in geographic ranges and species richness may be directly connected. According to Brown (1995), interspecific competition restricts geographic range size. However, such a suggestion would assume that some of the species are exploiting the same niches (Gaston *et al.*, 1998a). Thus interspecific competition is unlikely to cause such a general pattern.

By contrast, Stevens (1989) has proposed that the greater ecological flexibility of high latitude organisms allows them to exist in ephemeral populations at lower latitudes. Such populations are dependent on other populations in areas where the species does well and are therefore immune to competition effects and their ranges can overlap with other species. Due to the effects of gene flow, peripheral populations tend to be adapted to conditions at the centre of the range (Kirkpatrick and Barton, 1997). However, historical factors may have contributed to species richness in these areas, as discussed below (Eeley and Lawes, 1999).

The gradient in species richness has been documented in African and South American primates, as discussed above. In addition there is a relationship between species richness and species range, and habitat and dietary breadth, that remains once latitudinal and longitudinal variations have been removed (Eeley and Foley, 1999). These results are consistent with Stevens' (1989) explanation of the pattern. However if species richness were functionally related to range size we would expect to find a gradient in species richness only south of the equator, in parallel with the Rapoport effect (Cowlishaw and Hacker, 1997).

Eeley and Foley (1999) observe that regions of high species richness are characterised by low average species range sizes and increased specialisation in species' habitat and dietary niche. However, the pattern in African primates is not simply linear: regions of low species richness display considerably more variation in terms of both range size and habitat and dietary niche breadth (ibid.). This variation indicates that the relationship between range size, species richness and niche breadth is complex, and that a combination of ecological and evolutionary factors may provide the best explanation.

Abundance, body size and range size

In most cases there are highly significant positive correlations between range size and both body mass and some measure of average population density (Brown *et al.*, 1996). There is usually, however, a great deal of residual variation (ibid.). In most taxa, body size and abundance show a strong negative association, as large bodied species tend to have large home ranges and exist at a lower abundance than small species.

This relationship is conventionally explained by the 'energetic equivalence rule': equal amounts of energy are available to each species in a community (Damuth, 1981). An alternative explanation is based on niche breadth. Species with a larger body mass, tend also to have a higher reproductive rate and to be generalists who can utilise a wide variety of resources (Cowlishaw and Dunbar, 2000).

The relationship between range size and population density has been explained partly with reference to the effects of niche breadth. As discussed above, there is a relationship between niche breadth and geographic range size (Eeley and Foley, 1999). A species that has a large geographic range may find more favourable habitats at its range centre, and thus have a higher population density (Hengeveld, 1990). By contrast, species with small ranges may only find a few favourable locations, not only at their range margins, but also at their range centres (ibid.). The pattern appears to be more complicated when studied across larger taxa, and may be better described as triangular than strictly linear (Gaston, 1994, Brown *et al.*, 1996). Species that occupy a small range have a low density, while widespread species may occur at either a high or low density (Gaston, 1994). Most African cercopithecoids conform to this general pattern (Eeley and Lawes, 1999).

There is some debate concerning the theoretical pattern likely to be presented by the relationship between range size and body mass. Differences concern the mechanism that is emphasised, whether energetic constraint, dispersal ability or intrinsic rate of population increase. Brown and Maurer (1989) argue that large species are constrained by their energetic requirements to have larger individual home ranges and therefore occur at lower population densities than smaller species. As a consequence, large species that occupy only a small geographic range will have a higher probability of extinction than those that are widespread. Both the influence of body size on dispersal ability and the tendency for smaller species to be more specialised serves to reinforce this pattern (Brown and Maurer, 1989). Gaston & Lawton (1988) suggest that, since generally smaller bodied species have a higher rate of population increase, the opposite pattern might also occur and smaller species may occupy larger ranges. On colonising a site, smaller species may attain a higher population density than larger species, and consequently would be less susceptible to stochastic extinction and would become widespread. However supporting evidence for either pattern is limited (Eeley and Lawes, 1999). There does not appear to be a linear relationship between body mass and range size in primates (Harcourt, 2000). Gaston (1994) suggests that the relationship between body size and range size, like that between range size and abundance, might be triangular. Those species that occupy small ranges have a small body size, while widespread species may be either small or large. The African anthropoids provide evidence of such a pattern: in general, larger species occur over large geographical areas, while small species may either be widespread or relatively range restricted (Eeley and Lawes, 1999).

The relationship between range size and abundance, expected on the basis of niche breadth, can be combined with that between body size and range size to make some general range size predictions, as described by Eeley and Lawes (1999). In general, small specialist species might be expected to occupy small ranges and occur at a lower abundance, small generalist species might be relatively abundant and occupy large ranges, while large species will be widespread and generalist but occur at a lower abundance (ibid.; see Figure 2.3). Large species occupying small ranges will rapidly become extinct (Eeley and Lawes, 1999, Purvis *et al.*, 2000). Such a pattern appears to occur among the African anthropoid primates (Eeley and Lawes, 1999). However there is considerable variation within this pattern, and species occupying the small range size/low abundance category are, unexpectedly, on average the largest (ibid.). These species are all either paleoendemic, or in danger of extinction (ibid.). Thus these exceptions could be seen as an artefact of the historical trajectories of species' ranges, as discussed below.

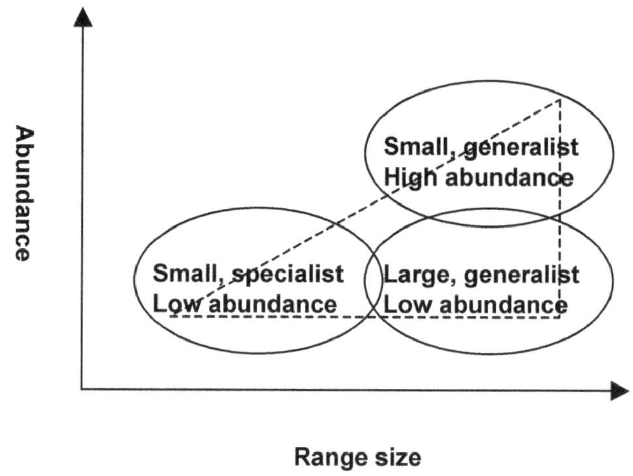

FIGURE 2.3
THE GENERAL THEORETICAL RELATIONSHIP BETWEEN SPECIES RANGE SIZE AND ABUNDANCE AND BODY MASS, COMBINED WITH THE GENERAL POSITIVE CORRELATION BETWEEN RANGE SIZE AND NICHE BREADTH, TO IDENTIFY SOME EXPECTED SPECIES CHARACTERISTICS IN RELATION TO RANGE SIZE
(AFTER EELEY AND LAWES, 1999)

Dispersal ability

> "A fundamental question in ecology is whether there are evolutionary characteristics of species that make some better than others at invading new communities"
>
> (Sol and Lefebvre, 2000, p.599).

Widely dispersed species tend to share certain characteristics, and it seems likely that these might have some part in invasion success. These include a broad ecological tolerance. Widely dispersed species also tend to have large home range sizes, and thus body sizes for their group, and are often gregarious (Anton *et al.*, 2001). Body mass may increase dispersal ability by allowing individuals to cross barriers more easily and disperse further (Cowlishaw and Dunbar, 2000). Relatively rapid life history characteristics may also be useful in dispersal (MacArthur and Wilson, 1967). In addition, widely dispersed species tend to be polytypic (including several subspecies or other variants) and to have a relatively long fossil record (Anton *et al.*, 2001). Evidence that some intrinsic traits of vertebrates can predispose them to be successful invaders comes from the fact that some species establish themselves in new areas while others fail to do so (Sol and Lefebvre, 2000). Sol & Lefebvre (2000) used data on Avian species introduced to New Zealand to test the link between forebrain size, feeding innovation frequency and invasion success. They demonstrated that species with larger relative brain size and a higher frequency of foraging innovations in their area of origin tended to be better invaders. This implies that behavioural flexibility improves invasion success.

Little is known about the distances dispersing primates travel, or their ability to establish new populations (Cowlishaw and Dunbar, 2000). Data on primate colonization ability, such as dispersal rates, tends to come from the fossil record (Anton *et al.*, 2001). For instance, it has been suggested that the initial spread of the macaques may have followed the origin and spread of agricultural systems over the last 10,000 years (Richard *et al.*, 1989). The main causes of primate dispersal are argued to be the search for better access to mates, and inbreeding avoidance, rather than to seek better or unexploited environmental resources (Pusey, 1991). However individual dispersal ability may be less important than the movement of groups after fission (Cowlishaw and Dunbar, 2000).

In addition to the characteristics of the species and the communities and habitats themselves, invasion success is also determined by stochastic factors. These include the numbers of individuals involved, number of invasions, the heterogeneity of their environment and gene pool, and the timing of their arrival relative to environmental conditions (Hengeveld, 1990). Such factors are randomly distributed and therefore make invasion success very difficult to predict.

Dispersal ability seems likely to influence invasion success and therefore the geographic range size of a species. However invasion is only one aspect of the evolution of a species range. For example, rapid dispersers may in the long run be out-competed by more slowly dispersing and behaviourally flexible species (MacArthur and Wilson, 1967). Thus species' adaptive responses to environmental or biotic change may be equally important in determining range size on the long term.

Environmental limitations

Physical and biological boundaries

The most important external limits on species range expansion are environmental barriers, such as mountain ranges and rivers. In addition, biotic barriers such as the boundary between the savannah and forest biomes, and the distribution of other species, may limit species' ranges. Eeley (1994) analysed the concurrence between species range boundaries and geographical barriers in Africa, and found that both riverine and coastal barriers were important in limiting the geographical distribution of catarrhine species: ecological boundaries such as the transition from predominantly forest to savannah habitats were less important.

Species may be able to expand their distributions more easily within than across biogeographic boundaries (Gaston *et al.*, 1998a). This might partly explain some exceptions to Rapoport's rule. The range size of New World birds is inflected at 17° north (ibid.), and that of carnivore species richness in South America at around 25-30° south (Ruggiero, 1994). In South America the number of habitats peaks at 15° south rather than the equator (Ruggiero, 1994).

As discussed above, Brown (1995) has argued that competition effects are higher at low latitudes. Positive evidence that interspecific interaction determines species range size is limited (Cowlishaw and Dunbar, 2000). Geographical overlap between taxa increases as more distantly related groups of primates are considered (Eeley, 1994). This may reflect the effect of competitive interaction between similar species: however it is difficult to distinguish the effects of competition from patterns of allopatric speciation (ibid.). Interspecific competition is not the only interaction between species that may limit distribution. Predators and parasites would also be expected to have an effect, however little is known about either (Cowlishaw and Dunbar, 2000).

Topographical barriers may also limit species ranges more effectively nearer the equator. For instance, the effect of mountains on species' distribution may be stronger at lower latitudes (Ruggiero *et al.*, 1998). A lowland tropical species that attempts to cross a mountain pass will face environmental conditions more extreme than it normally experiences, whereas this is less likely to be true for lowland species resident at higher latitudes. The distribution of geographic range distortion suggests that this is indeed the case in South America (Ruggiero *et al.*, 1998).

Historical processes

Introduction

The environmental and phylogenetic history of a particular biome or region may be crucial to understanding modern species distribution. The influences of both environmental and phylogenetic history are reflected in many characteristics of species and clades, including the size, shape and internal structures of their geographic ranges (Brown *et al.*, 1996). The history of environmental change in a particular biome or region is likely to have influenced past colonization, speciation and extinction events in ways that may affect present geographic distributions (Brown *et al.*, 1996). In addition, species characteristics are dynamic through time and are affected by environmental change through natural selection. The characteristics of past environments have acted as selective agents to influence the environmental requirements and tolerances, and the demographic, dispersal and life history characteristics of contemporary organisms, and these characteristics in turn affect the geographic range (Brown *et al.*, 1996). These interacting temporal factors may explain many of the exceptions to ecological rules, and deviations from theoretical relationships, described above.

Environmental history

Environmental change in the past can be shown to have influenced modern distribution patterns. For example, as the majority of primates are restricted to tropical forest habitats, present day patterns of species richness or range size are strongly influenced by historical and regional processes occurring within this biome (Eeley and Lawes, 1999). Climatic fluctuations in the Pleistocene period were characterised by the expansion and contraction of glacial ice sheets at high latitudes. Refuge theory suggests that these climatic fluctuations were reflected at lower latitudes by successive periods of cooler and drier, and warmer, wetter conditions, and the associated contraction and re-expansion of tropical forest (Hamilton, 1976, Hamilton, 1989, Haffer, 1982).

Multiple cycles of forest expansion and contraction would have provided ample opportunity for the isolation and divergence of populations and speciation within the modern primate community (Eeley and Lawes, 1999). High levels of speciation over the Pleistocene period provide an explanation for the high species richness and endemism found in the forest primates today. There is some evidence that periods of primate speciation and radiation coincide with cycles of forest contraction and expansion, for example for the genus *Cercopithecus* in Africa, and marmosets and tamarins in South America (Eeley and Lawes, 1999).

Stevens (1989) has attributed high species richness in the tropics to the Rapoport's rescue hypothesis: species at higher latitudes with wider ranges have higher tolerance and can exist at low latitudes in peripheral populations. However the environmental history of areas with high species richness, such as the African tropical and sub-tropical forest, provide another level of explanation.

Lineage evolution

Phylogenetic history, particularly the date of significant radiations, can shape the distribution of species ranges. At any one time, the ranges of different species may be in phases of expansion, decline or stasis. In the context of evolutionary time, species range size distributions can be seen as summarising a slice across the temporal trajectories of the sizes of the geographic ranges of the species in an assemblage (Gaston, 1996).

For example, the relatively recent radiation of the African cercopithecines introduced a large number of species with ranges at an early phase of evolution (Eeley and Lawes, 1999). As this genus is forest adapted, this recent radiation may have skewed the range size distribution in such areas towards smaller sizes. Thus phylogenetic history may have influenced the current distribution of the African cercopithecoids (ibid.). By contrast, species such as *Theropithecus gelada* or *Macaca sylvanus* are examples of a late period in the history of a lineage. These species are each the sole examples of their genus in Africa, and have very small geographic ranges.

An increased understanding of these processes could explain the characteristic frequency distribution of species geographic range area (Gaston, 1996). In addition, the temporal factor in species distribution may explain why species do not all adhere equally well to ecological rules. Ruggiero (1994, 1998) suggests that a taxon's evolutionary history and ecological characteristics are a significant determinant of how well it conforms to Rapoport's rule. The relatively recent radiation of the cercopithecines may be a factor behind species richness at low latitudes in Africa, while the relict species described above present exceptions to Rapoport's rule.

Despite its importance for fully understanding spatial patterns, Gaston (1996) highlights the lack of any generally accepted model for the long-term (speciation to extinction) temporal dynamics of species' geographic range sizes.

Geography and evolution

The discussion above highlights the strong connection between geographical distribution and processes of evolution and speciation. The environmental history of an area may affect geographic range and population density, both of which influence gene flow. The appearance of new geographical barriers (for instance through habitat fragmentation) may isolate populations, causing them to diverge faster. The disappearance of such barriers may lead to speciation through adaptive radiation. In an adaptive radiation, species arriving in environments or ecosystems that are relatively 'empty' often undergo rapid diversification into many new species, which are able to capitalise on some existing adaptation and, by adding new adaptations, are able to expand into many new niches (Harrison *et al.*, 1988). Such adaptive radiation may occur when species enter new areas, as geographical barriers disappear, or as new species capable of immigrating evolve: it may also occur if species become extinct, allowing others to move into 'vacated' ecological space (Harrison *et al.*, 1988). As discussed above, a number of adaptive radiations in the primates may have occurred in response to cycles of expansion and contraction in the African and South American rainforests.

In addition, geographic range size may influence the likelihood of speciation: however there is some debate as to whether species with large or small geographic ranges are more likely to speciate (Gaston, 1998). Species with small geographical ranges are more prone to extinction (Purvis *et al.*, 2000).

The connection between evolution and geography provides a basis for integrating long term processes of evolution and climate change, with short-term patterns of geographical distribution. However it should noted that other mechanisms for speciation may also exist, for instance reproductive isolation through differences in behaviour, particularly behaviour associated with mate choice and reproduction.

Conclusion

In this Chapter I have summarised current knowledge regarding the spatial distribution of species, focussing on the area of species geographic ranges. I reviewed general patterns in species geographic range area, including frequency distribution, the latitudinal gradient, and variation between higher taxonomic groups. I then summarised our current understanding of the processes that lead to these patterns. I described the effect of species' characteristics including niche breadth, body mass and abundance and dispersal ability. I outlined important environmental factors that act as a limit to species ranges, from mountain ranges to other species. Finally I discussed the role of historical factors of environmental change and lineage evolution in shaping modern spatial distribution.

In this review I have highlighted some of the remaining questions in biogeography, and the relevance of the primates and of comparison of taxonomic or functional groups. The analyses described in Chapters 4 and 5 will increase our knowledge of these processes. There is a scarcity of models that address the entire process of speciation, range expansion, range contraction and extinction. In the following Chapter, I will present some models of hominin range expansion and evolution. Understanding how temporal trends produce spatial patterns, and the connection between geographical and evolutionary processes, provides the basis for the integration of theories of human evolution with spatial biogeography described in Chapter 3.

CHAPTER 3

MODELS OF HOMININ EVOLUTION AND RANGE EXPANSION

Abstract

The fossil evidence suggests a number of expansions of the hominin range over time, from first appearance to the emergence of the *Homo* lineage and the first move out of Africa. I will begin this Chapter by summarising this evidence. I will then present a number of models of hominin range expansion. The arguments for these models are based on evolutionary and ecological theory, with particular reference to the literature on human origins and biogeography. Where possible these models are backed up with reference to trends in the fossil record.

Hominin range expansion

The distribution of fossil and to a lesser extent archaeological sites, and taxonomic identification of fossils, suggests a number of expansions of the hominin geographic range in the course of human evolution. The restricted range of the earliest hominins is suggested in Figure 3.1. Early expansions (such as the first appearance of australopithecines in southern Africa) are accompanied by speciation. *H.habilis* is the first hominin species to have expanded its range to include both East and Southern Africa. The emergence of early *H.erectus* in Africa is followed by the occupation of new areas of Africa (notably North Africa) and new habitats within East Africa. The largest range expansion discussed here is the first move of hominin species into Eurasia.

The earliest hominin fossils have all been found in East or Central Africa (Figure 3.1). These include a 6-7 my old cranium and other specimens from Chad, in Central Africa, which have been attributed to a new species, *Sahelanthropus tchadensis* (Brunet *et al.*, 2002). Postcranial fossils from 6

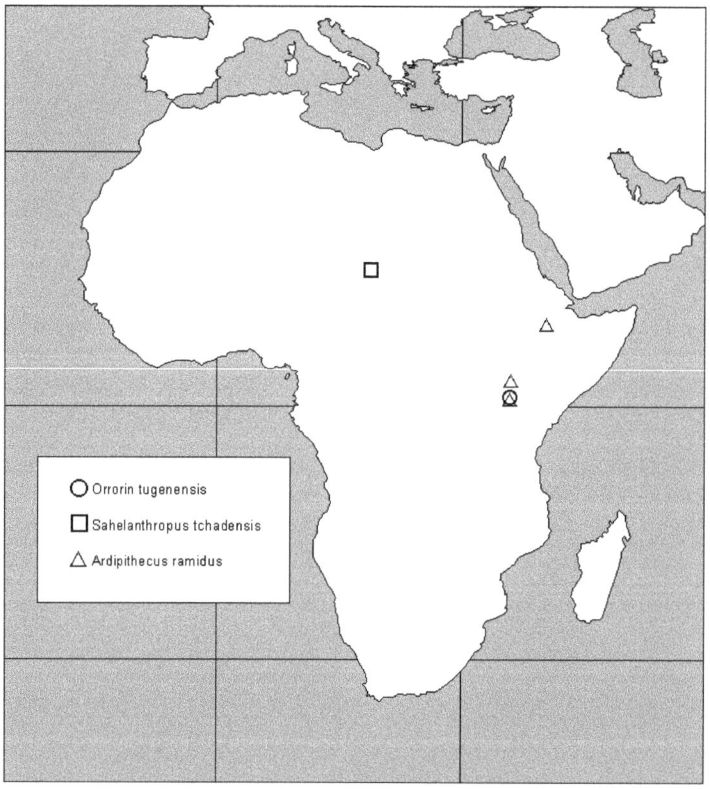

FIGURE 3.1
EARLIEST FOSSIL HOMININ SITES

my old sediments in the Lukeino Formation, Tugen Hills, Kenya are attributed to another species, *Orrorin tugenensis* (Senut *et al.*, 2001). Specimens from the Middle Awash dated to 5.54-5.77 my ago are attributed to a subspecies of *Ardipithecus ramidus* (Wolde-Gabriel *et al.*, 2001, Haile-Selassie, 2001).

A.ramidus has also been found at Aramis, Middle Awash, and As Duma, Ethiopia, and possibly Lothagam, Baringo and Tabarin (Chemeron Formation), Kenya (White *et al.*, 1994, Hill *et al.*, 1992, Hill, 1985, Semaw *et al.*, 2005). This species is dated to about 4.4 my ago (by $^{40}Ar/^{39}Ar$ dating) at Aramis (White *et al.*, 1994).

A.anamensis (*c.* 4.2-3.9 my ago) and *A.afarensis* (*c.* 3.9-2.8 my ago) are likewise found in East Africa (Figure 3.2). Fossils of *A.anamensis* have been found at Aramis and Asa Issie, Middle Awash (White *et al.*, 2006), Allia Bay, East Turkana (Coffing *et al.*, 1994, Leakey *et al.*, 1995), and Kanapoi (Patterson and Howells, 1967, Leakey *et al.*, 1995, Ward *et al.*, 1997). Fossils of *A.afarensis* have been found at Hadar, Belohdelie, Maka and Dikika in the Awash River Valley and Laetoli, and possibly also Fejej, Omo, Koobi Fora and West Turkana (Johanson *et al.*, 1982, Kimbel *et al.*, 1994, Asfaw, 1987, White, 1984, White *et al.*, 1993, Johanson *et al.*, 1978, Fleagle *et al.*, 1991, Suwa *et al.*, 1996, Klein, 1999, Alemseged *et al.*, 2005). A fossil from West Turkana, called *Kanyanthropus platyops* by its discoverers and dating to 3.5-3.2 my ago, probably represents a new species (Leakey *et al.*, 2001, Liebermann, 2001). Another australopithecine fossil, currently attributed to a species known as *A.bahrelghazali*, has been found at Chad in central Africa, dated to between 3.4-3 my based on associated fauna (Brunet *et al.*, 1997). This is one of only two hominin fossils found outside the east African rift valley and southern Africa, and shows that the range of the australopithecine genus included central Africa.

The earliest australopithecine fossils to be found in southern Africa are those of *A.africanus* (Figure 3.3). *A.africanus* has been found at Taung, Sterkfontein (Member 4 and possibly Member 2), Makapansgat (Members 3 and 4) and probably Gladysvale (Clarke and Tobias, 1995, McKee *et al.*, 1996, Clarke, 1988, Clarke, 1994a, Klein, 1999, Berger and Tobias, 1994). This species may have existed between 2.7 and 2.3 my ago, based on faunal dating at the sites. The australopithecine fossil from Member 2 Sterkfontein may be a member of this species, and is dated prior to 3 my ago (McKee *et al.*, 1996).

A number of hominin species are characterised by the development of cranial structures that allowed great force to be applied between huge upper and lower cheek teeth. These are referred to as the 'robust' australopithecines. Fossils of robust australopithecines have been found in East and South Africa. Specimens attributed to the species *A.aethiopicus*

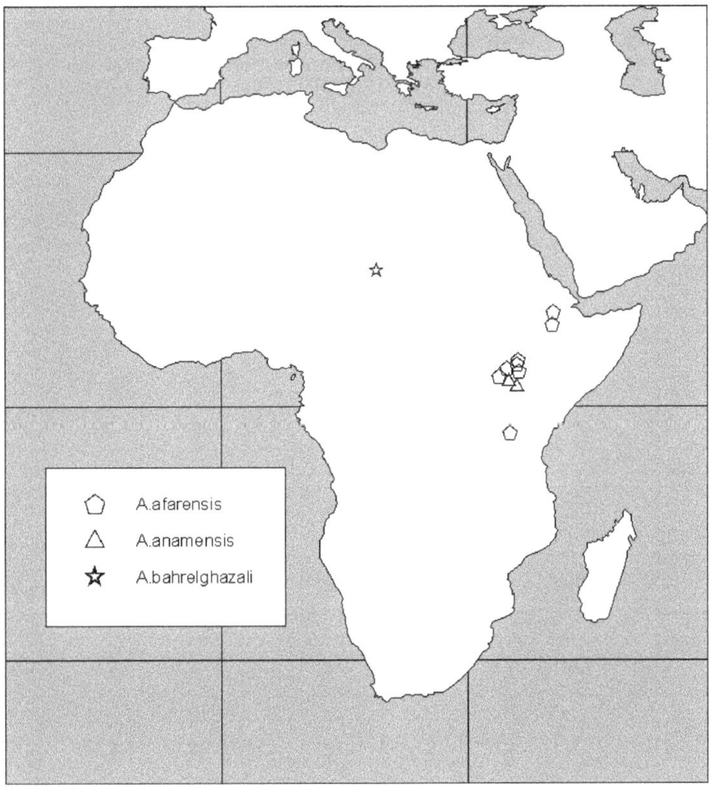

FIGURE 3.2
DISTRIBUTION OF THE AUSTRALOPITHECINES *A.AFARENSIS*, *A.ANAMENSIS* AND *A.BAHRELGHAZALI* (4.2-3.8 MY AGO)

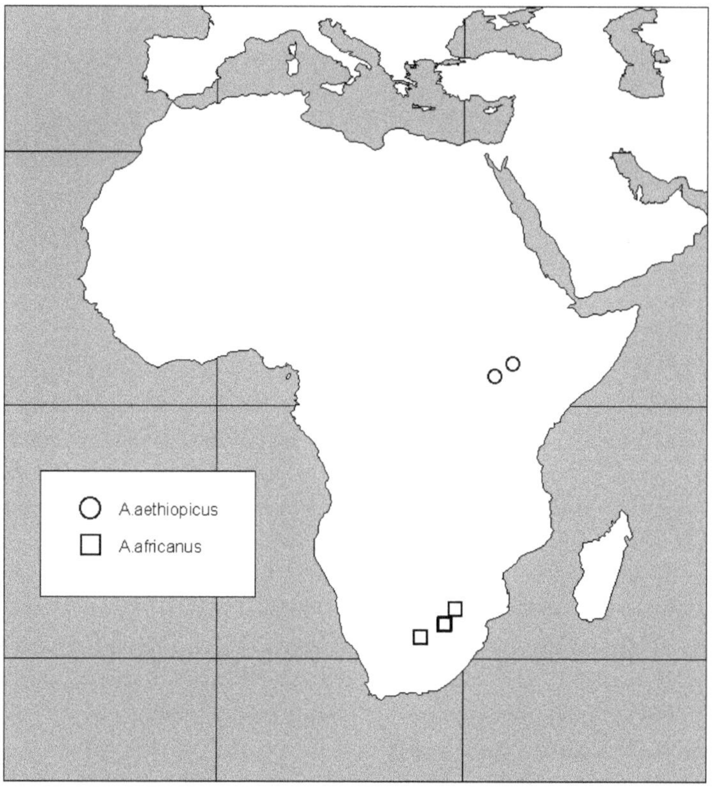

FIGURE 3.3
DISTRIBUTION OF *A.AETHIOPICUS* AND *A.AFRICANUS* (2.7-2.3 MY AGO)

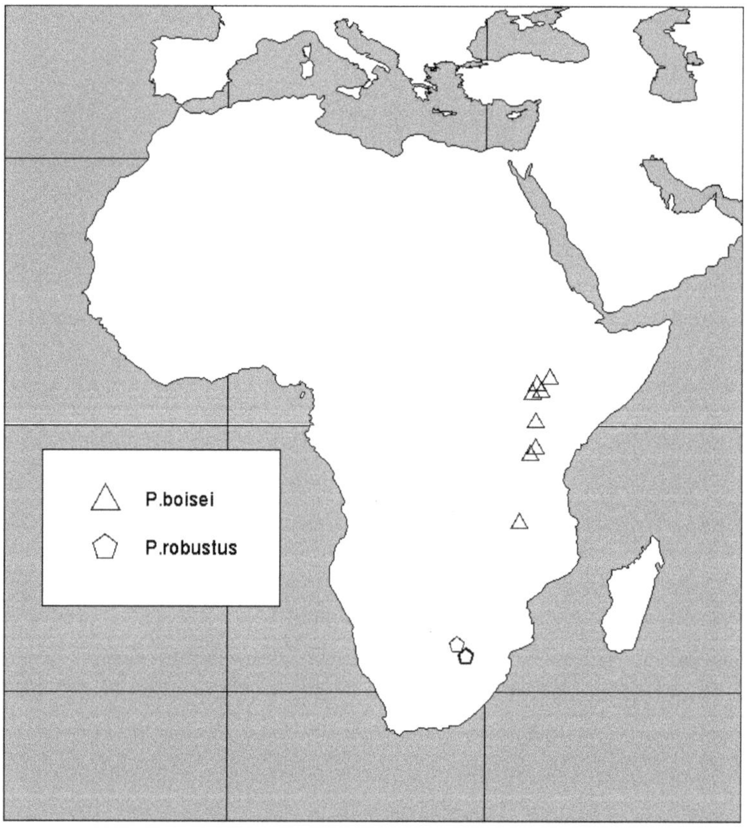

FIGURE 3.4
DISTRIBUTION OF *PARANTHROPUS* SPECIES

have been found at a number of sites in the Turkana Basin, dating to between 2.7 and 2.3 my ago (Walker *et al.*, 1986, Leakey and Walker, 1988, Suwa *et al.*, 1997) (Figure 3.3). *P.robustus* has been found only in the Sterkfontein Valley, at Kromdraai, Swartkrans, Drimolen and possibly Sterkfontein, from roughly 1.8-1 my ago (Broom, 1938, Brain, 1981, Grine, 1993, Keyser *et al.*, 2000, Kuman and Clarke, 2000) (Figure 3.4). *P.boisei* has been found at a large number of East African sites, dating from about 2.3 my ago to 1.2/0.7 my ago (Suwa *et al.*, 1997, Suwa *et al.*, 1996, Klein, 1999, Wood, 1991, Carney *et al.*, 1971, Leakey, 1971, Kullmer *et al.*, 1999).

Early *Homo* appears during the same period as *Paranthropus* (Figure 3.5). Fossils of early *Homo* are dated to 2.3 my ago at Hadar (Kimbel *et al.*, 1996), about 1.9 my ago at East and 2.3-2.4 my ago at West Turkana (Feibel *et al.*, 1989, Klein, 1999, Prat *et al.*, 2005), and about 2.3 my ago at Omo (Suwa *et al.*, 1996). Many authorities argue that there are two species of early *Homo*, with *H.rudolfensis* represented at Uraha, Chemeron and Koobi Fora (Chamberlain, 1989, Groves, 1989, Lieberman *et al.*, 1988, Rightmire, 1993, Stringer, 1986, Walker, 1981, Walker and Leakey, 1978). *H.habilis* may have been the first hominin species whose geographic range included both South and East Africa: fossils of this species may be present in Sterkfontein Member 5, dated to around 1.9 my ago (Hughes and Tobias, 1977): however Clarke considers this to be *Paranthropus* (Clarke, 1995).

Fossils of a hominin designated *A.garhi* have been found at Bouri, Middle Awash, dating to 2.5 my ago (Asfaw *et al.*, 1999). The earliest archaeological sites also appear between 2.5-2.0 my ago, and are situated in East Africa, including Hadar (Afar Locality 666), Kada Gona and Omo (Member F, Shungura Formation) in Ethiopia, and Kanjera South, West Turkana (Lokalalei) in Kenya (Figure 3.5) (Kimbel *et al.*, 1996, Semaw *et al.*, 1997, Semaw, 2000, Howell *et al.*, 1987, Kibunjia, 1994, Roche *et al.*, 1999).

Early African *H.erectus* emerged about 1.7 my ago (Figure 3.6). The best dated early specimens come from the Turkana basin: a nearly complete skull (KNM-ER 3733) and a partial skeleton (KNM-ER 1808) from Koobi Fora, are roughly 1.8-1.7 my old; a second skull (KNM-ER 3883) is only slightly younger; and a skull and associated skeleton (KNM-WT 15000) from Nariokotome III, West Turkana, is about 1.5 my old (Brown, 1994, Brown *et al.*, 1985, Feibel *et al.*, 1989). Possible *H.erectus* fragments from the lower Omo Valley (Shungura Formation) date to 1.4-1.3 my ago (Klein, 1999).

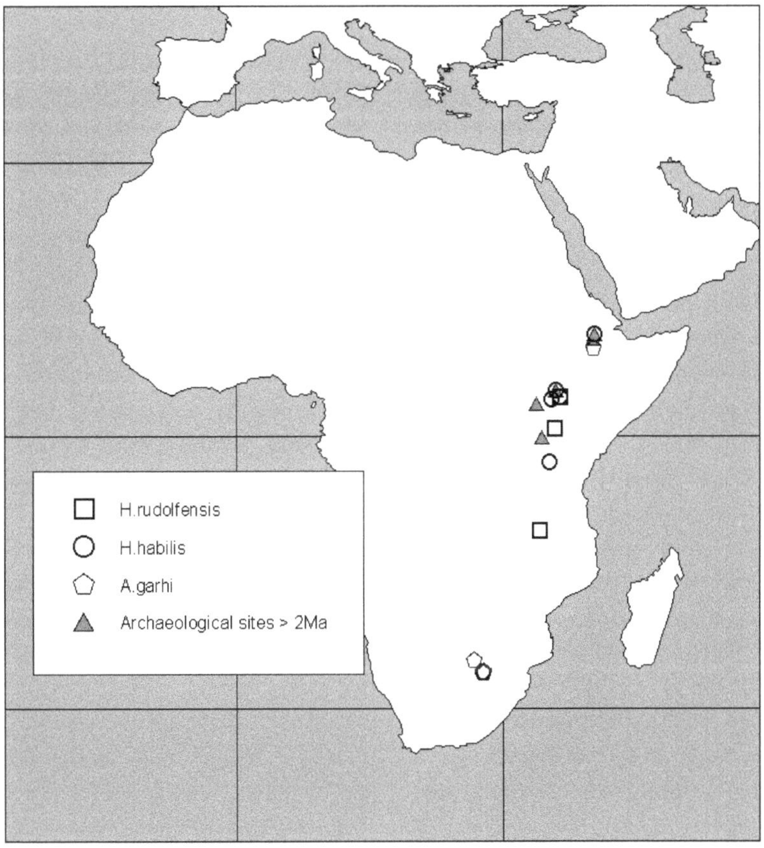

Figure 3.5
Distribution of early *Homo*, *A.garhi* and the earliest stone tools (>2my)

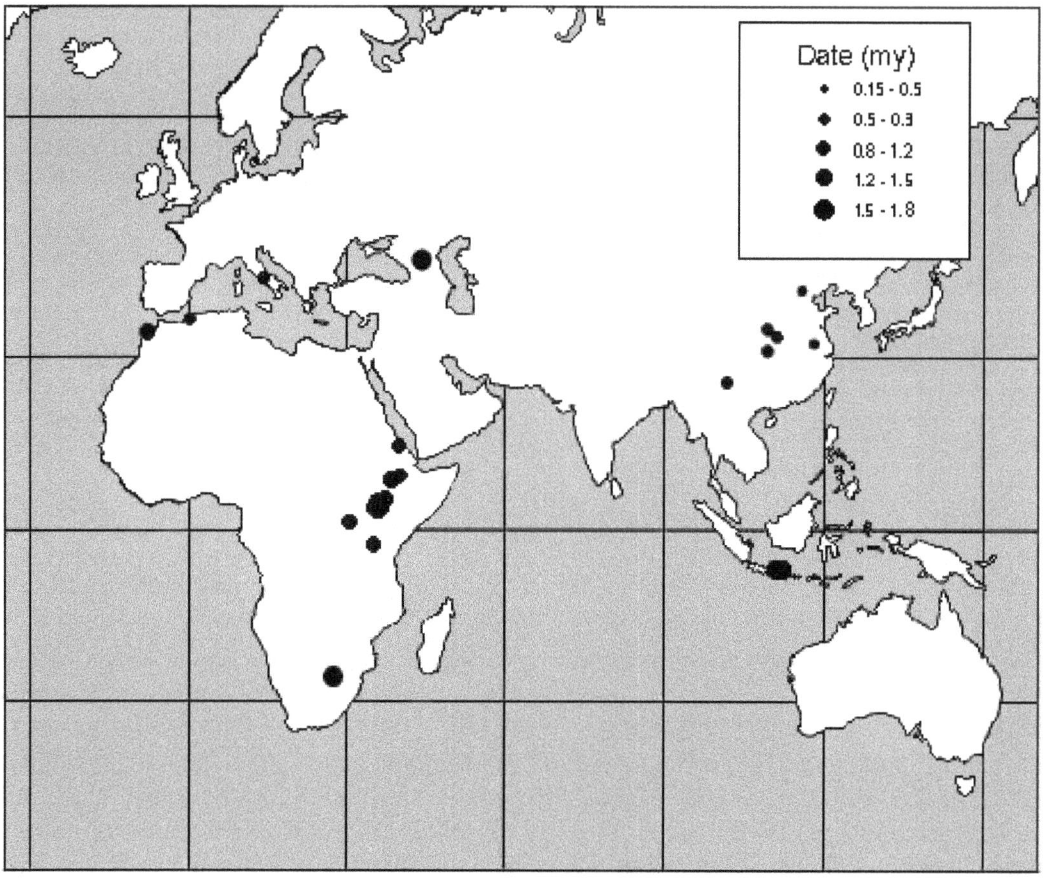

FIGURE 3.6
DATES OF FOSSILS ATTRIBUTED TO ***H. ERGASTER/ ERECTUS***

The distribution of fossil and archaeological sites indicates a major increase in range size in Africa during this period (Figure 3.6). Sites appear in both the southern and eastern regions of sub-Saharan Africa, and in new areas in Africa, notably the far north of the continent. In addition environmental reconstruction suggests that a number of sites within east Africa represent colonization of novel habitats. Hominins occupied the drier peripheries of sedimentary basins on the floor of the eastern Rift Valley by about 1.7-1.6 my ago (Harris, 1983), and reached the high plateau of Ethiopia by 1.5 my (Clark and Kurashina, 1979).

The largest range expansion thus far occurred when hominins first moved out of Africa (Figure 3.6). At Dmanisi, human fossils including two skulls and several crania and mandibles and Oldowan-like pebble tools and flakes have been dated to *c.* 1.75 my ago (based on a combination of sedimentary polarity, fauna and radiometric dates of an underlying layer) (Gabunia *et al.*, 2001, Vekua *et al.*, 2002, Lordkipanidze *et al.*, 2005). Dmanisi represents a major range expansion at 44°N on the southern slope of the Caucasus mountains. At 'Ubeidiyah, lake and river deposits contain bifaces and other Acheulean or Developed Oldowan artifacts that probably accumulated between 1.4 and 1.0 my ago, and a fauna including African species (Bar-Yosef and Belfer-Cohen, 2001, Bar-Yosef and Goren-Inbar, 1993). A number of sites further outside Africa have been given very early dates, notably the fossil hominins from Mojokerto and Sangiran in Java. The Mojokerto cranium now seems to have been found in context (Huffman *et al.* 2005) and has been given a date of 1.8 my old (Swisher *et al.*, 1994). The specimens from Sangiran have been dated to about 1.6-1.7 my ago (Swisher *et al.* 1994, Larick *et al.* 2001). The teeth from Longuppo, China are probably not hominin (Huang *et al.*, 1995, Schwartz and Tattersall, 1996, Wolpoff, 1999). In addition, there are a number of archaeological instances of Early Pleistocene artefacts from Asia, including dates of 1.66 mya from the Nihewan basin in north China (Zhu *et al.* 2004).

As Dennell and Roebroeks (2005) have recently pointed out, we have very little evidence for hominin range expansion into Eurasia, especially considering the area under discussion. In addition, we have limited evidence for earlier absence of hominins, and there is considerable discussion about the taxonomic attribution of the key fossils (Dennell and Roebroeks 2005). Given these uncertainties, we cannot currently be sure where, when and which hominins first appeared in Asia.

There may have been limits to the distribution of Early and early Middle Pleistocene hominins. In China, hominin distribution appears to have been limited to lower latitudes: archaeological sites in the Nihewan Basin (Schick and Dong, 1993, Zhu *et al.* 2004) (approximately 41°N) are the northernmost well-established instances. In addition, it has been argued that many early sites in Europe north of the Alps prior to 0.5 my ago are not valid (Roebroeks and van Kolfschoten, 1994). Currently it seems that people were present before 500 ky ago, but only sporadically, and that to begin with they were restricted to warmer climes south of the Pyrenees and the Alps (Roebroeks, 2001). Such an earlier presence is indicated at Trinchera Dolina (Aurora Stratum, Layer 6) Atapuerca, northern Spain (Carbonell *et al.*, 1995) and stone tools from Orce in southern Spain may be older still (Martinez-Navarro *et al.*, 1997, Oms *et al.*, 2000). Recent discovery of stone tools in Britain dated to *c.* 700 kya (Parfitt *et al* 2005) are accompanied by paleofaunal remains indicating that hominins were present in a warm interglacial period, confirming this view of hominin occupation.

Models of hominin range expansion

1.1. Environmental variability may act as a selective pressure. Generalism or species structures and characteristics which allow a species to respond flexibly to change, including a large brain that is effective in processing data and generating complex cognitive responses, would be adaptive under conditions of increasing environmental fluctuation. These characteristics would be useful in buffering the effects of future changes in an organism's environment, allowing it to stay in one place. In addition, these characteristics are likely to be useful in coping with the novel environments encountered in range expansion, and in dealing with the variable conditions within a larger geographic range. This will affect the future selective conditions faced by the organism, which will probably experience a wider range of variation than other animals. A capacity for behaviour which protects a species from the effects of changes in its environment, may inhibit genetic change in response to changes in the environment, and may also trigger genetic change in response to major behavioural changes. Thus two feedback mechanisms would cause continued selection for increasing behavioural flexibility within a lineage.

1.2. I argue that social learning is the most important aspect of cognitive complexity for increasing hominin habitat tolerance and range expansion. Human social learning can be interpreted as part of general variation in primate social learning. A greater capacity for social learning may enhance a species' ability to respond flexibly to environmental change, as a part of general behavioural flexibility. In addition, social learning may be critical in accurately assessing the risk involved in range expansion, and coping with increased seasonality in a larger geographic range. An increased capacity for social learning could have coevolved with environmental variability and range expansion: however it may also have been a by-product of social processes.

2. Wide-ranging primates are characterised by a strategy involving a relatively rapid reproductive rate and dietary niche breadth. Dietary niche breadth allows species to exploit a wide range of habitats, while a relatively rapid reproductive rate would help populations to recover rapidly if environmental change or migration into new environments led to high levels of mortality. The early hominins appear to have matured slowly, and been relatively long-lived. *H.erectus* may have had a relatively flexible reproductive rate due to care-taking behaviour (O'Connell *et al.*, 1999). Based on the primate pattern, I suggest that reproductive flexibility combined with dietary niche breadth was crucial in hominin range expansion.

3. Dietary constraints on the distribution of large brained species mean that the ability of these species to expand their ranges into less productive environments may depend on an increase in dietary quality. The occupation of new, more open habitats by *H.erectus* suggests the relaxation of these constraints. In addition, dietary quality, and meat eating, can be linked to spatial patterns at a range of geographical scales from local to regional and of units from the individual organism to the species or higher taxonomic group. A number of environmental changes may have relaxed environmental constraints on carnivore distribution during the period in which hominin species first moved out of Africa. Thus the range expansion of *H.erectus* may have been dependent on a change in dietary niche, and in the environmental factors limiting distribution.

Discussion

Niche breadth, behavioural flexibility and environmental change

Species' niche breadth may affect their response to environmental change and hence their distribution. In general, organisms can evolve two ways of responding to major shifts in habitat (Potts, 1998a). The first is by mobility or dispersal, allowing a species to track a preferred habitat or a key resource. The second approach is to broaden the range of conditions under which an organism can live (ibid.). Generalist species are likely to adopt the second strategy, specialists the first (Vrba, 1985a). Thus species' niche breadth will affect their regional distribution over time. As alternative environments recur, specialist species will ebb and flow out of an area, while generalists stay in one place (Vrba, 1985a). These geographical patterns may be visible in the regional record for the hominin lineage, when some hominin species have more specialised dietary niches than others.

An organism's response to changes in its environment will affect the selective pressures it faces. Populations that tend to track particular habitats and resources will experience similar selection pressures despite local habitat change (Potts, 1998a). Populations that stay in one place will experience more diverse conditions of reproductive success and survival over time (ibid.).

Explanations of hominin evolution have tended to stress the consistent selective effects associated with specific habitats or directional trends (Potts, 1998b). However varying environmental conditions may also act as a selective pressure. Vrba (1995) has suggested that cyclical changes in climate may have macroevolutionary effects. Potts (1998a) has suggested that environmental variability may act as a selective force via variability selection. According to the VS hypothesis, wide fluctuations over time created growing disparity in adaptive conditions. Inconsistency in selection eventually caused habitat-specific adaptations to be replaced by structures and behaviours responsive to complex environmental change (Potts, 1998a).

Characteristics that afford a flexible response to change are likely to be adaptive in conditions of increasing environmental fluctuation. Vrba (1995, 1999) argues that cyclical changes in climate select for generalist ecological characteristics above the species level. Another form of flexibility is afforded by complex structures or behaviours that are designed to respond to novel and unpredictable adaptive settings (Potts, 1998a). These might include dental structures or foraging strategies allowing a shift to newly available food types (ibid.). To this extent the predictions of the variability selection hypothesis conform well to Vrba's suggestion of selection for generalist ecological characteristics (Vrba, 1985b). However Potts (1998a) suggests a range of other structural and behavioural characteristics which might also be useful in weathering environmental variability. For instance, a locomotor system designed to allow a wide repertory of movement, or a large brain that is effective in processing external data and generating complex cognitive responses (Potts, 1998a). Boyd and Richerson (2000) also argue that increasing climatic variability during the Plio-Pleistocene may have selected for more complex cognition, although they focus on variation on a shorter time scale.

These evolutionary processes can be related to the spatial patterns highlighted in Chapter 1. According to Stevens' (1989) environmental variability hypothesis, increased climatic variability at higher latitudes gives organisms with greater tolerance a selective advantage. This corresponds to suggestions of climatic fluctuation selecting for generalism or behavioural flexibility in hominin evolution (Potts, 1998a, Vrba, 1985b). Such wide climatic tolerances are likely to allow species to occupy a variety of habitats and thus expand their range (Stevens, 1989). Patterns in modern species distribution confirm this: in addition to the widespread latitudinal trend, geographic range size increases with niche breadth in primates (Eeley and Foley, 1999). Niche breadth could represent species' generalism or more flexible behaviour especially in foraging strategies. Integration of the evolutionary model with ecological patterns in spatial distribution suggests that species with a broader niche might over time not only remain in areas experiencing climate change, but also expand their geographic ranges beyond those areas.

The theoretical relationship described here may create a feedback mechanism leading to an evolutionary trend. Environmental variability over time may select for an increased ability to respond to environmental change through flexible behaviour, whether through increased niche breadth or more complex cognition. These adaptations would allow animals to remain in a place throughout environmental changes, and would also allow expansion of the species geographic range. Range expansion into new and more varied habitats would continue selection for increasingly flexible behaviour. Furthermore, the characteristics described above – high niche breadth, and structures and behaviours allowing flexible responses to change – alter the effects of the environment on an organism, by acting as a buffer to environmental changes.

The effects of organisms' behaviour on their own evolution have been explored in niche construction theory. Evolution by natural selection, in combination with genetic inheritance, can plausibly explain the evolution of organisms as a function of their changing environment. Thus strong directional climate change, or environmental variability, may be used to explain the emergence of particular hominin adaptations, as described above. However animals affect their environments, often via behaviour, in what may be evolutionarily significant ways (Odling-Smee, 1988). Understanding the mechanism by which animals effect their own naturally selecting environments is of particular importance in a discussion of hominin species with a high level of technological skill and perhaps also of learning and communication.

Odling-Smee (1988) argues that developing and behaving organisms routinely select and perturb their own relative environments. Thus, at least some of the selection forces that ultimately select genes are themselves selected and modified by the behaviours, and activities, of phenotypes. This implies the existence of two routes to fitness: active organisms may bequeath either 'better (or worse) genes' for 'anticipated environments', or they may bequeath 'better (or worse) environments' for 'anticipated genes' (Odling-Smee, 1988). This concept allows behaviour a larger role in evolution;

> "In the presence of niche construction, adaptation ceases to be a one-way process, exclusively a response to environmentally imposed problems; it becomes instead a two way process, with populations of organisms setting as well as solving problems."
>
> (Laland *et al.*, 2000, p.135)

The ways in which organisms can modify their own naturally selecting inputs by their niche constructing outputs vary from simple to complex. These include the selection of an environment, for instance by dispersal, or aspect of the environment such as a mate or a breeding perch (Odling-Smee, 1988). Another way is by inflicting qualitative changes on their environments, whether by depleting resources or building a nest. Finally, 'prediction', for example by learning, does not alter the environment in which the organism exists, but can change the problems posed by that environment. For instance, learning can change an objectively complex and capricious environment into one that is subjectively stable and ordered, simply because the organism can predict what is coming next (Odling-Smee, 1988).

Hominin species may have been particularly good at buffering the effects of environmental change due to a number of characteristics. These include a capacity for social learning that in humans is facilitated by the additional processes of culture and language (Laland *et al.*, 2000). Culture can be seen as one of a number of processes by which populations of complex organisms acquire adaptive information that can be expressed in niche construction (ibid.).

'Every species is informed by naturally selected genes, and many are also informed by complex, information acquiring developmental processes, such as learning or the immune system, while hominins, and perhaps a few other species, are also informed by culture.' (Laland *et al.*, 2000, p.137)

In humans the ability to learn from other individuals as well as from personal experience is facilitated by the additional process of language, which underlies culture (Laland *et al.*, 2000). In addition, more technologically advanced cultures may have a greater capacity for counteractive niche construction (ibid.). These abilities (innovation, learning, technological skill) are likely to be based on a large brain capable of complex cognitive responses.

What effect would an increase in hominin capacities for niche construction have on hominin geographic ranges? The effects of inceptive and counteractive niche construction can be distinguished (Laland *et al.*, 2000). Inceptive niche construction (a novel form of niche construction, whether the discovery of a new habitat or an innovation) opens up new possibilities for exploitation for a species, which may allow range expansion. Counteractive niche construction acts as a buffer against selection pressures in an animal's environment, and is more likely to help species to stay in one area, and to occupy more niches within that area (Reader and MacDonald, 2003). However, I argue that an increased ability to buffer the effects of environmental change in one area would be useful in the occupation of novel habitats and therefore in range expansion. Thus both forms of niche construction are likely to contribute to range expansion.

This hypothesis can be linked to modern spatial patterns. Lee (1991) hypothesised that behavioural flexibility may radically affect the survival chances of animals under new conditions. Behavioural flexibility, in the form of learning, cognition, and/or rapid adjustment to new conditions, allows animals to respond more readily to changes in the environment and can therefore be an advantage when invading novel habitats (Sol and Lefebvre, 2000). As discussed in Chapter 2, Sol & Lefebvre (2000) used data on Avian species introduced to New Zealand to test the link between forebrain size, feeding innovation frequency and invasion success. They demonstrated that species with larger relative brain size and a higher frequency of foraging innovations in their area of origin tended to be better invaders.

I argue that behavioural flexibility, based on a large brain that is effective in processing external data and generating complex cognitive responses, may have been a key characteristic in hominin tolerance of environmental change and in the expansion of the hominin range.

This has significance for the processes of human evolution. Laland *et al* (2000) argue that as unusually potent niche constructors, hominins should be particularly resistant to genetic evolution in response to changing environments and at the same time, capable of dramatic evolutionary change following major innovations. Thus rather than finding that strong habitat changes are causally associated with significant evolutionary changes, hominin speciation and dispersal may be linked to behavioural innovations (technological, social) that altered the subjective environment of subsequent generations. We might also predict that more examples of new adaptations occurring out of sync with significant environmental change would have occurred in more technologically advanced hominin lineages.

Niche construction may also have affected the rate of human evolution. Wilson (1985) has argued that changes in niche, resulting from complex social behaviour and cultural (or protocultural) transmission, might generate a 'behavioural drive' which accelerates morphological evolution by fixing a greater proportion of genetic mutations (Laland *et al.*, 2000). This effect would be stronger in species with a larger relative brain size, which tend to have more frequent episodes of innovation or cultural transmission, thus exposing themselves to novel selection pressures. Theoretical analyses suggest that cultural transmission may both speed up and slow down evolution (Feldman and Cavalli-Sforza, 1976). This is consistent with the predictions of niche construction theory, which can slow down evolutionary change by modifying external selection pressures, or speed it up in response to behavioural innovations. For example, species with a broad dietary niche or more flexible social, technological or foraging behaviour may be less affected by local environmental changes (Vrba, 1985b, Potts, 1998a). Laland *et al* (2000) conclude that the rate of cultural change may be rapid in relation to environmental change, resulting in a net increase in evolutionary rate with the development of complex culture. A recent comparative analysis of birds has demonstrated that behavioural flexibility predicts species richness, which is consistent with the behavioural drive hypothesis (Nicolakakis *et al.*, 2003).

I have identified a number of feedback mechanisms by which a trend towards increasing behavioural flexibility might develop in the hominin subfamily. An increased ability to buffer environmental change by behaviour may result in a geographical distribution exposing species to more environmental variability. In addition, behavioural innovation may itself act as a selective pressure. What would the implications of this trend be for hominin geographical distribution? According to my model, as behavioural flexibility increased, so would the size of hominin geographic ranges.

Social learning and transmission and range expansion

I have suggested that cognitive complexity has an important role in human evolution and in hominin range expansion. More specifically, I argue here that a capacity for social learning is the key factor. Social learning commonly refers to the social transfer of information and skill among individuals (Box and Gibson, 1999, abstract). According to an alternative definition, social learning is 'learning that is influenced by observation of or interaction with another animal or its products' (Box, 1984, Galef, 1988).

Are the processes operating in human social learning analogous to those in other primates? Humans differ from other animals in the complexity of socially learned behaviours, and in the mode of social learning (Boyd and Richerson, 2000). Humans transmit far more information from parents to offspring than any other species (Laland *et al.*, 2000). Social learning is important in human culture[1]: cultural inheritance depends on the transmission of learned 'knowledge' among individuals by one or more kinds of social learning (Laland *et al.*, 2000, p.141). The effects of the characteristic complexity and mode of human social learning can be seen in the time depth and intricacy of human culture.

The bulk of animal social learning seems to be dependent mostly on the relatively simple mechanisms used in individual learning (Boyd and Richerson, 2000). Imitation (learning a behaviour by seeing it done) is presumably more complex than using conspecifics' behaviour as a source of cues to stimuli that might be interesting to experience (ibid.). Likely instances of imitation have been documented in feral monkeys and chimpanzees: good examples are placer mining of wheat in macaques, and fishing for termites and ants in chimpanzees (Nishida, 1987). In addition, the existence of cultural variation in behaviour has been documented on a much larger scale in chimpanzees and orangutans (Whiten *et al.*, 1999, Schaik *et al.*, 2003). The difference between communities of chimpanzees represents contrasts between different versions of an otherwise simple pattern, and this is best explained by more complex learning processes such as imitation (Whiten *et al.*, 1999).

Learning by human children is likely to involve a combination of imitation, other forms of social learning, and individual learning (Nishida, 1987, Whiten *et al.*, 1999, Whiten *et al.*, 1996). Studies of primates in the wild provide convincing evidence that complex forms of social

1. Culture is defined in very different ways in different academic disciplines. Some cultural anthropologists insist on linguistic mediation, so that culture is constrained to be a uniquely human phenomenon (Bloch, 1991). In the biological sciences, a more inclusive definition is accepted: a cultural behaviour is one that is transmitted repeatedly through social or observational learning, to become a population-level characteristic (Nishida, 1987).

learning are present, if rare. I would argue that human social learning is best interpreted as part of a continuum of social learning. Within this continuum, humans, like other animals, employ a mix of imitation, other forms of social learning, and individual learning: and both vertical and horizontal transmission. A convincing hypothesis relating degree of social learning to range expansion will need to be phrased in terms of general evolutionary theory, and should apply to the primates as a whole.

As discussed in the previous section, learning and intelligence may allow more flexible responses to environmental change (Boyd and Richerson, 2000, Potts, 1998a). The ability to gather and share information would be useful in weathering environmental variability at a local level (Gamble, 1993, Boyd and Richerson, 2000). However, the link between social learning and environmental variability is complicated. We would expect to see learning processes in environments that are both variable but to some extent predictable (Bergman and Feldman, 1995, Stephens, 1991).

Compared with genetic determination or automatic responses to certain physical changes, learning is costly and prone to error (Sibly, 1999). Where there is no environmental variation, or environmental variation is regular and predictable by the animal, then learning is unlikely to be the way to achieve an optimal response. At the other extreme, learning is of no use in a totally unpredictable environment, since in such conditions past and present states provide no useful information about the future (ibid.). The adaptiveness of social learning relative to individual learning is determined by the chance that transmitters (the animal providing skill or information) experienced the same environment as receivers, and the accuracy of individual learning (Boyd and Richerson, 1988). These observations suggest that there may not be a simple linear relationship between learning frequencies, particularly social learning frequencies, and environmental variability.

Furthermore, social transmission of information and skills may be vertical from parent to offspring, oblique from adults to young other than their own offspring, and horizontal among members of a single generation (Sibly, 1999). Vertical transmission is conservative in that it can support traditions within populations, so that they extend over generations of animals (ibid.). Horizontal transmission is rapid and ephemeral, and is therefore most appropriate for the transmission of information pertaining to the most transient elements of the environment, such as transient food resources (ibid.).

These mechanisms of social transmission might also be useful in different environmental conditions. A variety of mathematical analyses have come up with quite consistent suggestions regarding when social learning is expected to be favoured (Aoki and Feldman, 1987, Bergman and Feldman, 1995, Boyd and Richerson, 1985, Cavalli-Sforza and Feldman, 1983, Feldman *et al.*, 1996, Laland *et al.*, 1996, Rogers, 1988). When environmental change is rapid, or when there are sudden environmental shifts, tracking by individual learning should be favoured, as should horizontally transmitted information (Laland *et al.*, 2000). In such conditions, the genetic system will change too slowly to cope, and social learning from the parental generation is likely to be too prone to error. It is when individuals encounter intermediate rates of environmental change that social learning from parents should be favoured (ibid.).

Interestingly, ethnographic analogy suggests that early hominin social transmission of craft skills was probably vertical and kin-based (Shennan and Steele, 1999). This would suggest a greater constancy in the hominin environment over time. However, we must also consider the possibility that hominin species constructed a niche through their behaviour in which vertical transmission was favoured (Laland *et al.*, 2000). This would suggest that, rather than experiencing a relatively constant external environment, hominins created a stable social environment.

Why might social learning be particularly useful in colonization? As suggested by Gamble (1993) such an ability to gather information may be particularly useful in aiding successful migration. Migration is based on an assessment of risk, which can be carried out through physiological or conscious processes (Baker, 1978). For all apes, characterised by long juvenile periods and parental investment, as well as large brain size, calculated forms of risk-assessment would be preferable. Social learning may be crucial in calculated risk assessment. The area an individual animal can assess is finite; social life provides the framework for the dissemination and evaluation of information from a wider area (Gamble, 1993, p.106). By increasing a population's knowledge of the environment, social learning will therefore increase chances of making an accurate assessment. According to Gamble (1993, p.142) the key to range expansion for primate species was the investment in behaviours that would not only select and alter the environments they and their offspring inhabit, but also gather knowledge about it.

Gamble (1993, p.241) characterises hominin range expansion as a decidedly human process: 'Through colonization we observe the evolution of our humanity', because 'Humans went everywhere in prehistory because humans have purpose'. However, the behavioural features highlighted here are not exclusive to hominins. The forms of social communication suggested by Gamble (1993, p.110-2), particularly at early stages of human evolution, are certainly not limited to the hominin lineage. The relatively slow life history characteristics and large brains of all of the apes mean that they are likely to carry out calculated forms of risk assessment. This suggests that if social communication is what allows hominins to expand their ranges beyond those of other primate species, it must be a matter of degree rather than absolute differences; social intensification rather than change is key.

A number of authors have suggested that social learning may have played a special role in brain evolution. Boyd and Richerson (2000) suggest that culture was crucial to the evolution of human cognitive complexity, and also discuss a possible role for social learning in other animals. According to Boyd and Richerson (2000), animals with complex cognition foot the costs of a larger brain by adapting more swiftly and accurately to variable environments. As discussed above, learning devices are favoured when environments are variable in time and space but to some extent predictable. However such systems are costly, with an individual spending time and energy in learning, incurring some risks in trial and error, and supporting the neurological machinery necessary to learn (ibid.). The authors suggest that social learning can economize on the trial and error part of learning. Thus social learning may make a general purpose learning system, and a larger brain, viable.

In a similar vein, Gamble (1993, p106-7) argues that the reduction of migration risk may have selected for increasing intelligence and brain size in human evolution. As discussed above, social communication may be a particularly useful form of risk assessment. Migration may be one way of gaining advantages, whether mates, food or predator avoidance, and these prizes are important in the selection of behaviour (Gamble, 1993). Thus we might propose a coevolutionary hypothesis linking climatic variability, increasing brain size and social learning, and range expansion.

Social learning may have become more advantageous as the hominin geographic range expanded. Expansion into higher latitudes would have exposed hominin populations to increased seasonality. Seasonal change in resource distribution would have stretched social frameworks (Gamble, 1993). Thus in order to sustain colonisation, and to continue the flow and storage of information crucial to weathering seasonal change, further social intensification was necessary (Gamble, 1993, p.122-3). Alternatively, according to Laland *et al* (2000), our ancestors constructed niches in which it 'paid' them to transmit more information to their offspring. The more an organism controls and regulates its environment, and the environment of its offspring, the greater should be the advantage of transmitting cultural information from parent to offspring (ibid.). Both of these processes would favour increasing amounts of social learning over time, and according to my model, further range expansion.

Alternatively, social intelligence hypotheses posit that complex social interaction was responsible for the selection pressures that favored enhanced primate intelligence (Jolly, 1966, Humphrey, 1976, Byrne and Whiten, 1988, Barton and Dunbar, 1997). Social learning is commonly regarded as a core aspect of social intelligence (Humphrey, 1976, Byrne and Whiten, 1988). This would suggest that if an increased capacity for social learning was important in hominin range expansion, it was a byproduct of social processes.

Trends in the fossil record

In this section, I will discuss trends in the fossil record in comparison with the predictions of model 1. This model will be most convincing where it is possible to identify fossil trends that back up the theoretical patterns.

Climatic variability

According to model 1, increasing environmental variability may have selected for certain characteristics in early hominin species. Periods of increased climatic variability can be identified in the global records. Environmental evidence suggests that adaptive conditions over the span of human evolution were highly inconsistent on a local to global scale (Potts, 1998a, Boyd and Richerson, 1985). While the isotopic record from deep-sea cores indicates a marked global cooling trend from six million years ago to the present, the overall climate curve was composed of many intricate reversals in the trend (Potts, 1998a). This indicates a two-to-three-fold increase in the amplitude of fluctuation over the span of hominin evolution (ibid.). Ocean cores also record the presence of wind-blown continental dust. This data indicates that, over time, longer and more extreme cycles of climatic oscillation were added to the preexisting variability, especially at 2.8 my, 1.7 my, and 1.0 my (de Menocal and Bloemendal, 1995).

This pattern is confirmed by the complex history of rainfall, temperature and vegetation on land (Potts, 1998a). Pollen sequences in Africa between 3 and 1 my ago document strong modifications and reorganization in the plant communities through time (Bonnefille, 1995). This included more humid conditions before 3.2 my ago in East Africa, colder conditions in the highlands and drier conditions in the lowlands at 2.35 my ago, and greatest aridity at 1.8 my ago (Bonnefille, 1995).

A spectrum of cycles is responsible for altering earth's climate, from seasonal variations in sunlight, rainfall and temperature to arid-wet oscillations corresponding to variation in the earth's orbit, to glacial cycles which occur every 100 ky (Potts, 1998a). With longer periodicities, the scale of environmental impact becomes more severe: while seasonal variation effects many organisms, entire landscapes may be reconstituted by arid-wet oscillations timed with the 19 ky and 23 ky precesssional variations in Earth's orbit (ibid.). While the idea of cycles might imply predictability, these cycles interact so that the impact of any given environmental variable may change considerably between one cycle and the next (Potts, 1998a). Such variation certainly influenced African mammals. After 2 my ago fluctuation in biomass in the Sterkfontein valley occurred within intervals of less than 100,000 years (Thackeray and Reynolds, 1997). Overall the environmental evidence suggests that large-scale environmental fluctuations were characteristic of the period of human evolution.

The evidence for unusual levels of short-term fluctuation is more equivocal. The Greenland ice cores produced evidence for extremely rapid oscillations during certain time intervals: during these intervals, fluctuation between interglacial warmth and severe cold take place over a century or even a decade (Broecker and Denton, 1990). This has led a number of authors to argue that such short-term variation was crucial in hominin evolution (Boyd and Richerson, 2000, Calvin, 2002). However evidence of such abrupt variation is absent from the Antarctic ice cores (Jouzel *et al.*, 1993), suggesting that these events may be largely a northern-hemisphere phenomenon (Potts, 1998b).

Hominin behavioural flexibility and climate

According to model 1, increased climatic variability selected for increasing behavioural flexibility in the hominin lineage. Potts (1998a) describes a number of key features that could be identified as variability selected adaptations, including the locomotor structure of early australopithecines, brain encephalization (and related cognitive sophistication) in Pleistocene *Homo*, and the complex social mechanisms that emerged in certain late Pleistocene hominins, particularly *H.sapiens*. There does seem to be some correspondence between these periods of increased climatic oscillation and the characteristics highlighted by Potts (1998a). For example, tool use appears in the archaeological record by 2.5 my ago; there is an increase in habitat breadth and the range of land use in early African *H.erectus* after 1.7 my ago; and the most prodigious pace of brain encephalization occurred between 0.6-0.15 my ago (Potts, 1998a). These adaptations occur after longer and more extreme cycles of climatic oscillation were added to the pre-existing variability over time at 2.8, 1.7 and 1.0 my ago (de Menocal and Bloemendal, 1995).

However the relationship between global climate and local environmental change is complex. Hominins would have experienced global scale climatic change or variability only as it influenced local climate and hence local habitats in terms of constitution or distribution. Thus the temporal association between changes in the climatic and the fossil or archaeological record are unlikely to be instantaneous. More direct evidence of the environmental variability experienced by hominins could be taken from the sites themselves.

The existing Pliocene fossil sites suggest that the australopithecines encountered a variable series of environments, from closed (e.g. at Sterkfontein and Aramis) to relatively open habitats (mosaic at Lothagam, Kanapoi and Tugen Hills in Kenya, savannah at Laetoli) (Potts, 1998b). This increased habitat diversity ties in with the appearance of the first VS adaptation, a locomotor structure that allowed a wide repertory of movement (Potts, 1998a).

Between 2.3-1.7 my ago the Turkana basin underwent a series of major paleogeographical changes (Feibel *et al.*, 1991). During the same time period, hominin toolmakers considerably altered their landscape interactions: this is evident in the archaeological record from the growing diversity of paleoenvironments in which stone tools were discovered, particularly after 1.7 my ago (Rogers *et al.*, 1994). A similar pattern is documented in the Olduvai Gorge. The sediments of Bed I and lower Bed II are considered to span the period between 1.9 and 1.7 my ago. During this time there were impressive shifts in climate and vegetation (Bonnefille, 1995, Cerling and Hay, 1986, Hay, 1976, Sikes, 1994). Stone tool sites occur throughout the sequence, suggesting that Oldowan toolmakers had the means to adjust to the changes they faced (Potts, 1998b). In addition, the toolmakers of middle Bed II expanded their range to include depositional environments both within and beyond the immediate lake margin area, and stone sources outside the lake margin were also newly exploited (Potts, 1998b).

Trends in other lineages

The predictions of model 1 should apply to other groups of species as well as the hominins. There is fossil evidence for the effects of climatic fluctuation on lineage evolution.

Potts (1998a) compares extinct large herbivores of southern Kenya and closely related surviving taxa. In the horse, baboon, hippo, elephant and warthog genera there exist species with characteristics which enhance flexibility that are related to highly specialised extinct taxa. For instance, *Papio anubis*, characterised by an omnivorous diet and variable social group size, was related to *Theropithecus oswaldi*, characterised by a very large body size and dietary specialization. All of these specialists last appeared in the record *c.* 600-700 ka, coinciding with extreme mid-Pleistocene fluctuation. This confirms that during periods of extreme environmental fluctuation, within a lineage species with greater adaptive flexibility emerge and are more successful than more specialised species.

We would also expect to find that species with a greater capacity for flexible responses to environmental change would be less sensitive to environmental fluctuation. The genus *Theropithecus* is characterised by complex cheek teeth, presumably related to dietary specialisation on grass blades, seeds and tubers (Fleagle, 1999). The only living representative of this genus is the gelada (*Theropithecus gelada*). Fossil remains of *Theropithecus* are known from the early Pliocene, and they extend throughout most of the Plio-Pleistocene. From an ecological model based on the living gelada, Dunbar (1992, 1993) has suggested that these large lowland grazers must have been dependent on standing water and especially vulnerable to local extinction through climatic fluctuation. As predicted, the relation between first and last appearance dates of primate and hominin species suggests that *Theropithecus* was more sensitive to climatic variability than hominins (Foley, 1993).

A lower sensitivity to environmental fluctuation would also affect the distribution of hominin species. I would expect to find that at some point their regional signature diverged from that of the majority of other mammals. Bromage and Schrenk (1995) have analysed major changes in the distribution of early hominins and have documented that such shifts occurred together with those of other mammals in consistent relation to physical changes in their environment. A number of studies (Turner and Wood, 1993, Bromage and Schrenk, 1995, Foley, 1999) have suggested that either two or three dispersal events took place and that in general the early hominins followed the continental dispersal patterns of other large bodied mammals. A further study comparing fossil distribution with hominin phylogeny suggested that one or more hominin species may have dispersed in a direction opposite to that of other mammals (Strait and Wood, 1999). If these species (*H.habilis* and *P.robustus*) departed from continental patterns of mammalian dispersal, this suggests that one or both of them possessed behavioural or anatomical adaptations that allowed them to do so (Strait and Wood, 1999). However these divergences are based on questions of dating and classification which have not been completely resolved.

Hominin evolution

If highly variable environments select for increasing cognitive complexity, we might expect to find that hominin brain expansion was part of a general trend. There has been a general increase in average encephalization (brain size relative to body size) among mammals during the Cenozoic (Boyd and Richerson, 2000). The diversity of brain size also increases, as mammals evolve under strong selective pressure to minimize brain size, and those that can cope effectively with climatic deterioration by range changes or non-cognitive adaptations do so (ibid.). This implies that a general selective process (such as variability selection) may have been operating.

According to model 1, very effective niche construction in hominins would affect the rate of evolution. Laland *et al* (2000) make a number of predictions for hominin evolution, assuming that the more technologically advanced a culture, the greater its capacity for counteractive niche construction. The first is that, considering Vrba's (1985b) hypothesis of turnover pulses we would expect more technologically advanced hominins to exhibit less of a response to fluctuating climates than less technologically advanced hominins. Secondly, more technologically advanced hominins would show less adherence to ecological rules such as Bergmann and Allen's rules (Gaston *et al.*, 1998a), which suggest respectively that populations in warmer climates will be smaller bodied and have larger extremities than those in cooler climates (Laland *et al.*, 2000). Reversing the inference, they argue that the greater the phenotypic response to environmental change by hominins (such as robustness), the more restricted their capacity for niche construction must have been (ibid.). For example, the less technologically advanced Neanderthals would adhere more strongly to these rules than the more technologically advanced moderns (ibid.). The greater robusticity of the Neanderthals in Europe is consistent with this prediction.

A comparison of first and last appearance dates with climatic trends indicates that the earlier part of human evolution is more sensitive than the later (Foley, 1993). This supports the suggestion that hominin responses to environmental change in the *Homo* lineage would increasingly have been behavioural rather than genetic.

Brain size increases exponentially during human evolution (from approximately 400 to 500 cc estimated for the australopithecines to 1,400 cc for modern humans). There is archaeological evidence for a trend towards increasing social learning during human evolution. Initial innovation and social learning are implicated in hominin technologies, as shown by the archaeological record. Changes in stone tool technology over time suggest an increasing investment in the acquisition of craft skills both by learners and teachers during human evolution (Pelegrin, 1991, Pelegrin, 1993, Shennan and Steele, 1999). These patterns are consistent with a prediction of feedback processes contributing to a lineage trend of increasing brain size or social learning.

Summary of fossil trends

There is a certain amount of support for models 1.1 and 1.2 in the fossil record. There is evidence that global environmental fluctuation was particularly strong in the course of human evolution, and that hominin populations experienced environmental fluctuation at a local level. In addition, there is some evidence that increasing behavioural flexibility in hominins proceeded in synchrony with increasing climatic variability. In other lineages, species with greater adaptive flexibility emerge during periods of high climatic fluctuation, replacing less flexible species, and such species are less susceptible to climate change. There is a general trend of increasing brain size, and brain size and the capacity for social learning increased over time in the hominin lineage. Hominin species appear to have been less influenced by climatic change over time, and we can predict the phylogenetic response to environmental pressures in hominin species based on their capacity for niche construction. In addition, as described at the beginning of this Chapter, early hominin geographic range size increased through time. This support for models 1.1 and 1.2 would be strengthened by increased evidence that similar processes take place in extant species.

Life history and dietary breadth

In an influential analysis of the feeding ecology of the living macaques, Richard *et al* (1989) proposed an ecological division between weed and non-weed species. This division was based on the differing ability of macaques to tolerate and even prosper in close association with human settlements (ibid.). They argue that 'weed' macaques possess characteristics that enable them to survive as human camp followers: these include such attributes as curiosity, behavioural adaptability, an aggressive and gregarious temperament, and speed and agility on the ground (ibid.). The 'weed' macaques are also relatively wide ranging.

'Weed' plants are characterised by rapid growth, early, frequent and rapid reproduction, short lifespan and adaptations for short- and long-distance dispersal. This can be linked to a general trend that also applies to animals. The suite of life history characteristics shown by a species can be characterised as *K*- or *r*-selected (MacArthur and Wilson, 1967, Pianka, 1970). Unpredictable environments favour individuals that produce large numbers of offspring early in life and have efficient dispersal mechanisms: this is known as *r*-selection. Unpredictable environments are often patchy and therefore many *r*-selected species have efficient ways of scattering their young over a large area. By contrast stable non-seasonal environments are predictable and therefore favour individuals that devote a lot of energy to a few offspring at a time. Rather than dispersing, *K*-selected species invest in behaviour to buffer the effects of environmental pressure (MacArthur and Wilson, 1967, Pianka, 1970).

This pattern is complicated by the effect of age specific patterns of mortality in determining life history traits (Pianka, 1970). Larger organisms are better buffered from changes in their physical environment. Any organism with a generation time longer than a year must be adapted to cope with the full range of physical and biotic conditions which prevail at a particular location. Because longer lived, larger organisms are better-buffered from environmental vicissitudes, their population sizes do not vary as much as those of smaller, short-lived organisms. Pianka (1970) suggests that the attainment of a generation time exceeding a year may well be a threshold event in the evolutionary history of a population, marked by a drastic shift from *r*- to *K*-selection. Large bodied, long-lived primates clearly fit this pattern. Thus, *r*- and *K*-selection theory cannot be used to predict a general relationship between climatic variability, life history parameters and range expansion in the primates.

Further understanding of the role of life history in primate environmental tolerance and dispersal can be gained from particular examples. Evidence that some macaques have higher reproductive rates than others is sparse and equivocal (Richard *et al*., 1989). In addition, the wide geographic ranges of macaque 'weed' species are specifically linked to modern human distribution. However, there is fossil evidence that niche breadth and short life history parameters are connected to large geographic range size in the primates. Jablonski *et al* (2000) have conducted a study of the patterns of spatial responses by catarrhine primates to changes in the distribution of the tropical zone during the Pleistocene of China. The spatial responses among these primates to environmental changes indicate that life history parameters and dietary selectivity are strong predictors of the type and magnitude of response to environmental change. Those animals (macaques, langurs) with shorter gestation times, shorter weaning periods, shorter interbirth intervals, higher intrinsic rates of increase of population, and abilities to survive on a wider range of vegetation in seasonal habitats were less adversely affected (Jablonski *et al.*, 2000).

This example suggests that a combination of dietary niche breadth and relatively rapid life history parameters might increase environmental tolerance and the probability of successful range expansion in primate species. As discussed in Chapter 2, primate species with a broader dietary niche and habitat tolerance tend to have larger geographic ranges (Eeley and Foley, 1999). The ability of primate species to switch foods in response to shortages of preferred items would allow them to meet their energetic requirements in times of environmental change. Food is an important determinant of fertility and mortality rates (Richard, 1985, Chapter 4): populations of species with a lower dietary niche breadth in similar circumstances might become stretched beyond their energetic limits and reduced to densities too low to sustain successful reproduction (Jablonski *et al.*, 2000, p.21).

Changes in average fertility, migration rates and mortality all contribute to changing rates of population growth (Richard, 1985). Species with an earlier age for onset of reproduction, shorter gestation time, shorter weaning period, and short interbirth interval have a higher intrinsic rate of increase of population (Jablonski *et al.*, 2000). In addition, cold winters in a seasonal environment have a strong influence on the survival of infants. Such species can time their breeding and birth schedules so that infants are born in time to take advantage of mid-spring weather conditions and can be semi-independent by the beginning of next winter (Jablonski *et al.*, 2000, p.20, Richard, 1985, p.269). These life history characteristics would reduce the chances of extinction: Purvis *et al* (2000) have demonstrated that extinction risk correlates with body size in primates. This result may reflect the fact that body mass is a better surrogate for reproductive rate than reproductive variables (Purvis *et al.*, 2000).

This pattern would to an extent contradict the former hypotheses. Species with higher levels of behavioural flexibility and social learning tend to have larger relative brain size (Reader and Laland, 1999). There is a close correlation between body mass, brain size and many life history variables including gestation length, lactation length, age at first reproduction and life span (Key, 2000, Smith and Tompkins, 1995). Differences in brain size might be expected to shape patterns in life history parameters. Brain tissue is metabolically expensive, so the cost to the mother of growing large-brained infants is expected to be high. However, a model of life history variation based on simple covariation is too simple to explain the patterns in primates (Key, 2000). In general, the species with the highest levels of behavioural flexibility or social learning are unlikely to have the most rapid reproductive rates. However it is possible that a wide ranging primate strategy might also involve intermediate levels of behavioural flexibility.

There is some indication that wide-ranging primates may be a good analogy for hominin distribution, at least at early stages. The geographic range of early hominins in Africa corresponded to that of Plio-Pleistocene cercopithecoid monkeys (Benefit, 1999). Even outside Africa, hominins seem to be ecologically associated with catarrhines in the early Pleistocene (Jablonski *et al.*, 2000). It has been suggested that *H.erectus* was able to expand out of Africa as a weed species, equivalent to *Macaca* (Cachel and Harris, 1997); however this expansion has also been compared to movement of prey and carnivore species (Turner, 1992, Arribas and Palmqvist, 1999). This comparison is complicated by evidence that the spread of macaques to relatively high latitudes and altitudes may have depended on the inadvertent preparation of the environment by people (Richard *et al.*, 1989). Anton *et al* (2002) argue that *H.erectus* was able to disperse with greater efficiency than monkeys have in Pliocene to recent times. However this analysis of dispersal rate depends heavily on early dates from particular sites, notably Mojokerto and Sangiran in Java. As discussed at the beginning of the Chapter, the earliest dates from Java are very controversial.

However hominin species appear to have matured slowly, and been relatively long-lived. The great apes have the longest lifespans and period of development of the non-human primates, while modern humans are remarkably long-lived. The evidence from hominin skeletal material suggests that the australopithecines had similar life histories to chimpanzees, and that longevity and slow maturation became more extreme with time (Smith and Tompkins, 1995). The first shift towards more human-like life histories occurred in *Homo*, probably *H.erectus* (Smith and Tompkins, 1995, O'Connell *et al.*, 1999). Australopithecine brains were about the same size as those of modern chimpanzees (400-500 cc) suggesting similar adult mortality rates and, based on this index, lifespans of about 50 years. Dental eruption schedules suggest that the australopithecines matured at about age 10, as do modern chimpanzees, while *H.erectus* reached that threshold at about age 15 (O'Connell *et al.*, 1999). Adult body size is also an index of age at maturity. Estimates of fossil hominin body size (McHenry, 1994) suggest that the australopithecines and *H.habilis* were within the range of modern chimpanzees. The origin of *H.erectus* marked a dramatic increase in body size, particularly in the female (McHenry, 1994), and this was probably accompanied by a lengthened period of maturation.

While early hominin species had decidedly ape-like life history parameters, they also had some unique features. It has been suggested that *H.erectus* may have been characterised by an extended post-reproductive life, allowing 'grandmothers' to look after and perhaps most crucially to provision juveniles who were not their own (O'Connell *et al.*, 1999). This would both enhance the survivorship of youngsters they helped and allow the mothers of those offspring to begin a new pregnancy sooner (ibid.). One of the effects of this pattern of behaviour would be to increase the reproductive rate. Interestingly, there is an association between growth rates and social interaction in non-human primates. Species such as colobines and callitrichids in which infants are cared for by individuals other than their mother show higher growth rates than species with only maternal care (Fleagle, 1999).

Secondly, in industrialised societies, humans have become 'secondarily *r*-selected' by substantially decreasing interbirth intervals (Wood, 1994). However, reproductive patterns in human hunter-gatherers (Howell, 1979, Blurton-Jones *et al.*, 1992) suggest that this strategy for promoting population growth and dispersal is a relatively recent phenomenon that significantly postdates the initial dispersal of *Homo* from Africa (Anton *et al.*, 2002).

O'Connell *et al* (1999) suggest that 'grandmothering' would have been particularly important in environments where foods that could be gathered by children are rare. Environmental evidence suggests that forested environments shrank between 2.5 and 1.5 my ago (Leonard and Robertson, 2000). Such a change in environment would have reduced the availability of resources that younger children could take on their own (O'Connell *et al.*, 1999). Increased provisioning, possibly focussing on new resources such as underground storage organs, would have allowed hominin populations to occupy a wider range of habitats (ibid.). Thus according to the 'grandmother' hypothesis, a change in hominin life history might be an important factor in range expansion.

Based on the primate pattern, I suggest that the change in life histories may have had important implications for hominin geographical distribution because of its effects on reproductive rate. An extension of caretaking behaviour in *H.erectus* may have given that species the potential to speed up (or slow down) reproductive rate in response to circumstances, and thus cope with environmental constraints and opportunities encountered in range expansion. Reproductive flexibility, in combination with dietary niche breadth, may have had the same effect as relatively rapid life history parameters on primate species geographic ranges. In addition, this behaviour would have provided a buffer for hominin infants and children against the climatic or nutritional effects of winter in a seasonal environment.

Dietary niche

According to Gamble (1993), primate dispersal is limited by behaviour rather than biology and genes:

> "Wildebeest could walk from the Serengeti to Stockholm. New York as well as Nairobi could have packs of Hyenas. Chimpanzees have the brains and the omnivorous tastes to accomplish similar journeys."
>
> (Gamble, 1993, p.95).

In model 1, I described a feedback system in which brain expansion and higher innovation frequencies put species under novel selection pressure, that may be increased by range expansion. I also discussed the possible benefits of a larger brain in increased migration success. However it is also necessary to consider the costs of large brain size. A large brain has relatively high metabolic requirements (Aiello and Wheeler, 1995, Leonard and Robertson, 1996). There are certain dietary demands on large brained animals, tending to require either a high quality diet or long time spent foraging (or both). Thus diet tends to put a constraint on the distribution of large brained species that is absent in small brained species. Biomass and diversity of resources are related to latitude, with tropical habitats the most productive. In order to extend their range into more northern and less productive latitudes, large-brained primates might require an alternative source of high quality nourishment, such as meat. I will discuss the role of diet, particularly meat eating, in hominin range expansion in this section.

Dietary quality and the demands of a larger brain may explain limitations to the geographic ranges of the great apes. The large brained great apes all live at relatively low latitudes, where high quality and/or abundant nutrition can be obtained year round from abundant fruits, nuts and foliage. As described in Chapter 1, primates in general display a strong relationship between range size and dietary breadth (Eeley and Foley, 1999). The main food of chimpanzees and orangutans is fruit. Gorillas feed mostly on leaves, shoots, and stems; however despite eating more foods that are difficult to digest, they do not have the specialisation of the gut necessary to survive on a diet of browse[2], the main sort of vegetable food available in winter in seasonal

2 . Young shoots, twigs and bark of trees, and the woody parts of shrubs and herbs (Richard, 1985).

habitats (Richard, 1985). Richard (1985) suggests that the energy turnover associated with such a diet is too low to meet the needs of animals with the highest brain size to body size ratio of any living mammal except ourselves. Given the lack of physiological or behavioural adaptations to cope with seasonal food shortages in temperate zones, it seems unlikely that chimpanzees could migrate far beyond their ranges even given strong behavioural reasons to do so. This might lead us to modify Gamble's (1993, p.95) assertion that behaviour alone limits primate ranges.

Diet has been linked to the expansion of the brain (Aiello and Wheeler, 1995). Absolute and relative brain size increases throughout human evolution, with the most prodigious pace of encephalization occurring between 0.6-0.15 my ago, with the emergence of *H.sapiens*. The expensive tissue hypothesis suggests that a high-quality diet permitted a relatively smaller gut and thereby relaxed a metabolic constraint on brain size. Leonard and Robertson (1996) question the dynamics of the hypothesis, suggesting that the relationship with gut size reflects covariation between brain size and dietary quality. Again, this would suggest that major increases in brain size could involve a major change in diet.

Brain expansion is a general trend in the primates, and the relative brain size of early australopithecines was not greatly larger than that of modern chimpanzees. There is evidence that the earliest australopithecines lived in relatively forested environments (White *et al.*, 1994, Wolde-Gabriel *et al.*, 1994). This suggests that the australopithecines may have had similar dietary constraints on their distribution to modern apes. There is an increase in absolute brain size in *H.erectus*: however when these are corrected for a proxy of body mass, values are more similar to the Australopithecines than *H.sapiens* (Wood and Collard, 1999a). This would not seem to support the idea that a major increase in brain size required a dietary shift in this species.

However there is evidence that unlike the early australopithecines this species preferred more open habitats, and occupied a wider range of habitats. This is discussed in detail in Chapter 6. In addition, between 2.5 and 1.5 my ago, there was a marked decline in forested areas throughout eastern and southern Africa. Stable isotope studies suggest that a shift to C_4 dominated environments (open, grassy, heat-adapted vegetation) occurred as late as 1.7 my ago, with little evidence of consistently open savannah until after 1.0 my ago (Cerling *et al.*, 1991, Cerling, 1992, Kingston *et al.*, 1994, Sikes *et al.*, 1999). This change in distribution suggests that some of the constraints acting on distribution in other large-brained primates had been removed in *H.erectus*. This is best explained by a shift in dietary niche in this species. Hominin diets are discussed in more detail in Chapter 5, however I will summarise the evidence for meat eating in *H.erectus* here. The teeth and jaws of this species are more similar to those of *H.sapiens* than to the australopithecines (Wood and Collard, 1999a). Craniofacial and dental patterns suggest that early *Homo* ate food requiring less chewing than the australopithecines (Teaford and Ungar, 2000). Changes in the postcranial skeleton suggest a more efficient adaptation to rapid locomotion, which could have been useful in hunting (Shipman and Walker, 1989).

There is evidence that one of the functions of the earliest stone tools was to process meat: bones with cutmarks and percussion marks on them have been recovered from Bouri, Middle Awash dated to 2.5 my ago (de Heinzelin *et al.*, 1999). Early Pleistocene archaeological sites tend to be larger in size, and are often associated with faunal assemblages containing the remains of one or more large animals, mainly ungulates. The degree of association between the bones and stones at most of these sites has not yet been satisfactorily explained. However at a number of sites, analysis of the bone damage presents a convincing case that hominin species had primary access to large quantities of meat (Dominguez-Rodrigo, 2002, Dominguez-Rodrigo *et al.*, 2002). Thus the archaeological evidence suggests that meat was a part of the hominin diet from an early period. The appearance of early African *H.erectus* is associated with a change in technology in the archaeological record, characterised by the production of large bifaces. Experiments suggest that while these tools could potentially be used for a range of purposes, their morphology is most consistent with butchery tools for large mammals (Schick and Toth, 1993). Microwear analysis provides results consistent with this interpretation (Keeley, 1977).

I have suggested that increased dietary quality may have removed the constraints to distribution acting on early hominin species. In addition, meat eating can be linked to a number of spatial patterns, as discussed in Chapter 2. The range of variation and frequency distribution of geographic range size varies between orders (as well as species), and some of this variation may be due to differences in trophic level. In addition, dietary quality in general and carnivory in particular has an effect on the spatial distribution and abundance of populations. Carnivores tend to have a lower population density and larger home range. Based on these ecological patterns, Shipman and Walker (1989) have argued that range expansion in *H.erectus* may have been a response to a dietary change to increased carnivory and associated lowering of population density.

The spatial distribution of *Homo* has been compared to that of fossil carnivores in a number of areas and periods

(Turner, 1992, Arribas and Palmqvist, 1999, Jablonski *et al.*, 2000, Stiner, 2002). Jablonski *et al.* (2000) suggest that, as *Homo* fell out of ecological phase with its close relatives, it fell closer into phase with members of the scavenger and predator communities throughout the Pleistocene. Successful colonization of Eurasia by African hominins may have hinged on their ability to obtain flesh, since access to plant foods is intensely seasonal throughout much of temperate and periglacial Eurasia (Stiner, 2002).

As discussed above, environmental evidence suggests that an expansion of more open, grassy, heat adapted environments occurred in Africa between 1.7-1.0 my ago. This transition from woodland to open savannah environments resulted in changes in both the abundance and distribution of food resources. An analysis of ungulate biomass over time in Southern Africa shows a gradual increase in overall biomass and variability between 3 my ago and the present day (Thackeray and Reynolds, 1997). According to Leonard & Robertson (2000), a key aspect of this change for hominins would have been the associated change in the energetic structure of these ecosystems, with plant productivity declining and animal foods becoming a much more attractive resource. These environmental changes would have relaxed the environmental constraints on carnivore distribution in Africa. A change in dietary niche during this period would have been beneficial in terms of dietary resources and would have allowed substantial range expansion.

A number of authors have suggested that hominin expansion out of Africa corresponded with the opening of a new, carnivorous dietary niche (Turner, 1992, Arribas and Palmqvist, 1999). As meat eating species, hominins would have formed part of the carnivore guild or group of carnivorous species (Turner, 1992). The resources available for hominins in an area would thus have been decided both by the prey species available and the structure of the carnivore guild (Turner, 1992). This suggests that the major range expansion of *H.erectus* depended on a change in dietary niche, and also in the environmental factors limiting carnivore distribution.

Conclusion

I have described three models explaining the evolution of hominin geographic ranges. Each of these models is based on processes that are not exclusive to human evolution, and can therefore be evaluated and refined using comparative data from other mammals. Models 1 and 2 focus on aspects of cognitive complexity, social communication and life history parameters. In these characteristics, hominins share similar features with the non-human primates. Therefore modern primates provide a suitable dataset for analysis of these models: such an analysis will be presented in Chapter 4. Model 3 focuses on aspects of the ecological niche, notably meat eating. While some primates do eat meat, it makes up a relatively minor part of their diet. Because of the similarity of their adaptive niche to humans in this respect, carnivores make a better source of analogy in order to test this model. In Chapter 5, I will describe the results of an analysis of modern African carnivore distribution compared to that of primates and ungulates.

CHAPTER 4

PRIMATE BIOGEOGRAPHY ANALYSIS

Abstract

A comparative study of primate distribution in relation to physical and behavioural characteristics and environmental factors was conducted using GIS and statistical techniques. Geographic ranges of primates were digitised from maps and their composition in terms of climatic variability analysed using GIS statistical capabilities. Relevant variables (life history, body mass, brain size) were synthesised from the literature and entered into a comprehensive database. A database of innovation frequencies was obtained from Reader (2000). Regression techniques were used to assess the relative predictive value of variables with respect to range size and distribution. This analysis of adaptations took account of the effects of phylogenetic relationships.

The principal findings of this study are that there is no strong correlation between innovation frequency and geographic range size. This conclusion is supported by analyses conducted using different evolutionary assumptions, and alternative measures of behavioural flexibility based on brain size. This result contradicts the hypothesis that more behaviourally flexible species can counter environmental change or invade new habitats more successfully than other species. In addition, there is a significant relationship between behavioural flexibility and spatial but not temporal climatic variability. These results bring into question the hypotheses that more behaviourally flexible species can tolerate greater climatic variability; that behavioural flexibility is a response to climatic change; or that climatic variability has selected for a propensity to behavioural flexibility or social learning. It may be that other time scales have a stronger effect on primates. There is a conclusive lack of correlation with social learning frequencies and both range size and climatic variability. This provides strong evidence against the hypothesis that a capacity for social learning would be useful in range expansion and weathering climatic change. At the same time, linear regression analysis of life history data does not support the alternative hypothesis that a combination of relatively rapid life history parameters and dietary niche breadth form a wide-ranging primate strategy. It is difficult to find a suitable evolutionary model for a number of the variables, including geographic range size. However consistency across a number of alternative statistical methods and forms of measurement lends support to these results.

Introduction

Predictions for primate distribution

In the previous Chapter, I presented three models intended to explain the evolution of hominin geographic ranges. Models 1 and 2 focussed on aspects of behaviour for which non-human primates provide a good analogy. Based on these models, I can make a number of predictions for primate distribution.

I wish to clarify the relationship between niche breadth and behavioural flexibility. Previous analyses of primate distribution have focussed on niche breadth (Eeley, 1994, Eeley and Foley, 1999, Eeley and Lawes, 1999). More flexible behaviour might allow species to exploit a wider range of habitats and types of food. Many instances of primate innovation or tool use occur in the foraging domain (Reader, 2000), and this leads to a possible criticism of observation frequencies as a measurement of behavioural flexibility. I will test the prediction that more behaviourally flexible species will use more different habitats and food types than less flexible species.

Based on model 1.1, I predict that measurements of behavioural flexibility will covary with geographic range size and the environmental variability within the range in primate species. I focus on climate as an aspect of environmental variability, because there is evidence that climate influences primate distribution, and because climatic variability is often focussed on in discussions of human evolution (Boyd and Richerson, 2000, Calvin, 2002, Potts, 1998a). This is discussed in more detail below. A positive relationship between species geographic range size and behavioural flexibility would support the hypothesis that more behaviourally flexible species can counter environmental change or invade new habitats more successfully than other species. Positive relationships with measurements of climatic variability would be consistent with the hypotheses that more behaviourally flexible species can tolerate greater climatic variability; that behavioural flexibility is a response to climatic change; or that climatic variability has selected for a propensity to behavioural flexibility or social learning.

Social learning frequencies might be expected to affect the environmental tolerance and hence range size of primate species as a part of general behavioural flexibility. In addition, in model 1.2 I argued that social learning is critical in accurately assessing the risks involved in range expansion. I predict that social learning will covary with geographic range size and environmental variability, independent of behavioural flexibility and brain size. A positive relationship would be consistent with the hypothesis that species with a greater capacity for social learning can tolerate greater environmental variability, and also assess risk in migration more effectively, leading to a larger geographic range size.

In addition, it could be argued that behavioural flexibility would have a stronger effect on distribution at a local level. For instance, behavioural flexibility may have more effect on improved abilities to counter environmental change through 'counteractive' niche construction (Laland *et al.*, 2000). Based on this suggestion, I predict that the relationship between home range size and behavioural flexibility may be stronger than that with geographic range size.

Human activities may have destroyed the natural relationship between behavioural flexibility and geographic range size, for instance by selective hunting of larger primates. It has been suggested that primate species might innovate more in response to stress from their environments: the 'necessity hypothesis' (Reader and Laland, 1999). This would suggest that more innovative species might also be those under threat. If this were so, it might complicate any relationship between behavioural flexibility and geographic range size. Both of these hypotheses can be tested by analysing covariation between threat status and geographic range size.

According to model 2, wide-ranging primates are characterised by niche breadth and relatively rapid life history characteristics. Dietary niche covaries with geographic range size in the primates (Eeley, 1994, Eeley and Foley, 1999, Eeley and Lawes, 1999). My alternative prediction therefore is that life history variables will also co-vary with geographic range size. A strong negative relationship would suggest that this combination of characteristics is indeed important in range expansion or coping with environmental variation in a large range.

Definition of variables

Behavioural flexibility

Behavioural flexibility can be defined as the capacity of a species to engage in novel behaviour. This characteristic is also referred to in the literature as 'behavioural plasticity'. According to Sol *et al.* (2000) behavioural flexibility may take the form of learning, cognition, and/or rapid adjustment to new conditions. However flexibility in behaviour could presumably be reached through several different routes: it is possible that the concept of behavioural flexibility will subsume a number of different processes (Reader and MacDonald, 2003).

Traditionally studies of primates have characterised the social and ecological preferences of species as fixed and adaptive (Lott, 1991). However a number of recent studies have highlighted the existence of intraspecific variation in social and ecological behaviour between species and suggested that this flexibility could be adaptive (Lott, 1991). Within-species variation in social and ecological behaviour has also been identified in primates (Lott, 1991, Beauchamp and Cabana, 1990, Chapman and Chapman, 1990). One of these studies demonstrated that primate populations make extensive shifts in their dietary behaviour, to the extent that would cause a species to be classified as a different type of dietary specialist, or as generalist as opposed to specialist, at different times of the year (Chapman and Chapman, 1990).

How can we measure behavioural flexibility so that it can be compared across species? There are serious problems with comparative experimental cognitive tests. Species differ widely in their reliance on different sensory modalities, in their neophobia, in their response to humans and in many other characteristics that make construction of a fair experimental test problematic (Reader, 2000). This makes the results hard to interpret (ibid.). Behavioural innovation has been suggested as an alternative measure of behavioural plasticity (Lefebvre *et al.*, 1998). It may be preferable to judge animals not on their relative performance in anthropocentric tests, but on opportunistic departures from their species norm (Lefebvre *et al.*, 1998).

Reader and Laland have conducted a comparative study of innovation, social learning and tool use in nonhuman primates by collecting reports of such behaviour from the primate and social learning literature (Reader, 2000, Reader and Laland, 2002). This method provides a test fair to all species and data on large numbers of species by measuring the tendency to discover or learn novel solutions to environmental or social problems relevant to the animal. Observation frequencies can be corrected for research effort in terms of the number of articles searched for each species (Reader, 2000, Reader and Laland, 2002).

There is considerable variation in innovation, social learning and tool use frequencies among primate species even when correcting for research effort (Reader, 2000). Lefebvre *et al.* (1998) estimated innovation frequencies in North American and British birds, and found that innovation frequency differed between orders in a similar fashion in both areas. Relative innovation frequencies (corrected for species number per order) were related to relative forebrain size (Lefebvre *et al.*, 1998). Reader and Laland (2002) have demonstrated that individual variance in the propensity to innovate and socially learn is widespread amongst primate species, and that this variation is related to the relative and absolute size of the executive brain. This co-variation suggests that innovation, tool use and social learning frequencies are adaptive and provide a suitable quantitative measure of behavioural plasticity for the following analyses.

In the studies described above a broad definition of innovation was used: innovation is the discovery of novel information or behaviour patterns, or the performance of established behaviour patterns in a novel context (Reader, 2000). Examples of keywords used to define innovation included 'never seen before', 'invention', 'novel', 'new', 'innovation', and 'opportunistic' (Reader, 2000, Reader and Laland, 2002). Many reported innovations are relatively simple behaviour patterns such as eating novel foods. Examples of innovation from the database include *Pan troglodytes* eating mangoes, *Cercopithecus mitis* eating a flying squirrel, *Alouatta caraya* eating marsh plant roots, and *Saimiri sciureus* eating a bat (Reader and MacDonald, 2003). More complex examples include lemurs (*Lemur catta*) immersing their tails in water and drinking from the wet tail, and 'stepping-stick' use in chimpanzees: sticks were used to walk on, as protection against a spiny tree that bears edible fruit and flowers (Reader and MacDonald, 2003).

Social learning

Social learning commonly refers to the social transfer of information and skill among individuals (Box and Gibson, 1999, abstract). Reader (2000) defines social learning as 'learning from another animal or its products'. Some examples of keywords used to define social learning included 'tradition', 'cultural transmission', 'imitation', 'observational learning', and 'socially learned'. Examples of behaviour classified as social learning include differences between matrilines in louse egg handling techniques in *Macaca fuscata* (Reader, 2000).

Brain size

The advantage of a large brain is always taken to be the benefits of intelligence, and the extreme example of our own brain size can illustrate this point. The human brain is uniquely large among primates, but this does not come without severe costs; a large brain is energetically expensive, birth may be difficult and even dangerous, and the human infant is helpless for a comparatively long period of time. The high costs of brain expansion indicate that such a large increase in brain size must have been driven by a strong selective advantage. The advantage of a larger brain is likely to have been a factor of an increase in intelligence. Corroborative evidence for this approach comes from the correlation between executive brain measures and innovation frequency, social learning and tool use (Reader, 2000, Reader and Laland, 2002).

Larger animals generally have larger brains, which is not surprising given that, for most mammals, much of the brain is taken up with sensory and motor processes which might be expected to need brain tissue in proportion to their size (Byrne, 1995). In addition, as the absolute size of living things changes, so the relative proportions of their parts are found to change (ibid.). These regular trends complicate the use of brain size as a measure of intelligence.

A technique that takes account of the way bodily form changes with size is called allometric scaling. A double logarithmic plot of brain size against body size for a given group of animals forces the points onto a straight line, demonstrating that the two masses are related by some form of power relationship. The value of this measure of relative brain size is taken from the distance of a given species' brain size above or below the predicted size given by the line of regression (Byrne, 1995). However the assumption that it is possible for an animal to have less than enough processing power to deal with body size requirements is problematic. Secondly, an underlying assumption of this method is that the extra volume over that expected is best viewed as a proportion of the total volume (Byrne, 1995).

The choice of brain size measurement used as an indicator of intelligence depends on 1) competing theories of brain evolution and 2) ideas about brain structure and role of different parts. There are two main ways of interpreting the brain. The first is as an 'on board computer' (Dawkins, 1976). The comparison with computational machines suggests that the brain is ultimately limited in power by the number of its elements, that is that the absolute number of neurones is more relevant than the number relative to body size (Byrne, 1995). The second interpretation is that animal brains function by making reflex responses to stimuli. In this model the number of input-output connections, or sensory and motor neurones, would determine how big the system that needs to handle them would have to be. In this case brain tissue size relative to body size will show the extent to which processing can be more flexible, justifying traditional allometry (Byrne, 1995).

These different interpretations are not necessarily mutually exclusive: it is possible that both are true of different systems within the brain (Byrne, 1995). If the brain is a device in which different parts serve very different functions, those parts used for bodily functions should increase in a regular way with body size while those parts used for computation should not (ibid.). Different structures within the brain may evolve at different rates (Barton and Harvey, 2000) suggesting that the areas of the brain of interest should be examined instead of the total brain size. This is the logic behind a third method of measuring brain size as indicative of intelligence, by comparing the relative sizes of structures within the brain. The structures measured are those most closely connected with intellectual functions such as the neocortex.

There are practical constraints on the choice of brain size measurements for this analysis. The ultimate aim of the analysis is to identify patterns in primates that can be compared with hominin species. While measurements of many structural elements of the brain are available for modern primate species, hominin fossils provide only absolute brain volumes. The measurements used in this analysis have been selected in consideration of this practical constraint, and also to represent the three main theories relating brain size to intelligence. All analyses have been carried out using absolute brain size, brain size relative to body size and the neocortex ratio.

Life history

Primates show considerable diversity in many aspects of life history. Purvis *et al.* (2000) argue that gestation length is probably our most reliable indicator of a species' position on the fast-slow continuum of life-history strategies. Factors related to reproduction are particularly highly correlated with each other (Fleagle, 1999), so this measurement should be quite representative of other aspects of reproductive scheduling. I used gestation length and maximum recorded lifespan as possible predictors in these analyses. As species' life history is strongly related to its body mass, female body mass was included as a control variable. In addition, body mass may be a better surrogate for reproductive rate than reproductive variables in the primates (Purvis *et al.*, 2000).

Environmental variability

An organism's environment is complex, involving a set of interrelated variables (Lee, 1991). These include the type, distribution and availability of food resources; the presence of refuges and places of shelter from heat or cold stress or potential predators; the skill, hunger and number of predators; the presence of competing species; and the relative skills of interspecific competitors within the community (Lee, 1991). In addition, animals alter the world around them (Odling-Smee, 1988). All of these variables can improve, deteriorate or change cyclically through time, and have consequences for the survivorship and reproduction of individuals, and thus pose problems of adaptation (ibid.). Innovation may provide benefits only at particular time scales or in response to particular environmental changes; thus the choice of environmental variability measure is important (Reader and MacDonald, 2003).

It can be argued that the central issue in considering primate species' tolerance of environmental change lies in the ability to locate, consume and process adequate food (Chivers, 1991). In a study of individual difference in behavioural flexibility, Reader and Laland (1999) found that the largest number of recorded observations of primate innovation, social transmission and tool use were in the context of foraging. Variation in climate can cause changes in food availability. In particular, variation in rainfall can be an important factor in causing changes in resource productivity (Janson and Chapman, 1999). In addition, the seasonal distribution of rain could influence plant communities (Chapman *et al.*, 1999). This might suggest that climate is a relevant proxy variable for aspects of environmental change likely to affect primates. In addition, the studies of biogeography described in Chapter 2 (Cowlishaw and Hacker, 1997, Stevens, 1989) specifically discuss the role of climatic variability in species adaptation and distribution.

As discussed in Chapter 3, the rate of environmental fluctuation has implications for the selective pressures affecting organisms. Thus it is important that the rate of environmental variation used in these analyses is appropriate to the characteristics under discussion. A frequent finding of theoretical models is that the evolution of both asocial and social learning are favoured at intermediate rates of environmental change (Boyd and Richerson, 1988, Sibly, 1999). According to Clark (1991), environmental fluctuations scaled to an animals' developmental cycle and expected lifetime provides the necessary characteristics of environmental variability under which phenotypic plasticity may evolve. In quite constant environments, but also in environments that are alternating more rapidly than an animal can produce and utilise a new phenotype, animals may be expected to develop and behave conservatively (Clark, 1991). Frequent novelty and rapid change may select for another attribute, 'flexibility' (Fagen, 1982). However, variability selection is expected to occur at a much slower rate of change (Potts, 1998a).

The length of a species' life history has important effects on how environmental fluctuation is experienced. Pianka (1970) pointed out that species with a generation time longer than a year encounter the full range of annual conditions. Clark (1991) points out that primates and other relatively long-lived animals will be exposed to change on more time-scales. For example, a primate such as a baboon, or Japanese macaque, experiences within its lifetime the variation in availability of specific foods among repeated wet, dry or cold periods (Clark, 1991). Learning about alternative ways to forage can be the key to survival in a particularly severe season (Clark, 1991). Thus we would expect primate species to respond to environmental change with behavioural flexibility at scales of temporal fluctuation shorter than the generation period for that species; it may be more appropriate to measure environmental change at seasonal or inter-annual scales than longer periods of time.

In this study I have focussed on climate as a part of environmental variability likely to be relevant to primate species. I have used two time scales of climatic variability relevant to primate life histories: seasonal and interannual variation. I also measured spatial variability, my aim being to measure variation across the species range rather than within a particular habitat. It is a valid question as to whether these scales of analysis are appropriate: perhaps animals respond to more fine-grained climatic changes, or to other environmental changes. I believe that the choice of climatic measures and measurement scales is appropriate to investigations of environmental variability, but accept the possibility that other measures at other scales may be informative.

Possible confounding variables

Human interference has strongly influenced the modern distribution of primates, through behaviour such as habitat destruction, hunting, and pest control. The degree of human interference is likely to vary geographically, with population density.

In addition, human interference may have had a selective effect on the distribution of geographic range size in modern primates. We might expect to find that large-brained, innovative species such as the apes are particularly vulnerable to human persecution because they are larger-bodied, less arboreal, diurnal, have longer inter-birth intervals, and rely more on patchily distributed foods. Slow life history correlates with extinction risk (Purvis *et al.*, 2000). Disproportional endangerment of species with a large body size can be a sign of hunting pressure and selective decimation of primate communities: hunting may eliminate larger species, but have a positive effect on the population densities of smaller species (Peres, 1999a). This would suggest that recent human activities could have destroyed a natural relationship between brain size or innovation and geographic range size.

Method and analysis

Overview

Geographic ranges of primate species were digitised from maps. The digital maps were rectified and projected into equal area format. The range composition in terms of land cover and climatic variability was analysed using the capabilities of a GIS. Three separate databases were produced based on the main areas in which primates are found, being South America, Asia, and Africa (excluding Madagascar). Life history, body mass, and brain size variables were synthesised from the literature and entered into a comprehensive database. A database of innovation, social learning and tool use frequencies was obtained from Reader (2000). Regression techniques were used to determine which factors have the greatest predictive value in explaining variation among living primates in the spatial extent of their distribution. This analysis took account of the need for phylogenetically independent contrasts. All analyses were conducted using both raw data and independent contrasts.

Data sources

Life history, body weight and brain size

Female body weight values are from Smith and Jungers (1997). The use of female body weight is based on the assumption that different selective pressures act on the different sexes, with female body mass selected for resource availability (due to the female reproductive role) and male body mass more open to sexual selection. According to this assumption female body mass will be more directly scaled towards optimal resource availability while male body mass is more likely to reflect mating competition. Three different measures of brain size were used (see above for theoretical and methodological considerations): absolute brain weight, relative brain weight controlling for body weight, and neocortex ratio. Relative brain weight was calculated by taking the residuals of a log-log plot of brain weight on female body mass. The use of the ratio of neocortex volume to the volume of the rest of the brain follows Dunbar (1996). Measurements based on volumes or relative volumes of different structural parts of the brain are from Stephan *et al.* (1981), absolute brain weight from Harvey *et al.* (1987). Habitat and dietary niche breadth data are from Eeley and Foley (1999). They collated the number of different habitat types occupied (from a total of 23 possible habitats identified) and the number of different food types used (from a possible total of 10), based on an extensive literature search. Life history data are from Harvey *et al.* (1987). Home range data is from Leonard and Robertson (2000) and Nunn and Barton (2001). Individual home range values were obtained by dividing home range size by group size (Milton and May, 1976). Threat status categories are from (Rowe, 1996), who collated IUCN (World Conservation Union), USESA (United States Endangered Species Act) and CITES (Convention on International Trade in Endangered Species) classifications. The data was reclassified to make it numerical: Lower risk = 1, Rare = 2, Vulnerable = 3, Threatened = 4, Endangered = 5 and Critically Endangered = 6.

Innovation, social learning and tool use

Simon Reader (2000) provided the database of instances of innovation, social learning and tool use in primate species. Reader and Laland (2002) have used counts of instances of innovation cited in the primate literature as a measure of behavioural plasticity. The dataset included counts of the numbers of articles searched for each species in order to correct for research effort (Reader, 2000). Frequencies were corrected for 'research effort' into each species by taking the residuals of a natural log plot of the relevant frequency against the number of journal articles searched, according to Reader's (2000) methodology. These corrections were carried out prior to taking independent contrasts (this assumes that research effort is not phylogenetically biased, which may be questioned – see Reader and MacDonald (2003)). This allows residuals to be taken based upon all the species for which data on innovation frequency and research effort are available.

Geographic ranges

Defining and mapping geographic ranges is inherently problematic. The real units of geographic ranges are the complex spatial and temporal patterns in which individual organisms are dispersed over the earth (Brown *et al.*, 1996). Any maps of the geographic ranges of species necessarily simplify such complex distributions. In addition, geographic ranges change over time, and only the most up to date range maps will show all locations where a species has recently been recorded.

A number of alternative methods have been employed for estimating species range size in the literature. Some studies use measures of latitudinal extent rather than area.

Gaston *et al.* (1998) note that the observation that the magnitudes of the latitudinal and longitudinal axes of species distributions tend to be positively correlated suggests that patterns revealed using areas and latitudinal extents should be broadly similar. However they suggest that measures of latitudinal extent may be rather misleading representations of the distribution of species as they tend to reflect the positions of populations that are marked outliers from the main body of occurrence (Gaston *et al.*, 1998b).

A more significant distinction is between alternative methods of mapping species distribution. A map of the extent of occurrence of a species represents the limits of that species' distribution (Wolfheim, 1983). Such maps represent considerable variation in density and even partial absence as a solid area. Only large areas of absence are represented. More detailed maps attempt to show the actual area of occupancy of a species: for instance, "Dot maps" plot each location where a species has been recorded, allowing greater precision. However such detailed maps are only available for a few species.

The maps of primate geographic ranges used in this analysis were derived from *Primates of the World* (Wolfheim, 1983). Wolfheim (1983) collated literature references and information from field biologists, recorded locations for each primate species, and produced maps based on the sum of location references for each species. The limits of the range were determined from published and unpublished range maps and collecting localities. These maps show primate species' extent of occurrence rather than area of occupancy, and have a low resolution. However I decided this was appropriate to the continental scale of the analysis. In addition, I considered it advantageous to have a single source for a large number of range maps.

Environmental variables

The global surface climate data comes from two datasets: CRU05 0.5° 1961-1990 Mean Monthly Climatology (New *et al.*, 1999); and CRU05 0.5° 1901-1995 Monthly Climate Time-Series (New *et al.*, 2000).

Phylogeny

The primate phylogeny used was a composite tree derived from 112 previously published phylogenies (Purvis, 1995).

I resolved differences in nomenclature in the literature using Corbet and Hill (1991).

GIS database

Range size database

Primate range maps were scanned and digitised as vectors in AutoCAD Map 2 (1997). They were then imported to GRASS version 4.2 (1997) as vectors and converted to raster images. As the source maps were in an unknown projection, the raster images were rectified using a vector map of known projection and co-ordinates, and projected into equal area format using ArcINFO's (1999) PROJECT function. The databases were compiled in GRASS (1997) format to take advantage of GRASS's capacity for querying and analysing raster maps. ERDAS Imagine version 8.4 (1997) was used to derive statistics from the climatic variability maps.

Three databases were produced for each of the main areas of primate distribution, Africa, South America and Asia. The procedure was kept as similar as possible. However some considerations of the data made different approaches desirable for the Asian database. Both the African and South American databases were created in Lambert's Azimuthal Equal Area projection. This projection is made upon a plane tangent to the globe at any point. It preserves areas of individual polygons while simultaneously maintaining a true sense of direction from the centre. This is a useful projection for the continents of Africa and South America because it can project large areas in any part of the world with comparatively little distortion of distance and direction.

However for the wider landmass of Asia, extending in an east west rather than a north south direction, Cylindrical Equal Area provided a more suitable projection. Cylindrical Equal Area projection has several varieties but the most useful was a normal, perspective projection onto a cylinder tangent at the equator. This projection is recommended for narrow areas extending along the central line: shape and scale would be severely distorted further from the central line. In addition, the African and South American maps were rectified into latitude longitude format and then projected using ArcINFO (1999); for the Asian maps one stage of this process was bypassed by rectifying the maps using a previously projected vector coastline map.

Climate variability maps

Each map was based on a thirty-year period (1961-90). My intention was to include the full variation induced by various more short-term climatic cycles such as El Niño. However climatic cycles of longer duration have been identified and the effects of these on the representativeness of the climate dataset have not been taken into account. The use of a thirty-year period to calculate variables such as mean annual rainfall is standard practice in meteorology.

Map calculations were carried out using ERDAS Imagine's (1997) Modeller facility. Maps of the annual mean of daily rainfall and temperature were produced using the CRU05 0.5° 1961-1990 Mean Monthly Climatology dataset (New *et al.*, 1999, New *et al.*, 2000). The yearly mean for 1961-90 was calculated based on monthly means, allowing for leap years. Secondly, the temperature range data from the same dataset was used as an indicator of seasonality. Minimum and maximum yearly temperatures were calculated from monthly values and the difference taken to produce a map of the annual temperature range. Finally, a map of interannual variability was based on variability in mean annual rainfall. The mean for each year from 1961-90 was calculated using data from the CRU05 0.5 Degree 1901-1995 Monthly Climate Time-Series. New maps were then produced based on the standard deviation and coefficient of variation in annual rainfall for each 0.5° cell. I used the coefficient of variation because standard deviations can correlate with the size of the mean. There are more detailed descriptions of the production of these climate maps in the appendix.

Further processing was carried out to calculate values for the climatic variation within each species range. Values of spatial variability tolerated by species were calculated as the coefficient of variation in mean rainfall and temperature values in those raster cells falling within the species range. The seasonality and interannual variability tolerated by species was calculated as mean values within the range. I obtained this data for the African and South American species only, due to time constraints.

Dataset composition

Range sizes are likely to be constrained by continent size, shape and habitat availability, and the historical patterns of environmental variability may have been quite different between different continents. There is thus a case to be made for separate analyses within each continent. However primate species' geographic distributions are clearly biased by phylogeny – for instance, all of the higher primates in South America belong to a different infraorder from those in Africa and Asia. This means that contrasts within the same clade will also tend to be made within the same continent, and very few contrasts will be made between continents, thus controlling for this potentially confounding variable.

Pooling data across different continents could be a problem if there were different relationships on different continents, in which case these relationships would be obscured by analysing all three continents together (Reader and MacDonald, 2003). However the models developed in Chapter 3 do not predict such differences. While the potential for range expansion may be much greater in Africa than say, the islands of south-east Asia, the same relationships are still predicted in each area. That is, more flexible species are predicted to have larger geographic ranges in both localities, even though absolute range sizes may be much smaller in south-east Asia than Africa.

I have also carried out analyses on datasets restricted to continents: the results of these analyses will tell us whether the predicted patterns apply to primates globally, or whether there are differences between the continents.

Analysis of comparative data in evolutionary biology

Introduction

Comparisons among species are frequently used to test hypotheses of how organisms are adapted to their environment. The comparative method is the most general technique for asking questions about common patterns of evolutionary change (Harvey and Pagel, 1991). Hypotheses of adaptation lead to predictions of correlated evolution between the presumed adaptive character and the proposed cause (Purvis and Rambaut, 1995). Comparative studies identify evolutionary trends by comparing the values of some variable or variables across a range of taxa. The aim of this project is to identify evolutionary trends in primate geographic ranges by comparing the values of a number of variables across the primate order.

The nature of comparative data presents problems for statistical analysis. Standard statistical tests like regression assume independence of data points and standard variance (Shennan, 1997). The problem arises because species are arranged hierarchically in a phylogeny: close relatives tend to be more similar than would be expected by chance (Harvey and Pagel, 1991). Related species often share traits by common descent rather than through independent adaptation. This may produce spurious correlations between variables and suggest an evolutionary trend where there is only a common inheritance (Harvey and Pagel, 1991). Non-independence can also sometimes cause important correlations to be missed (Purvis and Rambaut, 1995). Modern comparative methods in evolutionary biology attempt to distinguish independent evolutionary origins of character traits from cases of identity by descent (Harvey and Pagel, 1991); however see Harvey and Rambaut (2000).

A number of techniques are available for the comparative analysis of variables. These methods are based on a set of assumptions that comprise a null hypothesis of evolutionary change, and on a set of statistical techniques that apply those assumptions to real data (Harvey and Pagel, 1991). The statistical techniques are designed to produce data points that can be treated as independent for statistical analysis. One class of methods attempts to define a set of mutually independent comparisons calculated from the phylogeny. These methods recognise that what is phylogenetic inheritance at one level of a hierarchy may be part of an adaptive difference at the next highest level (Harvey and Pagel, 1991).

Felsenstein (1985) developed a method based on the logic of comparing pairs of species or higher nodes that share a common ancestor. He assumes that the evolution of characters along branches of a phylogeny can be modelled as a Brownian motion process (a random walk), with variance accumulating linearly with time. Comparisons are made among the direct descendants of each node.

Figure 4.1 shows a branching phylogeny for four species, a, b, c, and d. Most of the variation among these four species in a typical phylogeny would already have been present between the two higher nodes. However, the pairwise differences between species a and b, and between c and d, and between nodes 1 and 2, are all independent of each other. This is because, for example, the difference between species a and b reflects only the evolutionary changes that have taken place since they split from their common ancestor at node 1. All similarity between species a and b that is due to their shared phylogenetic history will be subtracted out. The same applies to species c and d. Finally, the differences

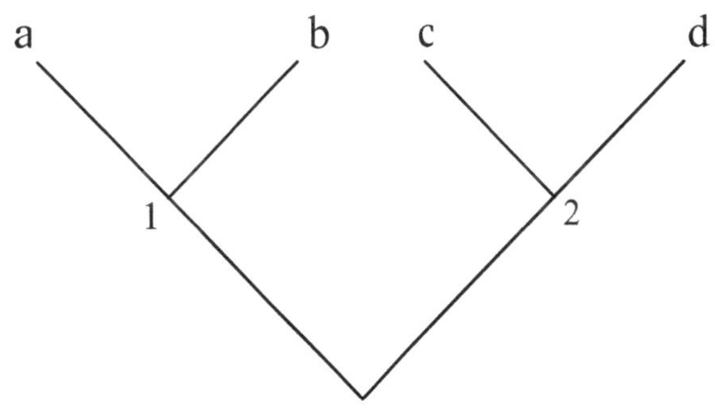

FIGURE 4.1
BRANCH OF PHYLOGENETIC TREE. FROM HARVEY (1991)

between the higher nodes, nodes 1 and 2, reflect only the evolutionary events that have occurred since they split from their common ancestor, and this difference will be independent of the other two (Harvey and Pagel, 1991). Thus three sets of independent contrasts (IC) are derived from this section of the taxonomy. If the comparisons for two variables are then plotted against each other, we can see whether evolutionary change in the two traits has been correlated.

Any comparative method must embody an assumption about how evolution has proceeded (Purvis *et al.*, 1994). Methods with more realistic and more reasonable assumptions will perform better. The IC method explicitly assumes a Brownian motion model of character change: the amount of trait variance expected along a segment of the working phylogeny is proportional to the square root of the length of the segment. Biologists using the method can opt for a punctuational view of evolution by setting all the segments to the same length, or can take a gradualist viewpoint by reflecting the ages of taxa in segment lengths (Purvis *et al.*, 1994).

Geographic range size and phylogeny

According to Garland *et al.* (1992), any continuous trait that is inherited from ancestors is appropriate for phylogenetic analysis, regardless of the mechanism of inheritance (e.g. genetic or cultural). There are strong phylogenetic patterns in distribution; at the continental level, the *Cercopithecinae*, *Colobinae*, *Hominoidea* and *Lorisoidea* are distributed in Africa and Asia, the *Lemuroidea* live in Madagascar, the *Tarsioidea* live in Asia and the *Ceboidea* in South Central America. In addition, relatively recent radiations of species with particular ecological characteristics and habitat preferences are likely to have a strong effect on trends shown in any species-based analysis.

Using independent contrasts

Independent contrasts were calculated using CAIC (Comparative Analysis by Independent Contrasts) version 2.6.9. This is a package for use on Apple Mackintosh, provided by Purvis and Rambaut (1995), and based on Felsenstein's method as described above. All ages and brain and body weights were natural log transformed before taking contrasts because CAIC assumes that different lineages are equally likely to make the same proportional change in size. Independent contrasts were regressed through the origin using least-squares regression (Purvis and Rambaut, 1995). All tests were carried out using both raw data and independent contrasts. Comparison aids interpretation of the graphs, and differences between the results may be informative.

The performance of the IC method varies depending partly on how well known the topology and branch lengths are, and how appropriate the evolutionary model. Phylogenies often contain polytomies – multiple branching nodes – which reflect our ignorance of the precise order of splitting. CAIC produces only one contrast for such groups. In addition, phylogenies are constantly being revised. Purvis used a method that produced a composite tree based on trends in a large sample of previous estimates (Purvis, 1995). Branch lengths are required for proper standardization of the contrasts. There are a number of possible ways of estimating branch lengths. For the primates, estimates of divergence times are available from Purvis' (1995) phylogeny. The timings for 90 nodes are based on molecular, fossil and karyotypic data, and the remaining branch lengths are interpolated. Alternatively equal branch lengths may be used. In this case contrasts are standardised by the number of speciation events, thus implying a punctuational model of evolution.

The adequacy for statistical purposes of any proposed branch lengths must be verified empirically for each phylogeny and for each character rather than assumed (Garland *et al.*, 1992). I carried out preliminary model tests (Garland *et al.*, 1992) to compare the adequacy for statistical purposes of Purvis' branch lengths and equal branch lengths, a transformation of Purvis' branch lengths, and different transformations of the data. In general the branch length models performed equally well, and in all of the analyses presented in this chapter equal branch lengths were used. Based on the results of the model tests, the geographic range and home range data is log transformed, as are some of the measurements of climatic variability. Since the values of observation frequency and journal research were already natural log transformed prior to correction, I do not think that further transformation would be helpful. In these analyses species for which no innovations are recorded are included. However it could be argued that there was something different in the way that primates with innovations recorded were studied (Reader and MacDonald, 2003). It was difficult to find a completely satisfactory model for analysis of the geographic range size, climatic variability and observation frequency data (see appendix for details).

Uncertainty about branch lengths or phylogeny are not sufficient justification for treating species as independent data points since simulation studies have shown that CAIC performs reasonably well under these conditions and out-performs species-level analyses (Purvis *et al.*, 1994). Thus use of independent contrasts is always preferable to treating species as independent data points, but it is important to test the assumptions of the IC method. Computer simulations have shown that incorrectly specifying the topology or branch lengths leads to higher type 1 error rates (the chance of rejecting the null hypothesis when it is in fact true) (Purvis *et al.*, 1994). However the chances of doing this are higher still when phylogeny is not taken into account. In my research, independent contrasts have been given the greatest confidence, but I have also taken a number of steps to ensure that no spurious results are accepted.

Based on the evolutionary assumptions of CAIC the absolute value of the standardised contrast should be independent of the estimated value of the character for the node at which the contrast was taken. In addition, according to the statistical assumptions of CAIC, the absolute values of scaled (standardised) contrasts should be independent of both the age of the node and the square root of the expected variance. Assumption tests were carried out for each analysis, and are presented in the results tables in the appendix.

It is a good idea to examine the scatter plot of contrasts to determine whether the overall relationship depends on only one or a few unusual points. This is important for a number of reasons. Violation of the branch length assumption will lead to some heteroscedasticity in the contrasts. In addition, data error can have a particularly strong effect on regression analyses using independent contrasts. Differences between close relatives, which may have arisen through data error rather than evolution, may be exagerated when they are converted into rates (Purvis and Webster, 1999). I have excluded contrasts with a Studentized residual greater than 3 (Jones and Purvis, 1997). In addition, in each analysis I have examined the outliers and their effects on the regression in order to see whether the result is strongly influenced by a few contrasts.

Another possible solution to persistent heterogeneity of contrasts would be to use a weighted regression technique: this would improve the precision of the estimate of slope. However this method is sensitive to outliers just as unweighted least-squares regression is. More work remains to be done on the use of regression models other than least-squares (Purvis and Webster, 1999).

In addition, confidence in the conclusions would be improved by carrying out parallel tests making different assumptions. It is possible to use the BRUNCH algorithm to generate contrasts when you are not happy with fitting any evolutionary models to your data (Purvis and Rambaut, 1995). The only assumption when this algorithm is used for continuous variables is that evolution in different branches is independent. CAIC sets all the contrasts in the independent variable to positive, so if evolution in the traits under investigation is not correlated, then roughly half the contrasts in the dependent variable will be positive, and the other half negative with a mean not significantly different from zero. The null hypothesis can be tested using a two-tailed sign test on the signs of the contrasts, or, more powerfully, a t-test on the mean of the contrasts or with a randomization test. In a number of the analyses below I carried out parallel tests using the BRUNCH algorithm.

Statistical considerations

A number of the analyses presented here have been repeated with several related variables. For instance, all analyses of observation frequency and distribution were carried out with the innovation, social learning and tool use variables. These variables are strongly correlated with each other (Reader and Laland, 1999). The reason for repeating very similar analyses is that measuring behavioural flexibility and intelligence is difficult and contentious. I did not want to miss an opportunity to understand the interaction between intelligence and geographic distribution through selecting an inappropriate measurement. I also wanted to include variables that might be less sensitive but provide a better comparison with the data available on early hominin distribution, such as tool use frequency and absolute brain size.

The use of multiple variables, and correlation between many of the variables, would seem to be a good argument for carrying out multiple regression analyses, and this is possible using independent contrasts. This would make it possible to establish which variables were the best predictors, investigate correlation between the independent variables, and perhaps produce the best multiple regression model based on a combination of the variables. This may be less relevant if none of the independent variables are good predictors in regression analyses examining each predictor separately. In addition, there are a number of drawbacks to multiple regression. The datasets vary in size, so inclusion of several variables will tend to reduce the sample size. A multiple regression analysis using several closely related independent variables is liable to suffer from problems of colinearity. I decided that multiple regression analysis in this case would not add anything to my interpretation.

A number of the variables used in different analyses here are likely to interact: for instance, the different measures of climatic variability, and the innovation, social learning and tool use frequencies. If you carry out multiple tests on one set of data, the chances of making a type 1 error are increased (rejecting the null hypothesis when it is in fact true). This can be corrected for by altering the significance level according to a formula (e.g. Bonferroni Corrections). I had assumed that since a number of the tests were carried out on not entirely independent variables, similar results in several cases would serve to strengthen the conclusions; however by the logic of Bonferroni corrections this is not the case. Bonferroni Corrections provide a method of allowing for the effects of repeated statistical tests by reducing the threshold significance level according to the number of tests. I did not use Bonferroni corrections in these analyses: instead I have used a standard level of significance of $\alpha = 0.05$ throughout, and have also presented actual p values, allowing for reassessment in the light of Bonferroni Corrections.

Results and discussion

Alternative measures of behavioural plasticity

Hypothesis: More behaviourally flexible species will use more different habitats and food types.

Habitat and dietary niche breadth data are from Eeley and Foley (1999). They collated the number of different habitat types occupied (from a total of 23 possible habitats identified) and the number of different food types used (from a possible total of 10) for African primate species, based on an extensive literature search. To test this hypothesis I carried out analyses with innovation frequency as the independent variable. Innovation frequency may be a particularly good indicator of behavioural flexibility (Reader, 2000).

There was a significant positive correlation between habitat niche breadth and innovation frequency using species as data points and also using independent contrasts (using independent contrasts $r^2 = 0.330$, $p = 0.005$, $n = 21$) (Figures 4.2-3). However there was a violation of the evolutionary assumptions of CAIC for the independent variable (observation frequency) in this analysis. There were two outliers with high values for contrasts in innovation frequency. These were the contrast between *Pan troglodytes* and *Pan paniscus*, and the contrast calculated at the node between *Papio anubis, Papio papio*, and *Papio cynocephalus*. The contrast between the two chimpanzee species may be due to error in the measurement of innovation frequency, particularly as *P.paniscus* is relatively rare. In addition, the latter contrast was calculated at a multiple node. Exclusion of these outlying contrasts did not alter the significance of the regression ($r^2 = 0.258$, $p = 0.022$, $n = 19$).

The results using independent contrasts for innovation frequency and dietary niche breadth were not significant ($r^2 = 0.107$, $p = 0.138$, $n = 21$). There was one outlier with a residual greater than ±3 (the contrast between *P.anubis* and *P.papio*), and two outliers with large values for innovation frequency (the same contrasts as those discussed above). Removal of these contrasts changed the significance of the relationship (using independent contrasts, $r^2 = 0.276$, $p = 0.021$, $n = 18$). The change in significance suggested that the significant result was not very robust.

Analyses carried out using measurements of brain size did not confirm the significant and positive results for habitat niche breadth described above (using independent contrasts, for absolute brain weight $r^2 = 0.000$, $p = 0.983$, $n = 27$, for relative brain weight $r^2 = 0.006$, $p = 0.699$, $n = 27$). There were assumption violations for the dependent variable (habitat tolerance) in these analyses. There was one contrast with very high values for the brain measurements, the contrast between *Cercocebus galeritus* and *Cercocebus torquatus*. Exclusion of this outlier did not alter the significance of the results.

The coefficient of determination indicates that species' habitat tolerance is partially explained by variation in innovation frequency. This suggests that variability in niche breadth is to a certain extent affected by species' behavioural flexibility, but that other factors are also important. Many instances of innovation are in the foraging context, and it could be suggested that species with broader dietary niches could be over-represented. The lack of a robust significant relationship with dietary breadth clears up this possible criticism of innovation frequency as a measurement of behavioural flexibility.

Range size and behavioural plasticity

Hypothesis 1: general behavioural flexibility will allow species to respond well to environmental change, is useful in range expansion and will lead to larger geographic range size.

Hypothesis 2: social learning will aid range expansion through more accurate assessment of risk, and will therefore lead to larger geographic range size.

Innovation, social learning and tool use frequencies and range size

There was no significant relationship between the natural log of geographic range size and innovation frequency for species ($r^2 = 0.033$, $p = 0.102$, $n = 83$) or independent contrasts ($r^2 = 0.028$, $p = 0.158$, $n = 71$) (Figures 4.4-5). I removed one contrast which had a studentised residual greater than ±3 (Jones and Purvis, 1997). This was the contrast between the pig-tailed langur (*Simias concolor*) and the proboscis monkey (*Nasalis larvatus*).

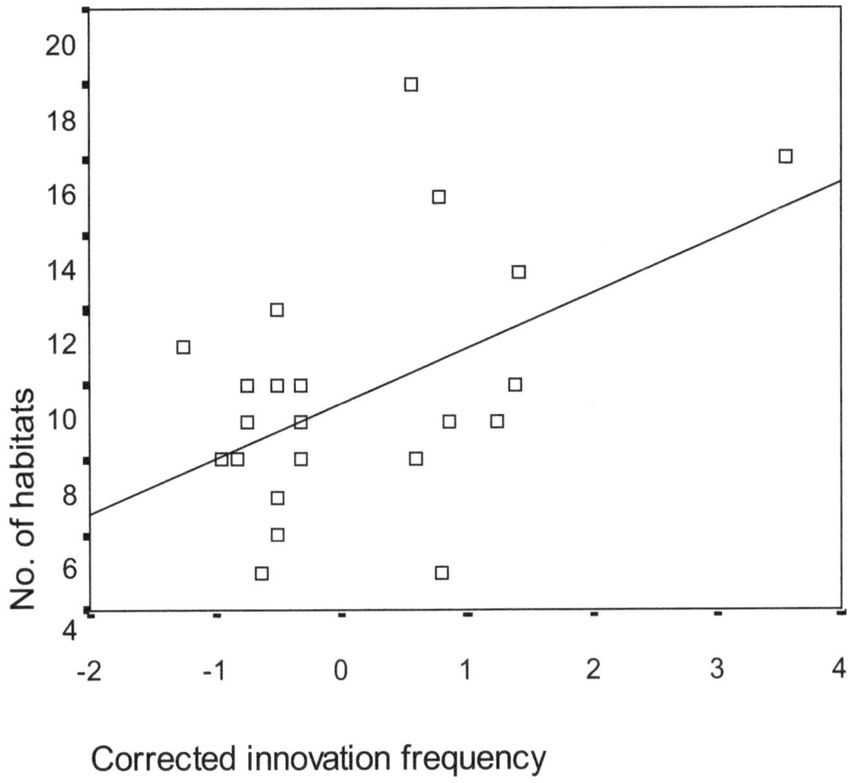

FIGURE 4.2

SCATTERPLOT OF HABITAT NICHE BREADTH AGAINST CORRECTED INNOVATION FREQUENCY. FREQUENCIES ARE CORRECTED FOR RESEARCH EFFORT BY TAKING THE RESIDUALS FROM A LN-LN PLOT THROUGH THE ORIGIN OF INNOVATION FREQUENCY AGAINST RESEARCH EFFORT. THE RAW DATA, WITH EACH POINT REPRESENTING ONE SPECIES.

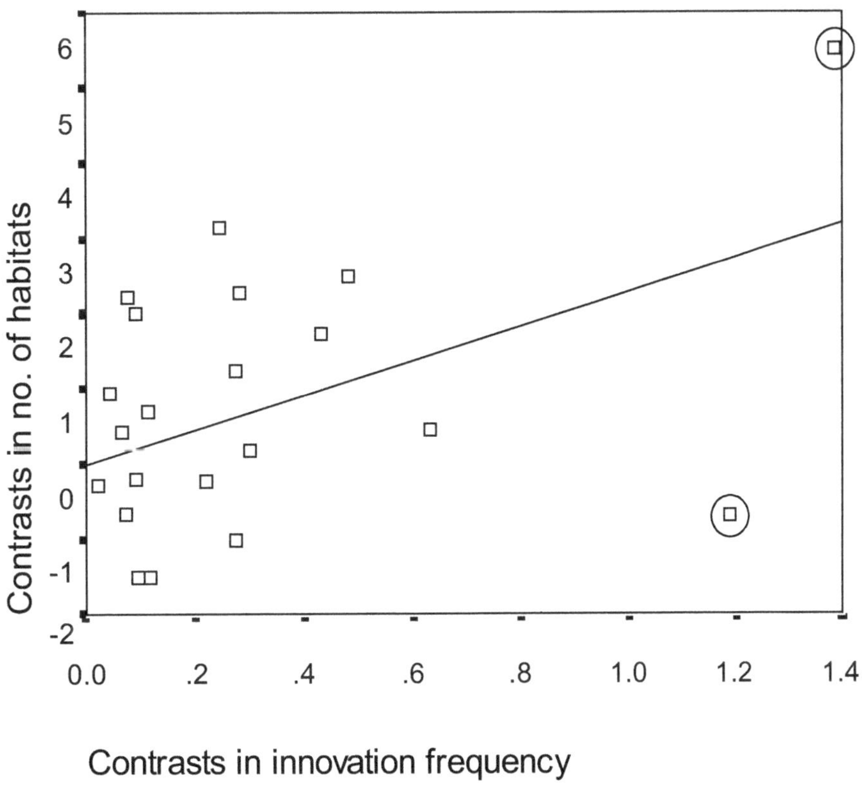

FIGURE 4.3

SCATTERPLOT OF HABITAT NICHE BREADTH AGAINST CORRECTED INNOVATION FREQUENCY. THE INDEPENDENT CONTRAST DATA. OUTLIERS CIRCLED.

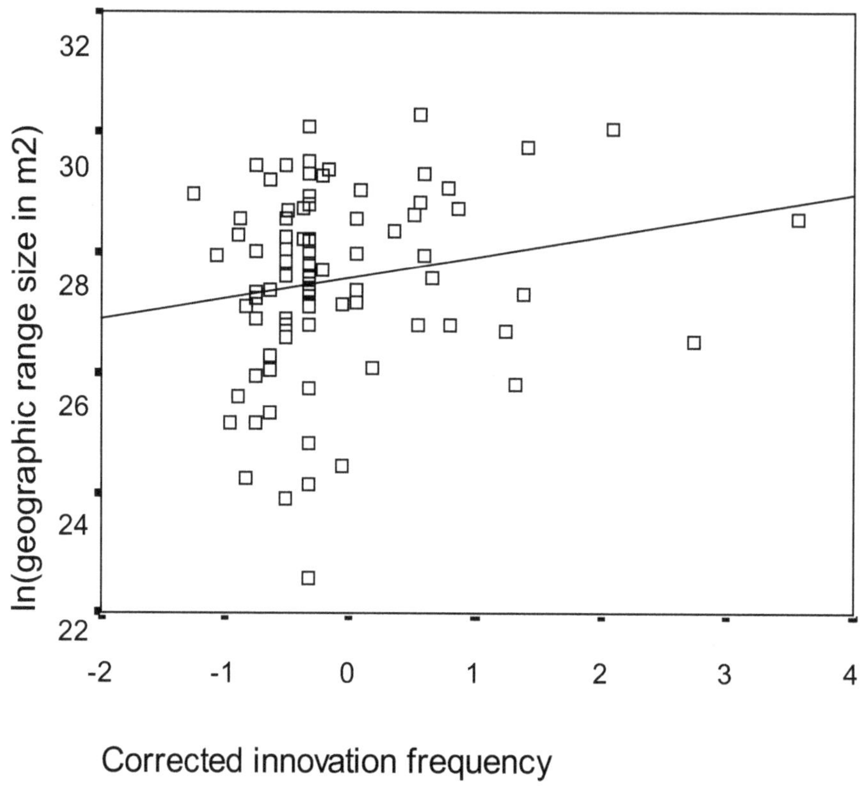

Figure 4.4

Scatterplots of geographic range size in m^2 (natural log transformed) against corrected innovation frequency. Frequencies are corrected for research effort by taking the residuals from a ln-ln plot through the origin of innovation frequency against research effort. The raw data, with each point representing one species

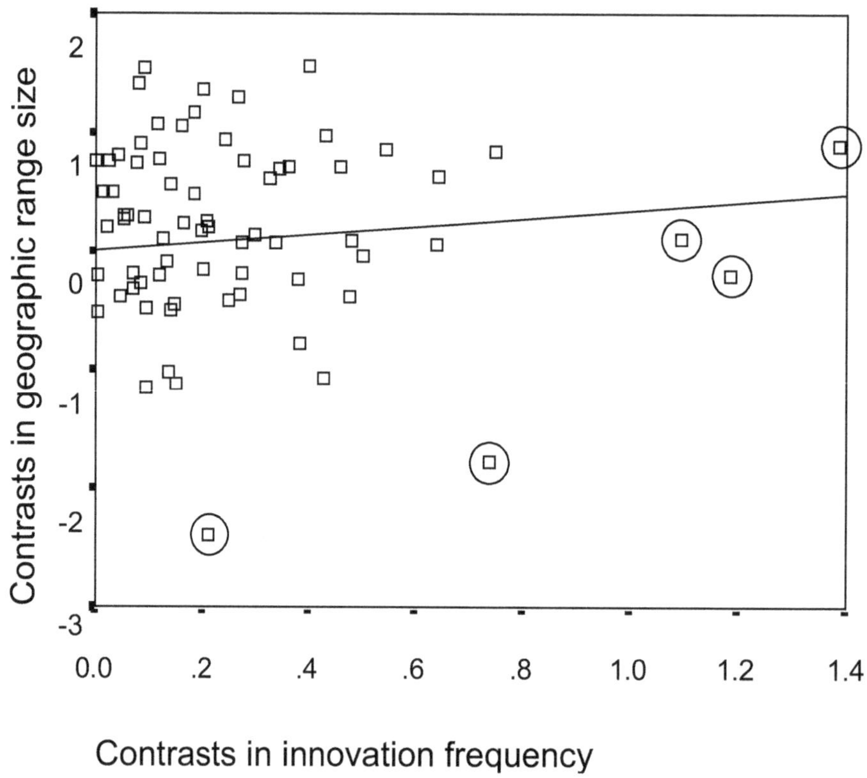

Figure 4.5

Scatterplot of contrasts in geographic range size in m^2 (natural log transformed) against corrected innovation frequency. Outliers circled

S.concolor is limited in distribution to the Mentawai Islands in Indonesia and therefore has a very small geographic range. Removal of this outlier improved the model fit for the geographic range data but did not meet all the assumptions of CAIC. The result remained non-significant (r^2 = 0.043, p = 0.083, n = 70). This result provides strong evidence to contradict the behavioural flexibility hypothesis, as innovation frequency may be a better measure of behavioural flexibility than tool use or social learning (Reader, 2000).

There were four other contrasts that were notable outliers. These included the contrast between *P.paniscus* and *P.troglodytes*, the contrast between *P.anubis*, *P. cynocephalus*, *P.papio* and *P.ursinus*, the contrast calculated at the node at which *Hylobates* diverges from the rest of the apes, and the contrast calculated at the node at which *Macaca cyclopis* diverges from *Macaca mulatta* and *Macaca fuscata*. Removal of all five outliers substantially decreased the significance of the assumption violations, and rendered the relationship significant (r^2 = 0.107, p = 0.007, n = 66). Because this relationship was only significant after the deletion of outliers, it cannot be considered to be robust. There may be a weak positive relationship between innovation frequency and geographic range size: an alternative method of analysis may be necessary to establish this for certain. In the meantime, these analyses show that any such relationship is not very strong, and suggest that other factors are likely to be more important.

There was no significant relationship between tool use frequency and geographic range size, using species or independent contrasts (Figures 4.6-7). The results using independent contrasts were r^2 = 0.000, p = 0.875, n = 70, with one outlier removed. The outlier was the contrast between *S.concolor* and *N.larvatus*, which again had a residual greater than ±3. The independent variable (tool use frequency) violated the evolutionary assumptions of CAIC in this analysis.

There was no significant relationship between social learning frequency and geographic range size, for species or independent contrasts (for raw data r^2 = 0.012, p = 0.331, n = 83) (Figure 4.8). After removal of one outlier with a studentised residual greater than ±3, the results using independent contrasts were r^2 = 0.017, p = 0.273, n = 70 (Figure 4.9). Removal of this outlier improved the meeting of assumptions for the dependent but not the independent variable. Removal of the two contrasts with the highest values for social learning frequency (the contrast between the two chimpanzee species, and the contrast between

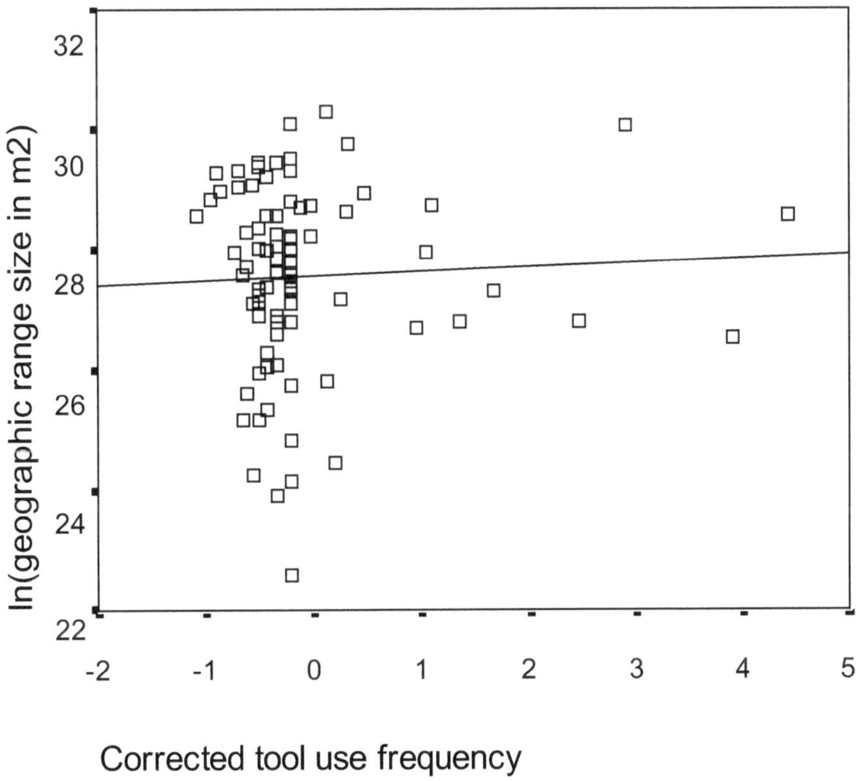

FIGURE 4.6.

GEOGRAPHIC RANGE SIZE IN M^2 (NATURAL LOG TRANSFORMED) AND CORRECTED TOOL USE FREQUENCY. THE RAW DATA, WITH EACH POINT REPRESENTING ONE SPECIES

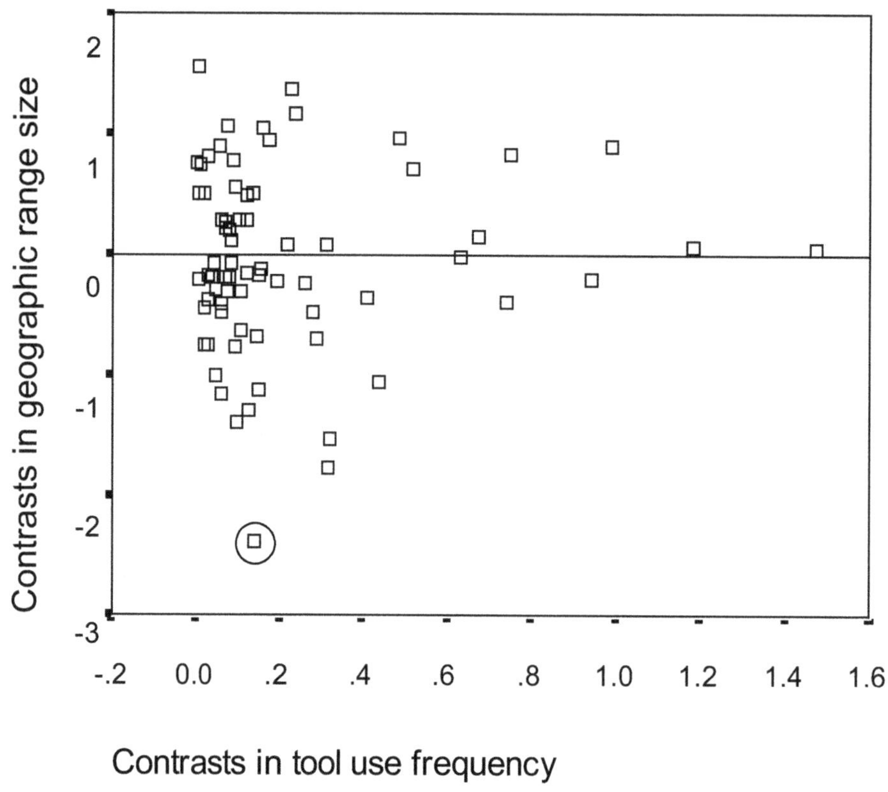

Figure 4.7

Geographic range size in m^2 (natural log transformed) and corrected tool use frequency. The independent contrast data. Outlier circled

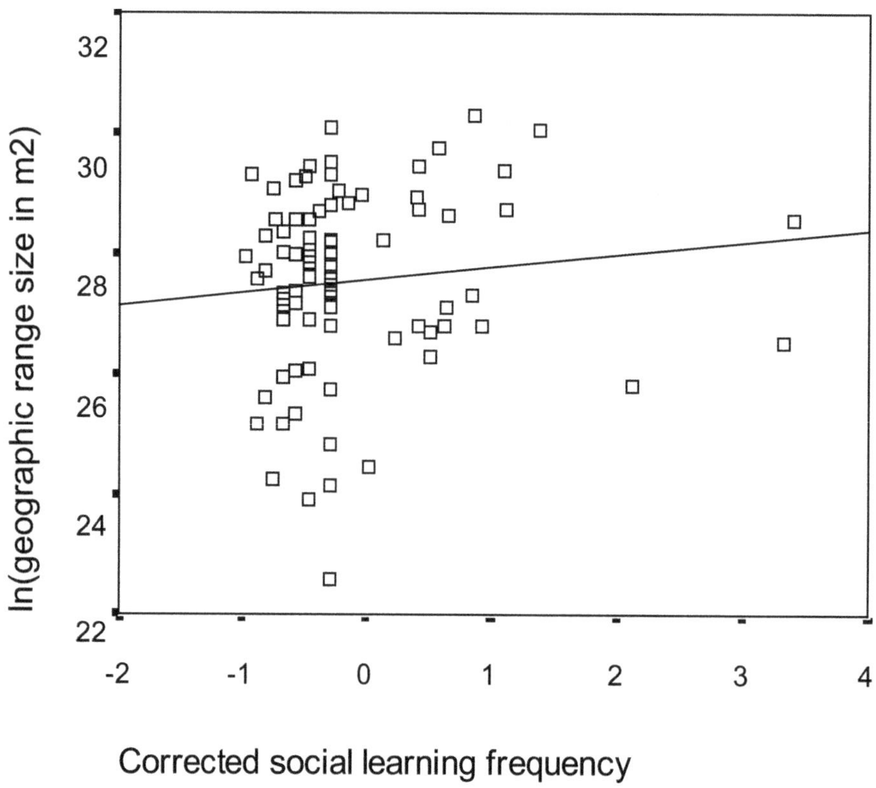

Figure 4.8

Geographic range size in m^2 (natural log transformed) and corrected social learning frequency. The raw data, with each point representing one species

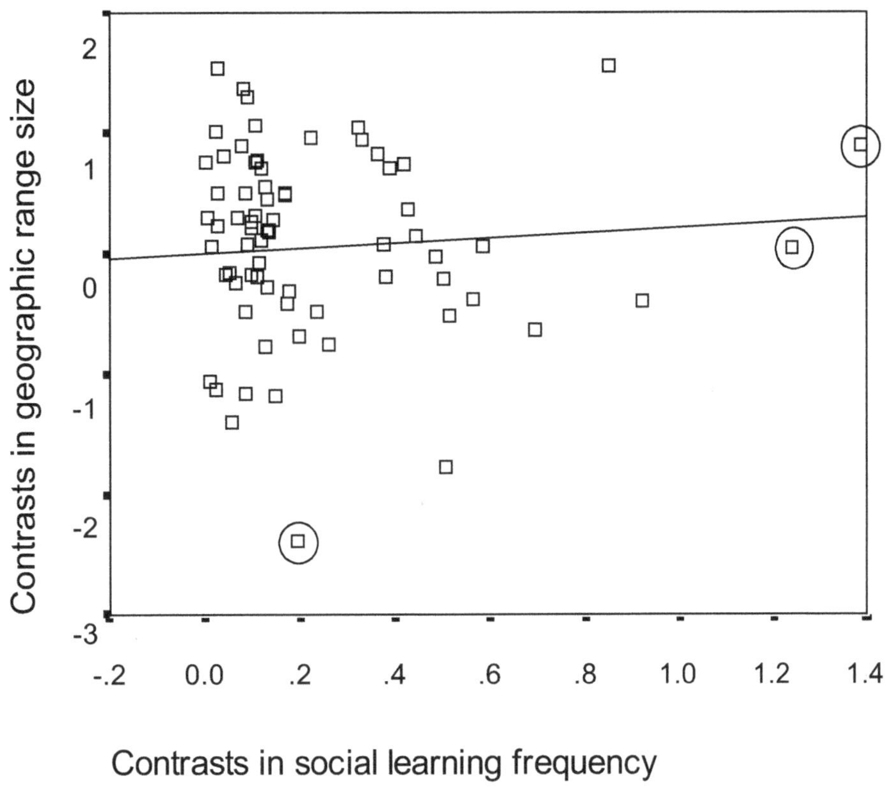

FIGURE 4.9
GEOGRAPHIC RANGE SIZE IN M^2 (NATURAL LOG TRANSFORMED) AND CORRECTED SOCIAL LEARNING FREQUENCY. THE INDEPENDENT CONTRAST DATA. OUTLIERS CIRCLED

Hylobates and the great apes) also did not change the significance of the regression. In addition, social learning was not a significant predictor of geographic range size when innovation frequency was controlled for, using raw data and independent contrasts (using independent contrasts $r^2 = 0.000$, $p = 0.897$, $n = 71$).

Because of the assumption violations in the above analyses I carried out equivalent tests using the BRUNCH algorithm. I carried out a one sample t-test to compare the mean of the dependent variable with zero. The results for innovation frequency were $t = -0.867$, $p = 0.392$, $n = 32$. I also used the Kolmogorov Smirnov test for normality, and found that the contrasts data meets this assumption ($Z = 0.569$, $p = 0.902$, $n = 32$).

The result of the t-test indicated that the dependent variable does not differ significantly from zero, confirming the null hypothesis that evolution in geographic range size has not been linked in any way to the evolution of behavioural flexibility as measured by innovation frequency. The results for tool use and geographic range size were $t = -0.867$, $p = 0.392$, $n = 32$, Kolmogorov Smirnov $Z = 0.650$, $p = 0.792$. The results for social learning frequency and geographic range size were $t = -0.532$, $p = 0.599$, $n = 32$, Kolmogorov Smirnov $Z = 0.403$, $p = 0.997$. Again these results confirm the null hypothesis.

In these analyses species for which no innovations are recorded were included. However it could be argued that there was something different in the way that primates with innovations recorded were studied (Reader and MacDonald, 2003). The results using a restricted innovation dataset give no significant relationship with geographic range size (Reader and MacDonald, 2003). This is consistent with the conclusions for innovation frequency described above. In addition, assumption violations related to the observation frequency dataset are a very good reason for repeating these tests using relative brain size as a proxy measure of behavioural flexibility.

I also carried out analyses using continental datasets (see Appendix for details). There were no significant relationships between observation frequencies and geographic range size for Africa and Asia. There were significant relationships between innovation and tool use frequency and geographic range size in South American primates, using species and independent contrasts (for the innovation analysis using ICs $r^2 = 0.290$, $p = 0.007$, $n = 23$). However the South American results were strongly affected by one outlying species, the tufted capuchin (*Cebus apella*), with very high observation frequencies of innovation, social learning and tool use and a large geographic range (Figure 4.10). Removal of this species, or log transformation, removed the significance of the results.

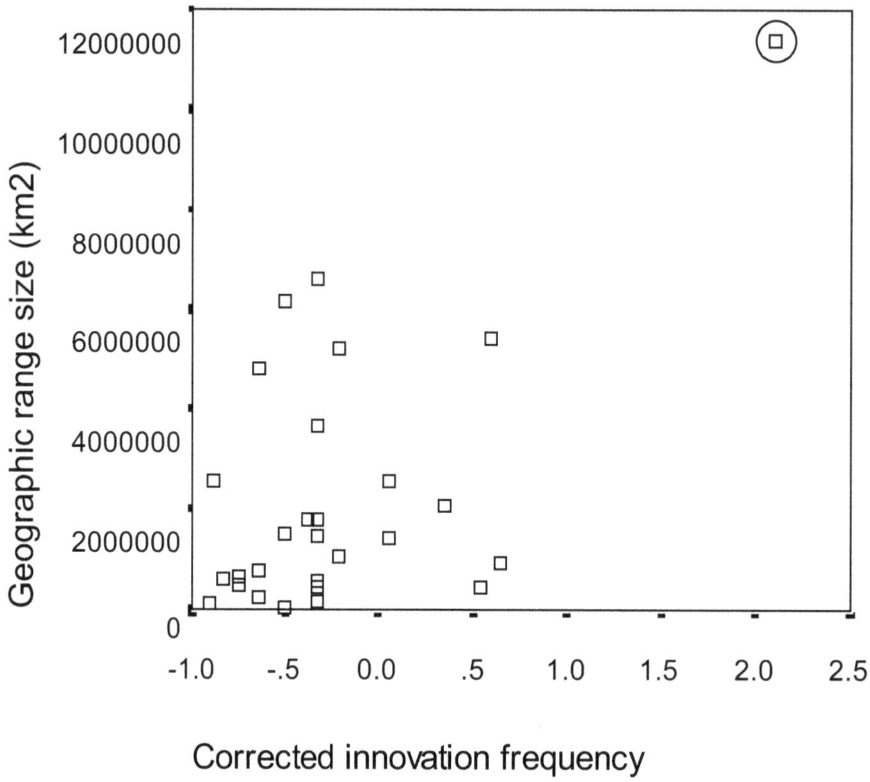

FIGURE 4.10
GEOGRAPHIC RANGE SIZE IN KM^2 AND CORRECTED INNOVATION FREQUENCY FOR SOUTH AMERICAN PRIMATES. THE RAW DATA, WITH EACH POINT REPRESENTING ONE SPECIES. *CEBUS APELLA* CIRCLED

Brain size and range size

The lack of the predicted positive relationship between brain measures and range size is very clear in the scatterplots (Figures 4.11-16). There was no significant relationship between geographic range size and either absolute brain weight or brain weight corrected for body mass using species as data points (absolute brain weight $r^2 = 0.000$, p = 0.875, n = 88, relative brain weight $r^2 = 0.029$, p = 0.109, n = 88, all variables natural log transformed) (Figures 4.11, 4.13). Interestingly, there was a slight negative trend in the species data for the natural logs of neocortex ratio and geographic range size, but this was not significant ($r^2 = 0.022$, p = 0.422, n = 31) (Figure 4.15). These results are liable to have been biased by phylogenetic relationships.

There was no significant relationship between geographic range size and absolute brain weight using independent contrasts ($r^2 = 0.002$, p = 0.659, n = 82) (Figure 4.12). There were relatively mild violations of the assumptions of CAIC for the dependent variable (geographic range size) in this analysis. None of the contrasts have studentised residuals greater than ±3. There were two contrasts with relatively high values for brain weight. These were the contrasts between *C.galeritus* and *C.torquatus*, and the contrast calculated from the node between the Tarsiers and the Anthropoids. Exclusion of these two contrasts did not change the significance of the regression. There was a slight positive trend in geographic range size and brain weight corrected for body weight using independent contrasts (Figure 4.14), however this was not significant ($r^2 = 0.019$, p = 0.210, n = 82). There was one violation of the assumptions of CAIC for the dependent variable in this analysis. There was no significant relationship between geographic range size and neocortex ratio using independent contrasts (both variables natural log transformed) (Figure 4.16). The results for independent contrasts were $r^2 = 0.000$, p = 0.968, n = 29. There was a relatively minor violation of the evolutionary assumptions of CAIC for the dependent variable in this analysis. There were no outlying contrasts with residuals greater than ±3. One outlier had a very high value for neocortex ratio, this was the contrast calculated at the root where the Strepsirhines diverge from all other primates. Removal of this contrast did not change the significance of the regression.

The lack of corroboration from any of the alternative measurements strengthens the interpretation that there is absolutely no linear relationship between brain size and range size. The brain size results back up the generally negative results for the observation frequency data discussed in the previous section.

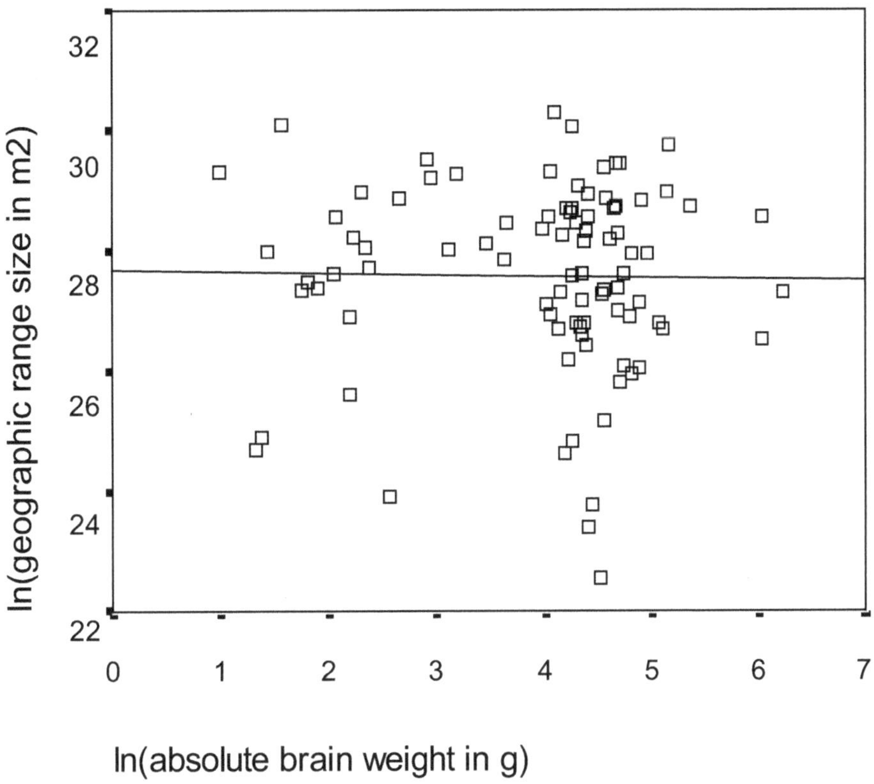

FIGURE 4.11

GEOGRAPHIC RANGE SIZE (M^2) AND ABSOLUTE BRAIN WEIGHT (G) (BOTH VARIABLES NATURAL LOG TRANSFORMED). THE RAW DATA, WITH EACH POINT REPRESENTING ONE SPECIES

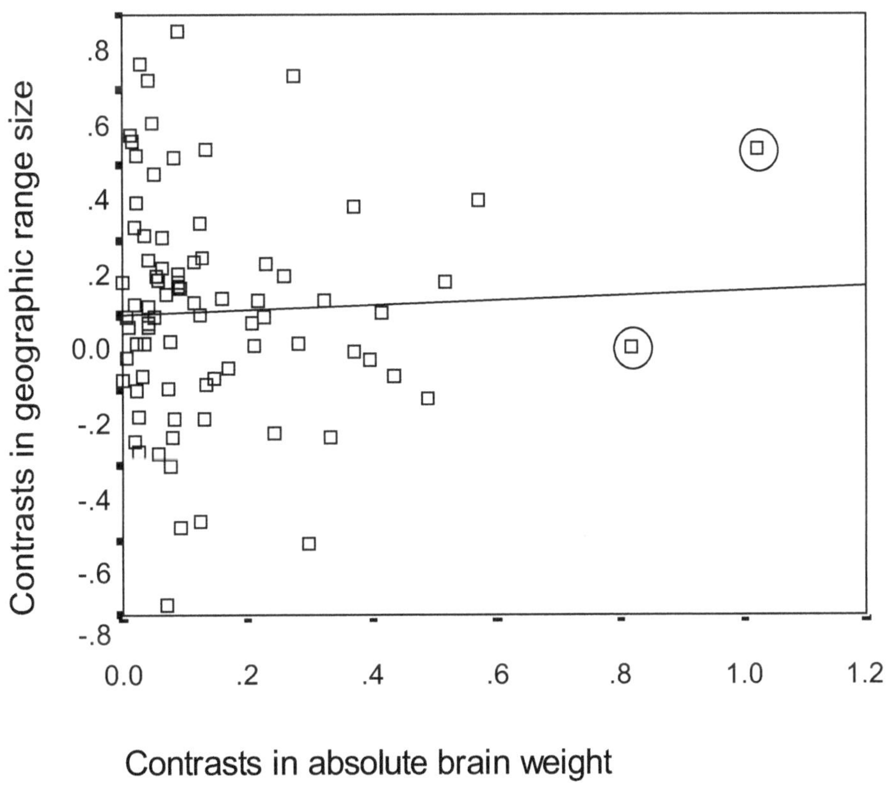

FIGURE 4.12.

GEOGRAPHIC RANGE SIZE (M^2) AND ABSOLUTE BRAIN WEIGHT (G) (BOTH VARIABLES NATURAL LOG TRASNFORMED). THE INDEPENDENT CONTRAST DATA. OUTLIERS CIRCLED

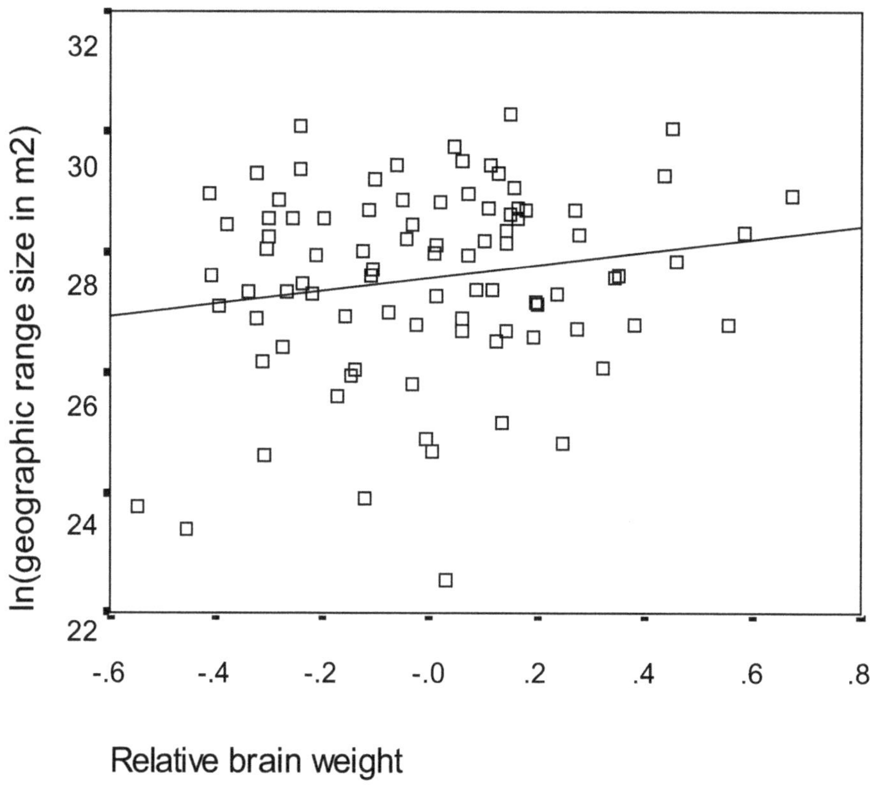

FIGURE 4.13. GEOGRAPHIC RANGE SIZE IN M^2 (NATURAL LOG TRANSFORMED) AGAINST RELATIVE BRAIN WEIGHT. RELATIVE BRAIN WEIGHTS ARE CALCULATED BY TAKING THE RESIDUALS OF A LOG-LOG PLOT OF BRAIN WEIGHT (G) AND FEMALE BODY MASS (KG). THE RAW DATA, WITH EACH POINT REPRESENTING ONE SPECIES

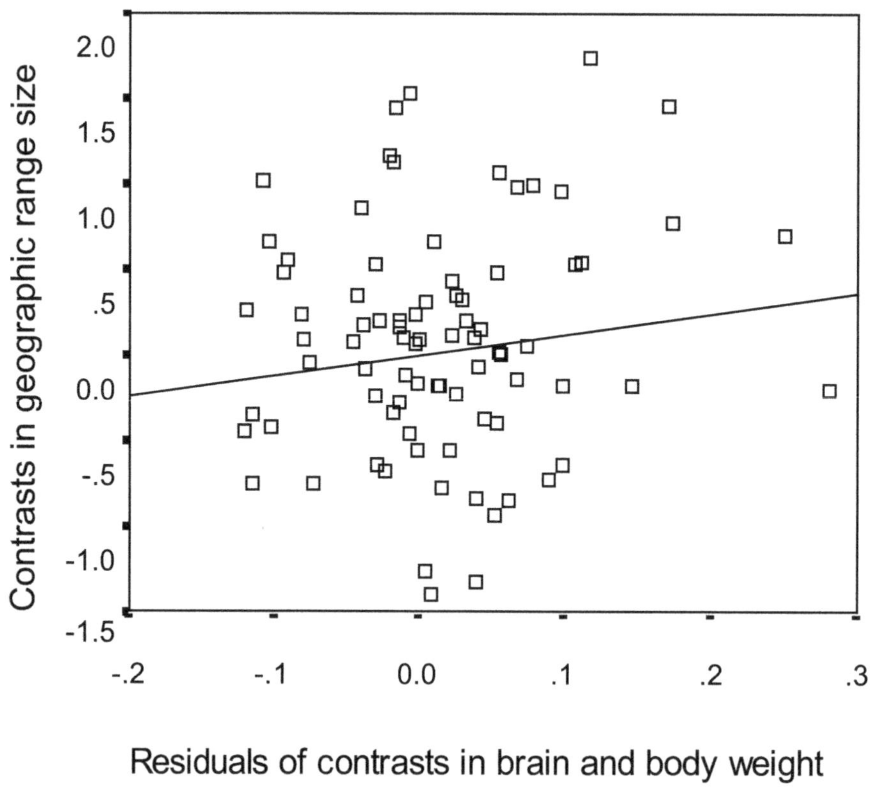

FIGURE 4.14. GEOGRAPHIC RANGE SIZE IN M^2 (NATURAL LOG TRANSFORMED) AGAINST RELATIVE BRAIN WEIGHT. RELATIVE BRAIN WEIGHTS ARE CALCULATED AS THE RESIDUALS OF A PLOT OF THE INDEPENDENT CONTRASTS OF ABSOLUTE BRAIN WEIGHT (G) AND FEMALE BODY MASS (KG), BOTH VARIABLES NATURAL LOG TRANSFORMED. THE INDEPENDENT CONTRAST DATA

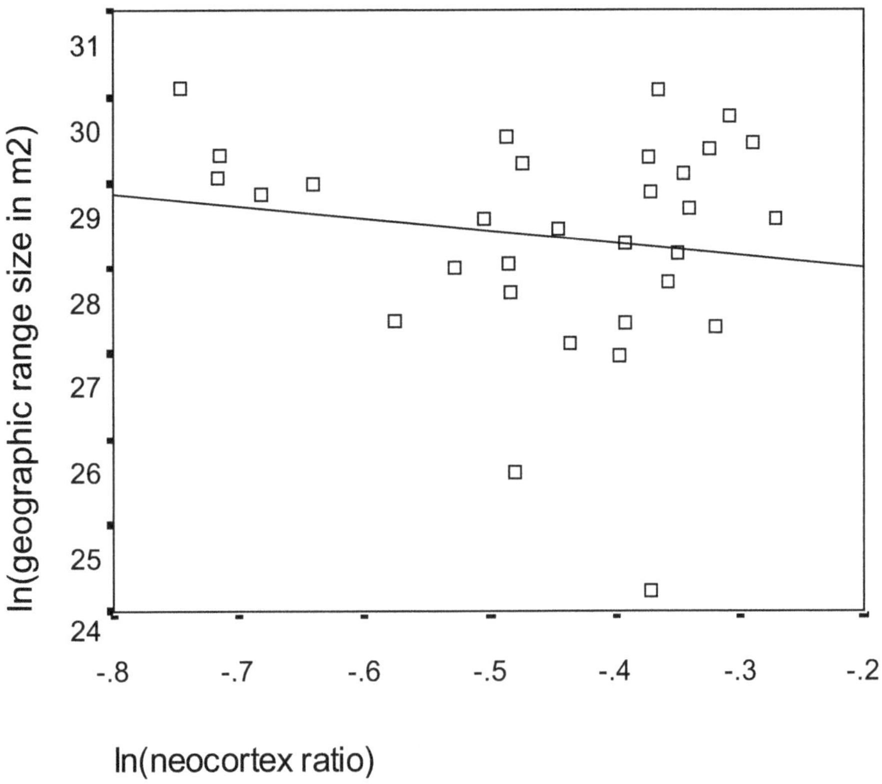

Figure 4.15. Geographic range size (m^2) and neocortex ratio, both variables natural log transformed. Neocortex ratio is calculated as the ratio of neocortex volume to the volume of the rest of the brain. The raw data, with each point representing one species

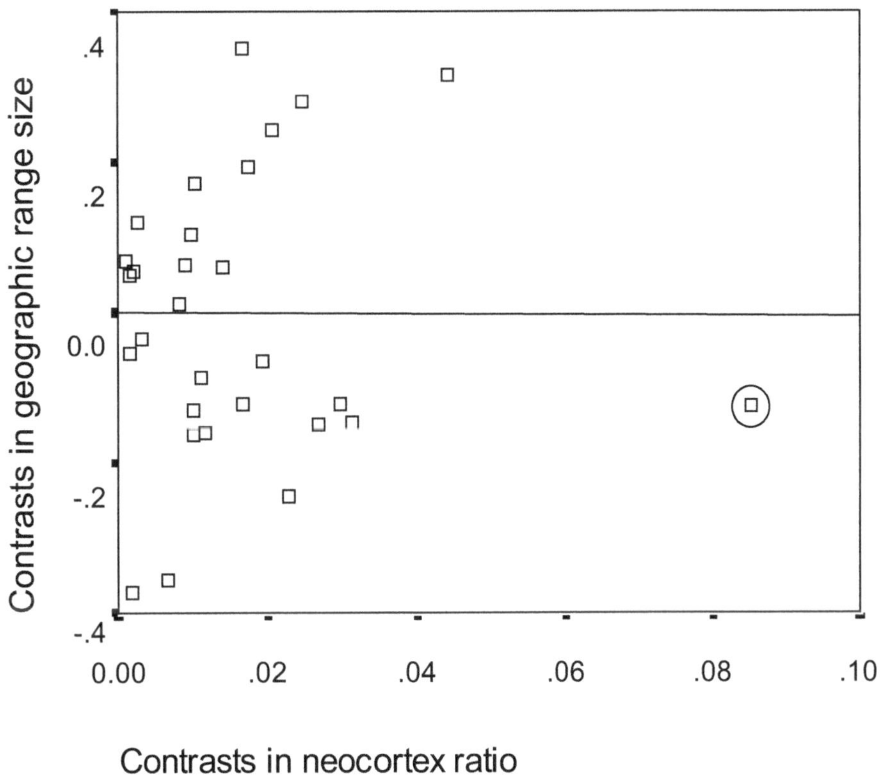

Figure 4.16. Geographic range size (m^2) and neocortex ratio, both variables natural log transformed. The independent contrast data. Outlier circled

Analysis of continental datasets also gave uniformly non-significant results when phylogenetic relationships were controlled for (see Appendix for details). In the raw data, there were negative relationships between the geographic range size and absolute brain size data in African and Asian primates, contrasting with South America. In addition, there was a negative trend using independent contrasts for African (and South American) primates, although this was not significant. There was a significant positive relationship between geographic range size and neocortex ratio for South American primates in the raw data: however this was not present in the independent contrast data. While these results were not very robust, it is interesting that the continental trends for brain size reflect the pattern in the innovation data described above. The pattern of contrasting trends in Asia and Africa compared with South America may be due to the presence of the great apes in Africa and Asia. These species have very large brains and are also under threat from human activity and therefore have small ranges.

Home range size

It has been suggested that behavioural flexibility may affect species' distribution more at a local than a regional level. Regression analysis of the raw data for primate species suggests that behavioural flexibility has a more direct effect on primate species home range size than geographic range size (Figure 4.17). There was a significant relationship between innovation, social learning and tool use frequency and individual home range size using raw data (for innovation frequency $r^2 = 0.121$, $p = 0.003$, $n = 72$). However analysis using independent contrasts produced no significant results, showing that this was a spurious relationship due to phylogenetic effects (for innovation frequency $r^2 = 0.000$, $p = 0.954$, $n = 67$) (Figure 4.18). There were violations of the assumptions of CAIC in the independent variable (innovation frequency) in this analysis. There were a number of contrasts with relatively high values for innovation frequency, these were the contrasts between *P.anubis*, *P.papio*, *P.cynocephalus* and *P.ursinus*, the contrast between the two chimpanzee species, and the contrast calculated at the node at which *Hylobates* diverges from the other apes. Removal of these contrasts did not change the significance of the regression.

This result was confirmed by analysis with relative brain weight (using independent contrasts, $r^2 = 0.001$, $p = 0.803$, $n = 66$). Incorporation of phylogeny removes the effects of unmeasured confounding variables (Nunn and Barton, 2001) – this may explain the contrasting results for the raw data and independent contrasts. This analysis

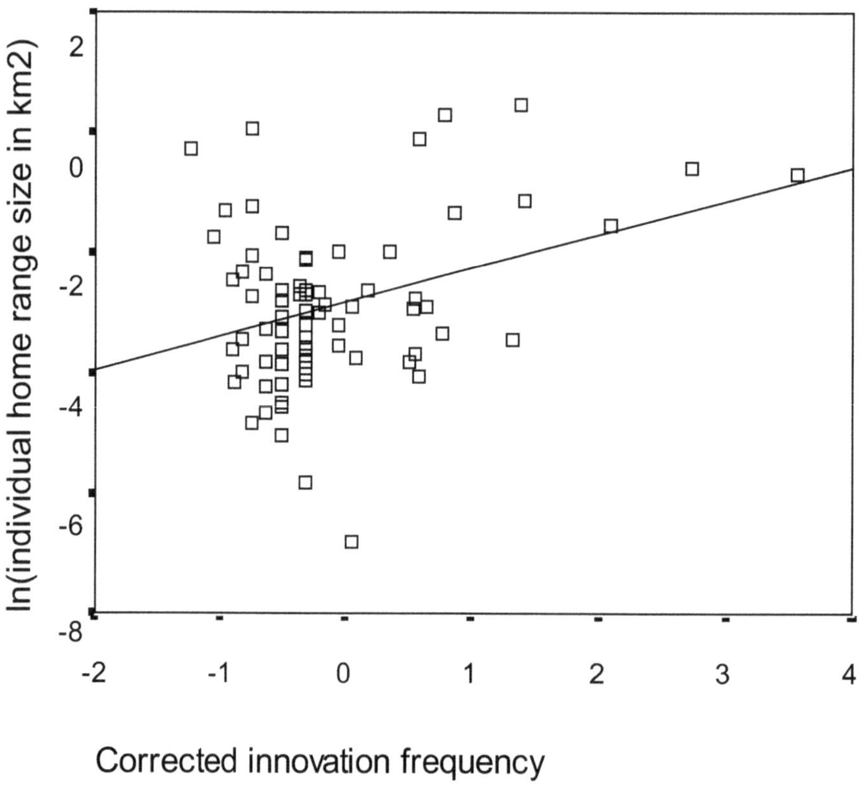

FIGURE 4.17

INDIVIDUAL HOME RANGE SIZE IN KM² (NATURAL LOG TRANSFORMED) AND CORRECTED INNOVATION FREQUENCY. THE RAW DATA, WITH EACH POINT REPRESENTING ONE SPECIES

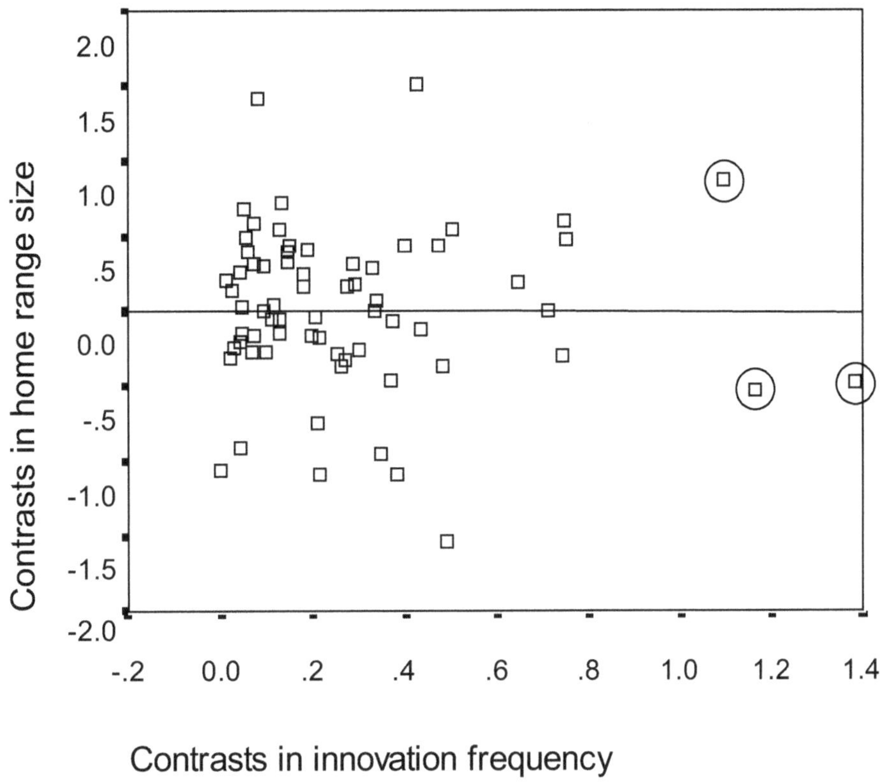

Figure 4.18. Individual home range size in km^2 (natural log transformed) and corrected innovation frequency. The independent contrast data. Outliers circled

demonstrates the importance of controlling for phylogeny in cross-species analysis.

Threat status

It has been suggested that primate species might innovate more in response to stress from their environments: the 'necessity hypothesis' (Reader and Laland, 1999). This would suggest that more innovative species might also be those under threat. If this were so, it might complicate any relationship between behavioural flexibility and geographic range size. However, a recent comparative analysis of taxonomic groups of birds indicated that behavioural flexibility was not associated with extinction risk (Nicolakakis *et al.*, 2003). Alteratively, more innovative species might be under threat from human activity as a by-product of other characteristics (such as large body size and relatively slow life histories). I have carried out an analysis of innovation frequency and threat status, using threat categories from the IUCN, USESA and CITES, collated in Rowe (1996).

There was no evidence from linear regression analysis of the raw data or independent contrasts that innovation, social learning or tool use frequency is dependent on threat status (Figure 4.19-20). There were two outliers with residuals greater than ±3 in the analysis of innovation frequency: the contrast between *P.troglodytes* and *P.paniscus*, and the contrast at the node at which *Hylobates* diverged from the other apes. However removal of these outliers did not alter the results (using independent contrasts, with two outliers removed, $r^2=0.008$, $p=0.462$, $n=68$). In the future, it would also be interesting to look at the geographic or temporal distribution of primate innovation. This would provide an alternative method of examining the 'necessity hypothesis' by comparing levels of innovation in populations living in conditions that are comparatively more or less stressful.

Opportunism and environmental variability

Hypothesis 1: More intelligent or behaviourally plastic species can tolerate greater climatic variability, both spatially and temporally.

Hypothesis 2: Species with an increased capacity for social learning can tolerate greater climatic variability.

Climatic variability patterns

South America has a relatively wide area with relatively high mean rainfall compared with Africa (Figure 4.21). Mean rainfall tends to be highest on the equator and on the coast, and in both Africa and South America is higher on the eastern side of the continent.

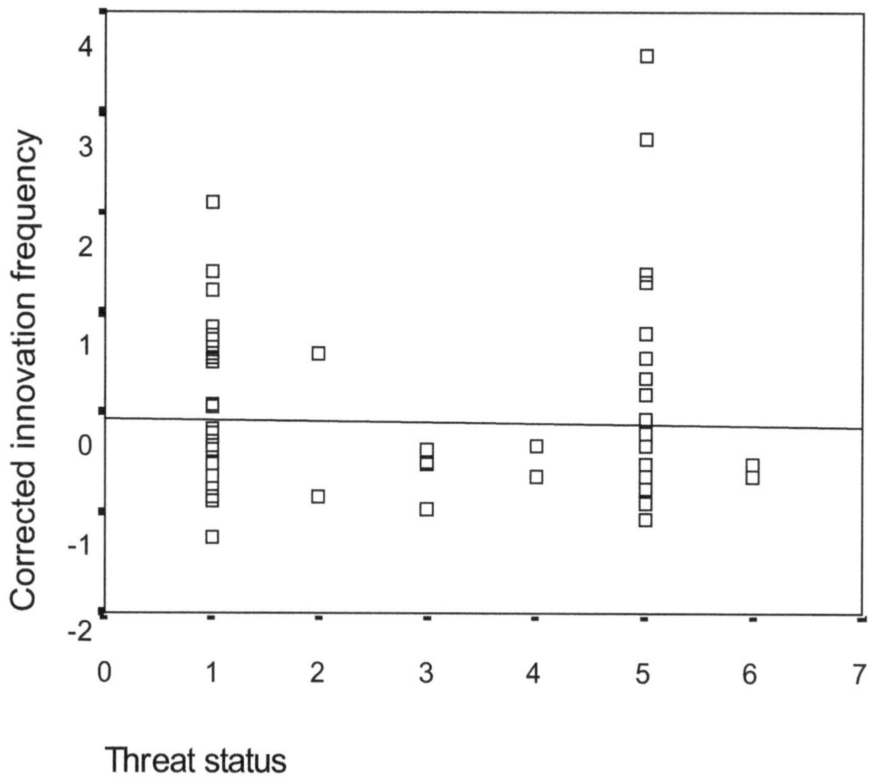

Figure 4.19

Corrected innovation frequency and threat status. The raw data, with each point representing one species

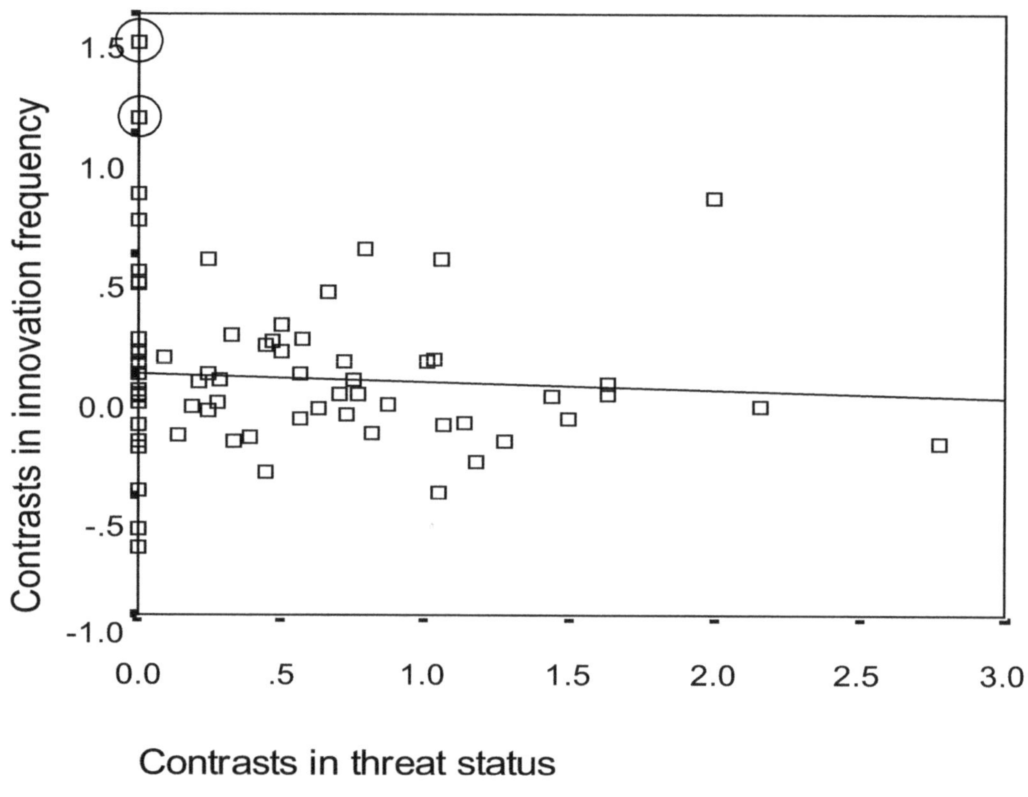

Figure 4.20

Corrected innovation frequency and threat status. The independent contrast data. Outliers circled

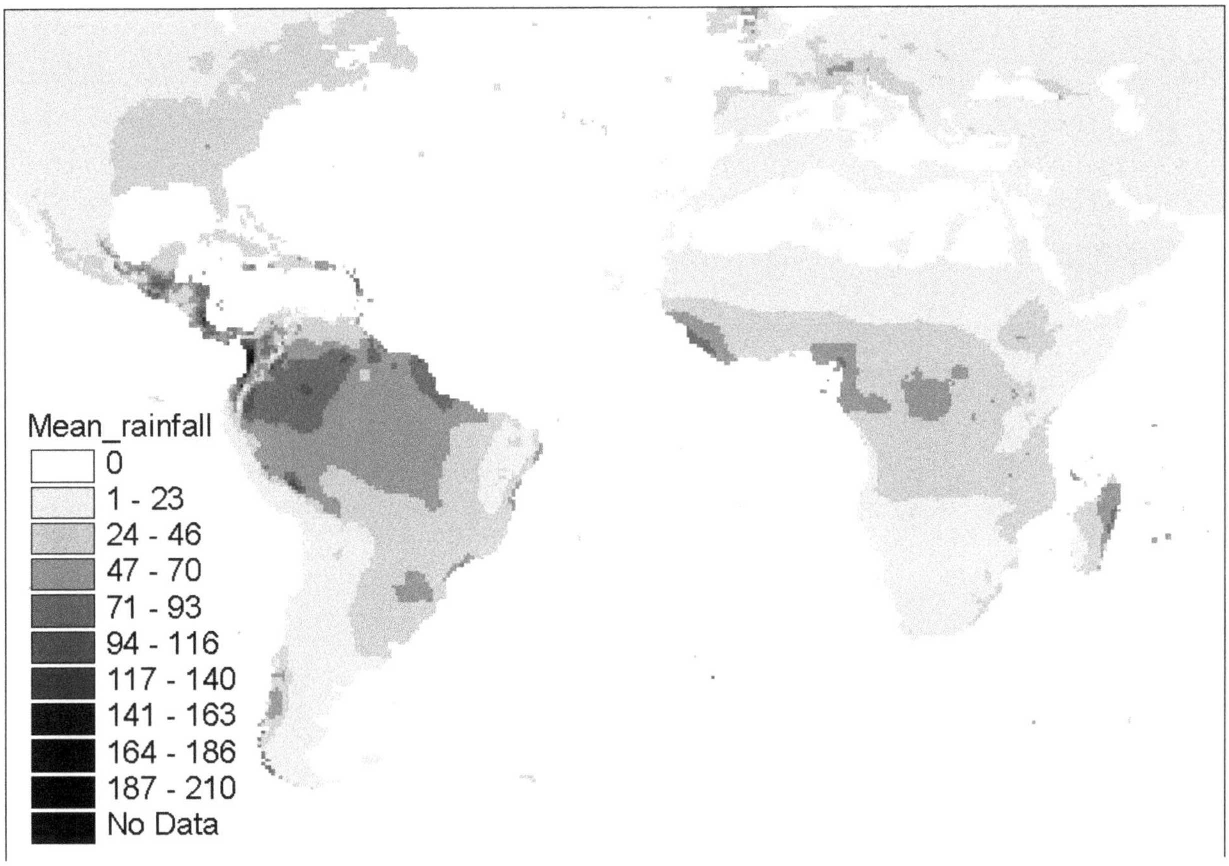

FIGURE 4.21
MEAN ANNUAL RAINFALL IN AFRICA AND SOUTH AMERICA (MM/DAY*10)

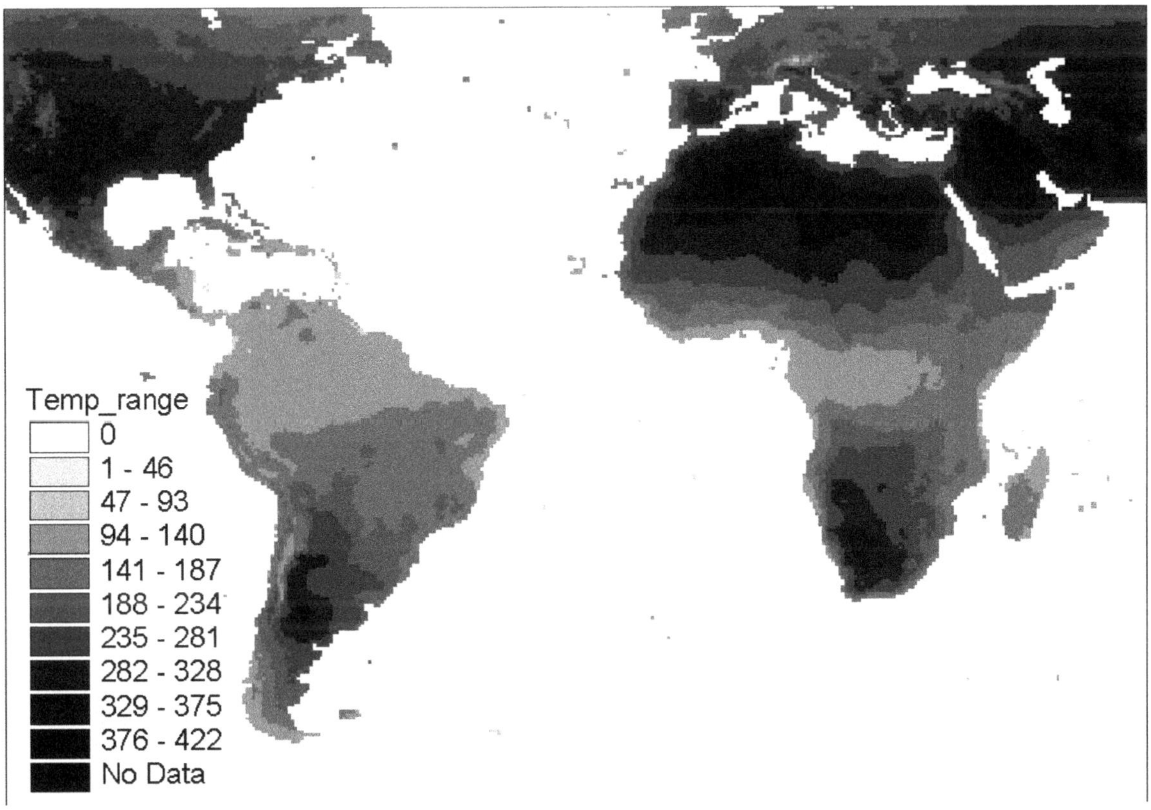

FIGURE 4.22.
ANNUAL TEMPERATURE RANGE IN AFRICA AND SOUTH AMERICA (°C*10)

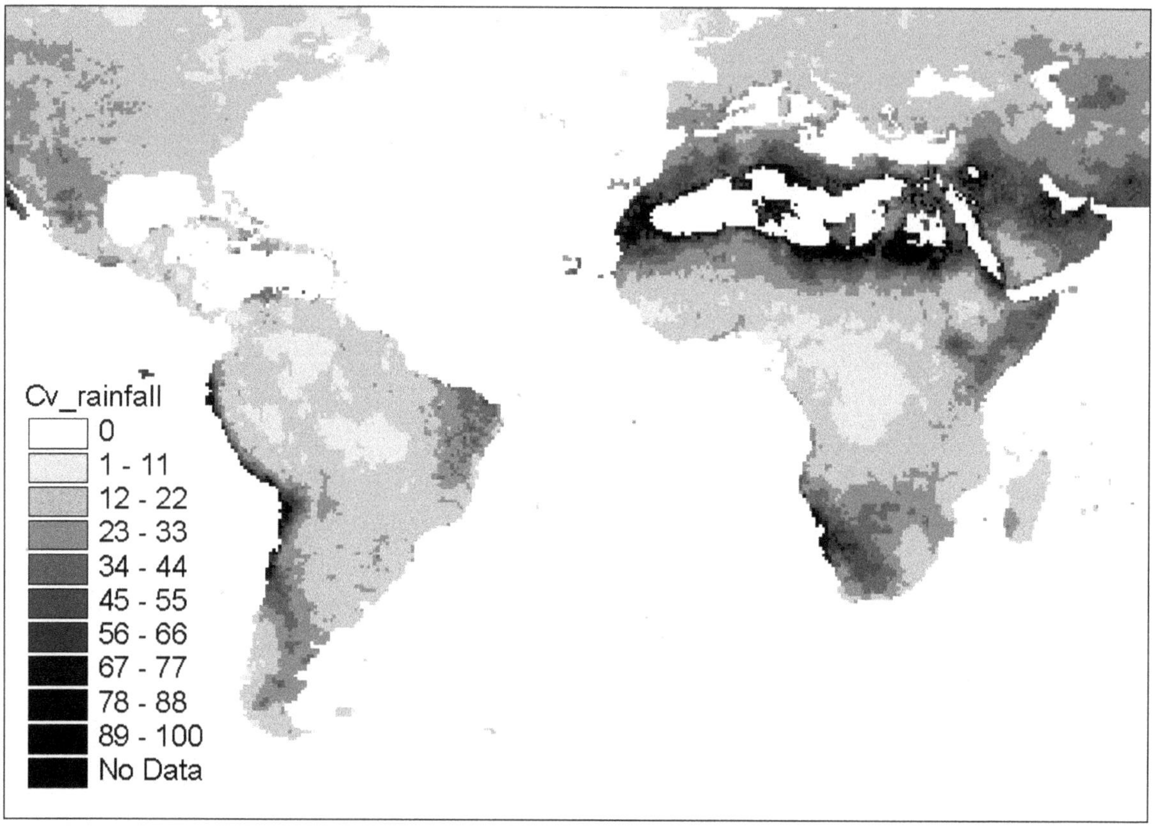

FIGURE 4.23
COEFFICIENT OF INTERANNUAL VARIATION IN RAINFALL IN AFRICA AND SOUTH AMERICA (%)

The gradient of temperature range is steeper in equatorial Africa than in South America (Figure 4.22). Temperature range generally increases north and south of the equator, and in the east of the continent in Africa. Differences in interannual variability in rainfall are less noticeable (Figure 4.23). High interannual variation occurs on the coast, in relatively arid areas (notably the Sahara desert), and in the south of both continents. There is a tendency for higher interannual variation in rainfall to occur where mean rainfall is low. Primate species richness in both continents occurs where rainfall is relatively high, the range of temperatures experienced relatively low and interannual variation also relatively low.

Spatial Variation

The relationship between spatial variation in rainfall and innovation frequency was not significant using raw data ($r^2 = 0.029$, $p = 0.208$, $n = 55$) (Figure 4.24). However there was a significant positive relationship using independent contrasts ($r^2 = 0.099$, $p = 0.027$, $n = 48$) (Figure 4.27). There were assumption violations for the dependent and independent variables in this analysis. There were a number of outliers among the independent contrast data. The contrast calculated at the node between *Saguinus bicolor* and *Saguinus midas*, and *Saguinus leucopus* and *Saguinus oedipus*, had a residual greater than ±3, so this was excluded without affecting the results. However there were also two contrasts with high values for innovation frequency. These were the contrasts between *P.troglodytes* and *P.paniscus*, and the contrast calculated at the node between *P.anubis* and *P.papio*, and *P.cynocephalus* and *P.ursinus*. Removal of these outliers removed the significance of the relationship, suggesting that it may have been caused by the strong influence of these contrasts, and is not very robust. This conclusion is reinforced by the non-significant results for an analysis carried out using the Brunch algorithm ($t = -0.753$, $p = 0.459$, $n = 24$, Kolmogorov Smirnov $Z = 0.488$, $p = 0.971$).

The relationship between spatial variation in rainfall and tool use frequencies is not significant using raw data. However there was a significant positive relationship with tool use using independent contrasts (with one outlier removed, $r^2 = 0.145$, $p = 0.008$, $n = 47$). I was not completely able to meet IC assumptions in this analysis, and the model fit was not improved by removal of a contrast with a residual greater than ±3. Examination of the residuals did not suggest that particular outliers with a high value for tool use frequency were having an undue effect on this pattern. However this conclusion was not supported by the results obtained using Brunch ($t = 0.035$,

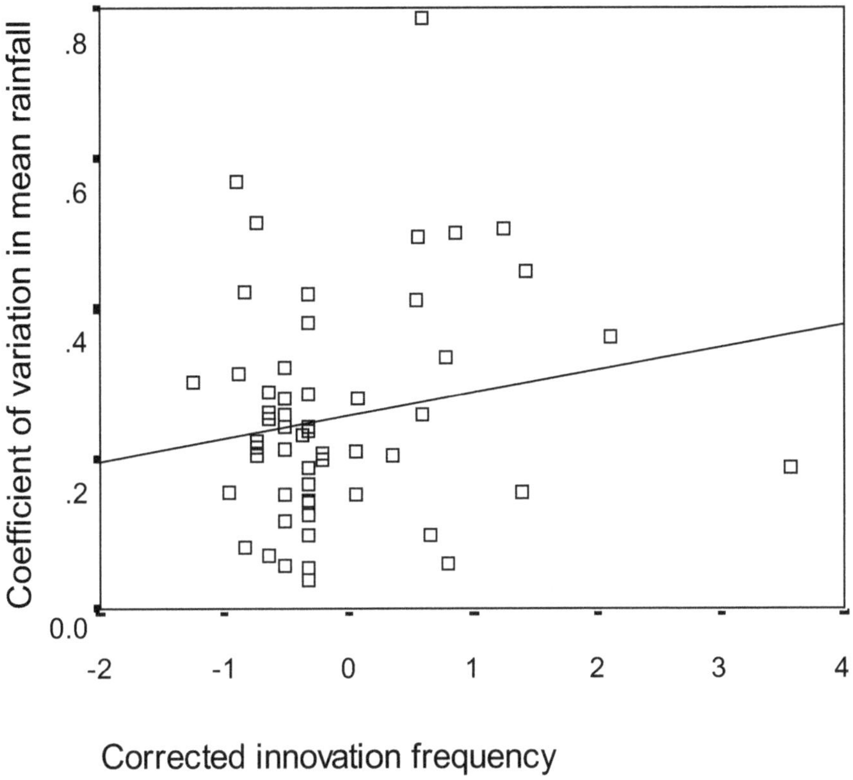

FIGURE 4.24.

SPATIAL VARIATION IN RAINFALL AND CORRECTED INNOVATION FREQUENCY. THE SPATIAL VARIATION WITHIN A SPECIES GEOGRAPHIC RANGE IS CALCULATED AS THE COEFFICIENT OF VARIATION IN MEAN DAILY RAINFALL IN MM/DAY*10 BETWEEN 0.5° CELLS. THE RAW DATA, WITH EACH POINT REPRESENTING ONE SPECIES

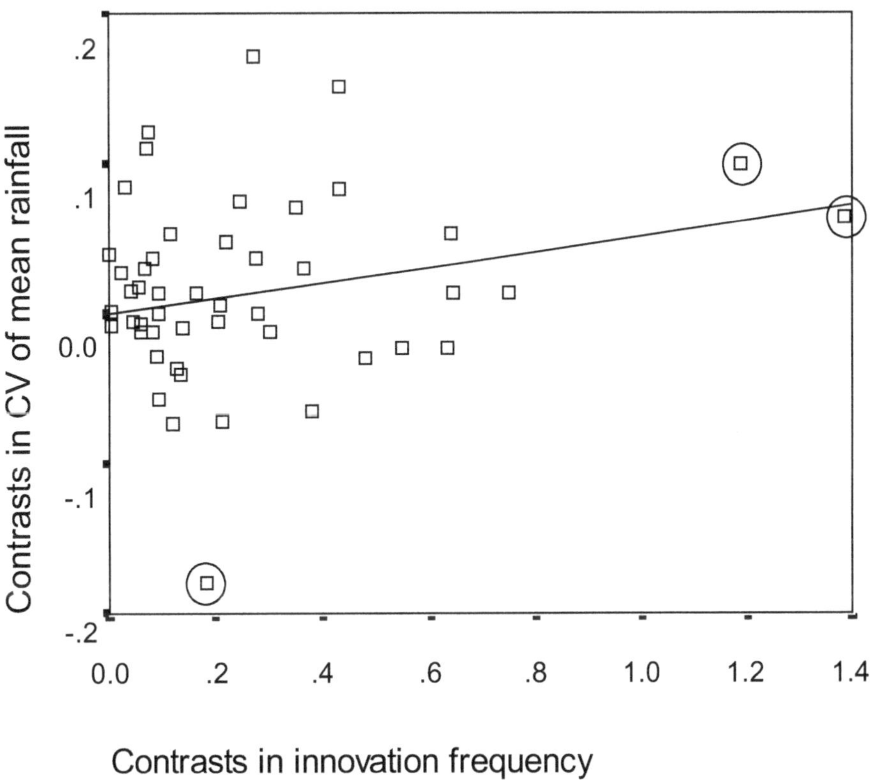

FIGURE 4.25.

SPATIAL VARIATION IN RAINFALL AND CORRECTED INNOVATION FREQUENCY. THE INDEPENDENT CONTRAST DATA. OUTLIERS CIRCLED

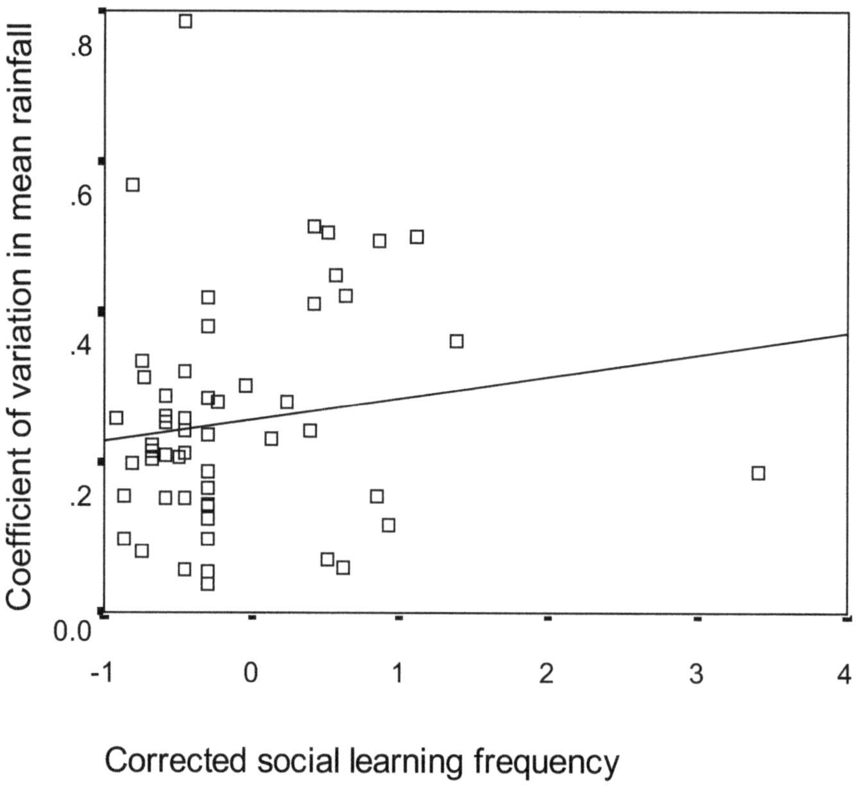

Figure 4.26

Spatial variation in rainfall and corrected social learning frequency.
The raw data, with each point representing one species

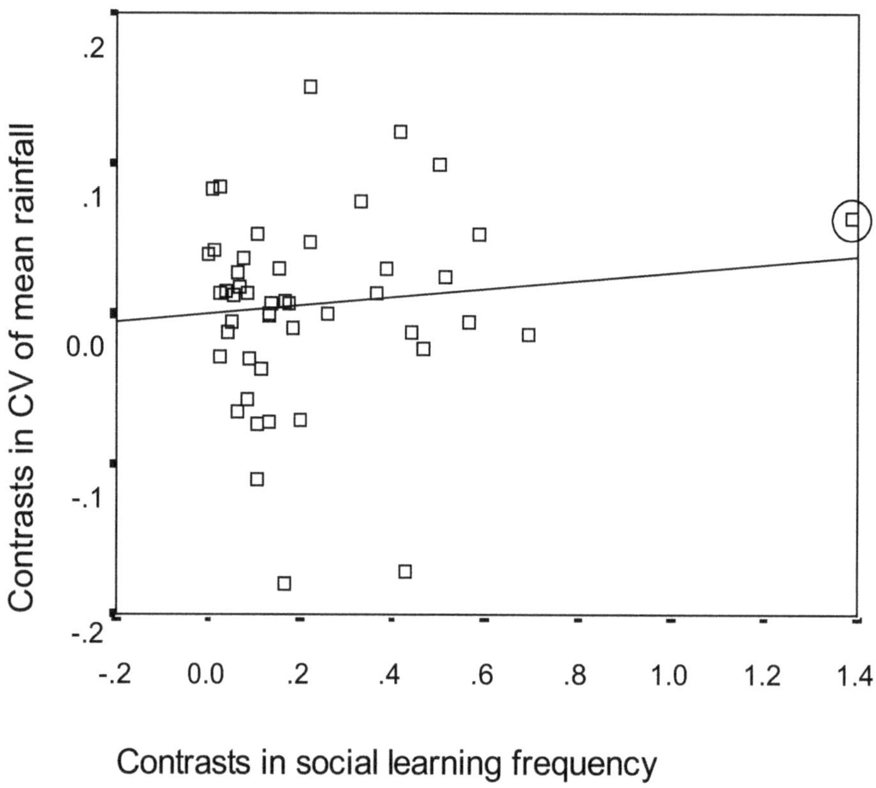

Figure 4.27

Spatial variation in rainfall and corrected social learning frequency.
The independent contrast data. Outlier circled

$p = 0.973$, $n = 24$, Kolmogorov Smirnov $Z = 0.424$, $p = 0.994$), which confirm the null hypothesis.

Regression of social learning frequency and the spatial variation in rainfall gave a non-significant result using both the raw data and independent contrasts (using independent contrasts, with one outlier removed $r^2 = 0.005$, $p = 0.635$, $n = 47$) (Figures 4.26-7); again this failed to meet all the assumptions of CAIC. The outlier was the contrast between the two chimpanzee species. There was also no relationship with social learning frequency when innovation frequency was controlled for ($r^2 = 0.001$, $p = 0.812$, $n = 48$). These results were confirmed by analyses using Brunch ($t = 0.965$, $p = 0.344$, $n = 24$, Kolmogorov Smirnov $Z = 0.570$, $p = 0.901$).

There was no significant relationship between spatial variation in mean rainfall and relative brain weight in the raw data or using independent contrasts (using independent contrasts $r^2 = 0.006$, $p = 0.573$, $n = 55$) (Figures 4.28-29). In addition, there was no significant relationship with absolute brain weight (using independent contrasts $r^2 = 0.000$, $p = 0.877$, $n = 58$). There were assumption violations for the dependent variable in these analyses. There was a relatively large value for brain weight for the contrast calculated between *C.galeritus* and *C.torquatus*. Removal of this outlier did not change the significance of the regression. There was no significant relationship with neocortex ratio (using independent contrasts $r^2 = 0.007$, $p = 0.701$, $n = 23$), and this analysis meets the assumptions of CAIC well. The results for the brain size data do not confirm the significant and positive results found in the observation frequency dataset.

I also carried out analyses using continental datasets. There was a significant positive relationship between innovation frequency and the variation in rainfall across a species range for African primates using independent contrasts ($r^2 = 0.212$, $p = 0.021$, $n = 24$). There were assumption violations for both variables in this analysis. This result was confirmed for tool use frequencies but not for social learning frequencies. This pattern is consistent with the hypothesis that behavioural flexibility and related intelligence allows species to cope with wider spatial variability in rainfall. It is interesting that this pattern is not shown in South American primates. A wider area of South America is characterised by relatively high rainfall than Africa (Figure 4.21). Spatial variability in climate may be less important in South America than Africa as a selective pressure and a factor in range expansion. Other factors such as competition may be relatively more important in primate adaptation in South America.

I carried out the observation frequency analyses using the coefficient of variation of temperature as well as rainfall (see appendix for details). None of the results were significant, suggesting that variation in rainfall is a more important aspect of climatic variability where primates are concerned. This is consistent with the results of earlier analyses (Cowlishaw and Hacker, 1997).

Seasonal variation

There was a mild positive trend in the relationship between temperature range and observation frequencies of innovation, tool use and social learning (Figure 4.30, 4.32). However this was not significant using the raw data or independent contrasts. Regression of innovation frequency and the natural log of temperature range, using independent contrasts, gave $r^2 = 0.027$, $p = 0.26$, $n = 48$ (Figure 4.31). Examination of the outliers indicated that there was one outlier with a residual greater than ± 3, and two outliers with large values for contrasts in innovation frequency. These latter contrasts were the contrasts between *P.troglodytes* and *P.paniscus*, and the contrast calculated at the node between *P.anubis* and *P.papio*, and *P.cynocephalus* and *P.ursinus*. Reanalysis after removal of these outliers did not change the significance of the results, and improved adherence to the assumptions.

Regression of tool use frequency and natural log of temperature range, using independent contrasts, was not significant ($r^2 = 0.023$, $p = 0.294$, $n = 48$). There were no obvious outliers in this analysis. The results for the analysis with social learning frequencies were also not significant ($r^2 = 0.030$, $p = 0.236$, $n = 48$). There was one outlier with a residual greater than ± 3, and one outlier with a large value for social learning frequency (the contrast between *P.troglodytes* and *P.paniscus*) (Figure 4.33). Removal of these outliers did not change the significance of the result. There were assumption violations for the dependent and independent variables in these analyses.

These conclusions were reinforced by the results obtained using Brunch (for innovation frequency $t = -0.279$, $p = 0.782$, $n = 24$, Kolmogorov Smirnov $Z = 0.668$, $p = 0.764$; tool use frequency $t = -0.148$, $p = 0.884$, $n = 24$, Kolmogorov Smirnov $Z = 0.475$, $p = 0.978$; social learning frequency $t = 0.307$, $p = 0.762$, $n = 24$, Kolmogorov Smirnov $Z = 0.577$, $p = 0.894$). These results confirm the null hypothesis that there is no relationship between behavioural flexibility and the seasonal variation within a primate species geographic range.

There was a significant positive relationship between absolute brain weight and temperature range in the raw data (both variables natural log transformed, $r^2 = 0.069$, $p = 0.036$, $n = 63$). This was absent in the independent contrast data ($r^2 = 0.015$, $p = 0.359$, $n = 58$), and in analyses using relative brain weight (Figure 4.34).

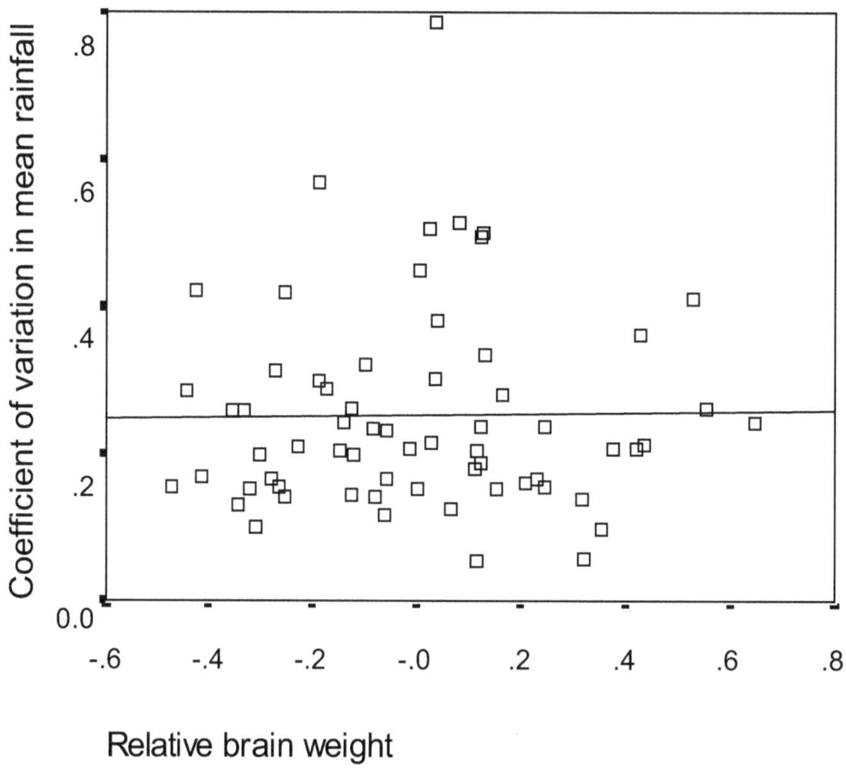

Figure 4.28

Spatial variation in mean rainfall and relative brain weight. Relative brain weights are calculated as the residuals of a log-log plot of absolute brain weight (g) against female body weight (kg). The raw data, with each point representing one species

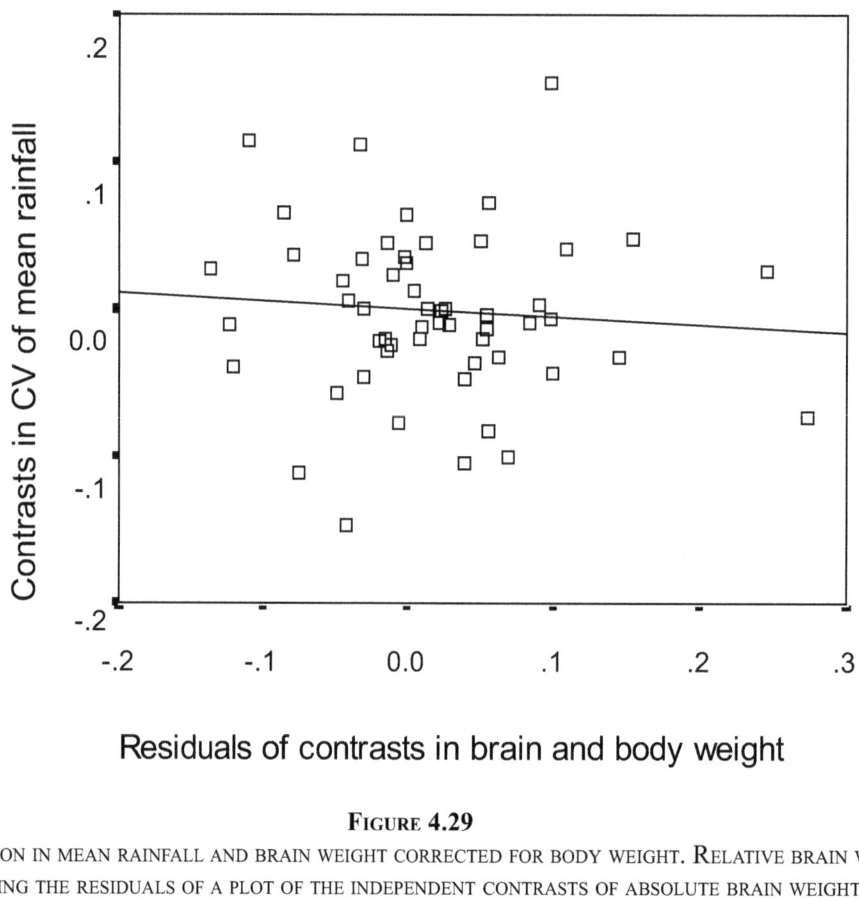

Figure 4.29

Spatial variation in mean rainfall and brain weight corrected for body weight. Relative brain weights are calculated by taking the residuals of a plot of the independent contrasts of absolute brain weight (g) and female body mass (kg), both variables natural log transformed. The independent contrast data

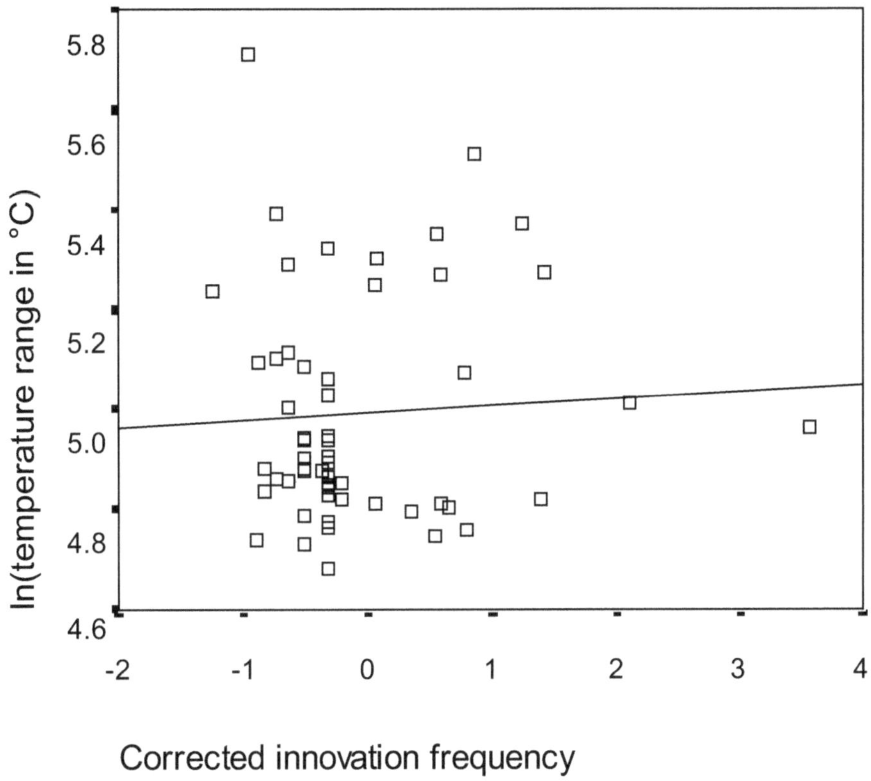

Figure 4.30

Temperature range in °C (natural log transformed) against corrected innovation frequency. Temperature range for each species is calculated as the mean across the range. The raw data, with each point representing one species

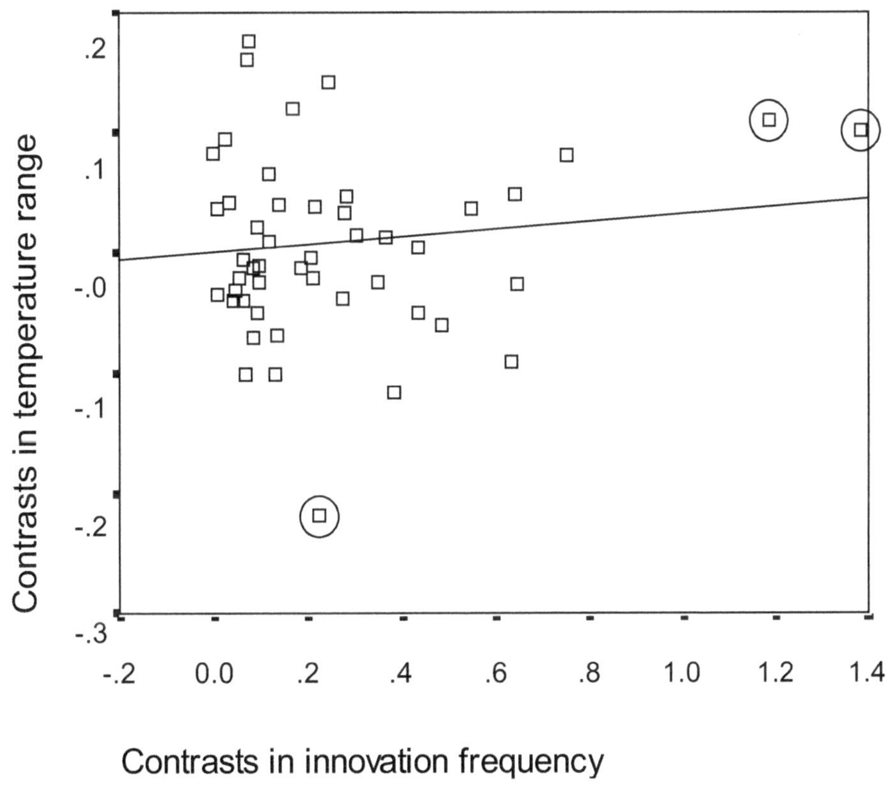

Figure 4.31

Temperature range in °C (natural log transformed) against corrected innovation frequency. The independent contrast data. Outliers circled.

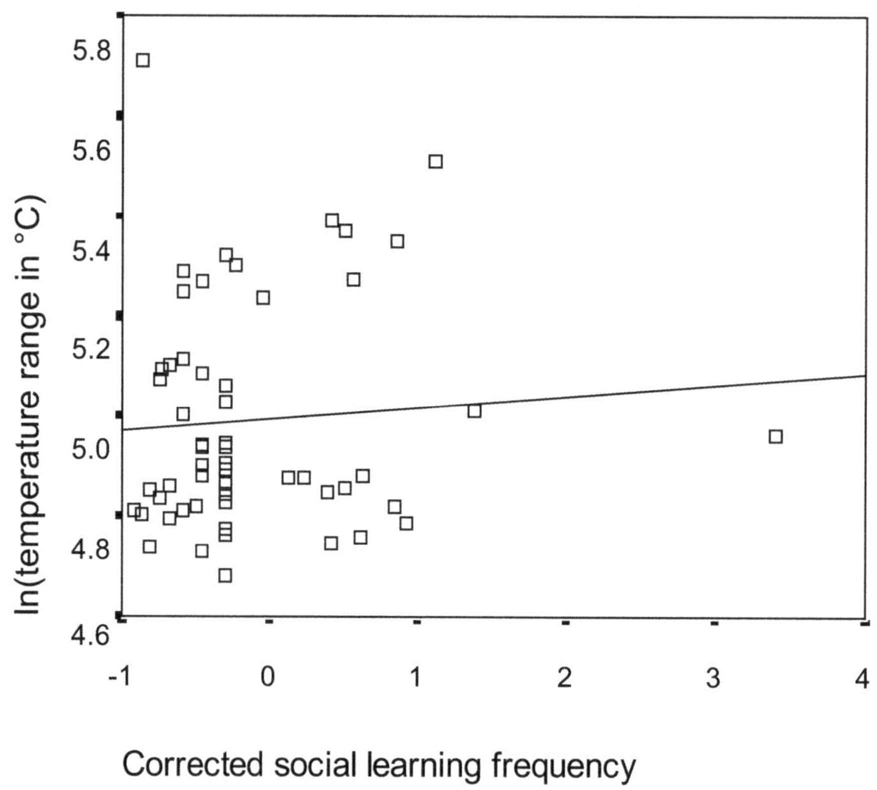

FIGURE 4.32

TEMPERATURE RANGE IN °C (NATURAL LOG TRANSFORMED) AGAINST CORRECTED SOCIAL LEARNING FREQUENCY. THE RAW DATA, WITH EACH POINT REPRESENTING ONE SPECIES.

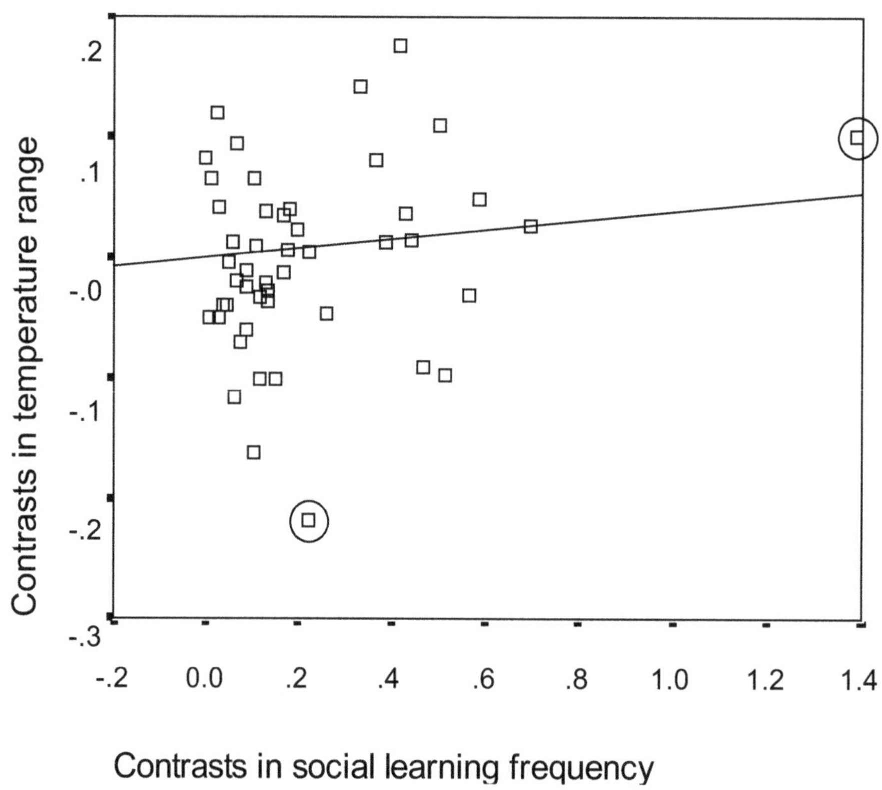

FIGURE 4.33

TEMPERATURE RANGE IN °C (NATURAL LOG TRANSFORMED) AGAINST CORRECTED SOCIAL LEARNING FREQUENCY. THE INDEPENDENT CONTRAST DATA. OUTLIERS CIRCLED.

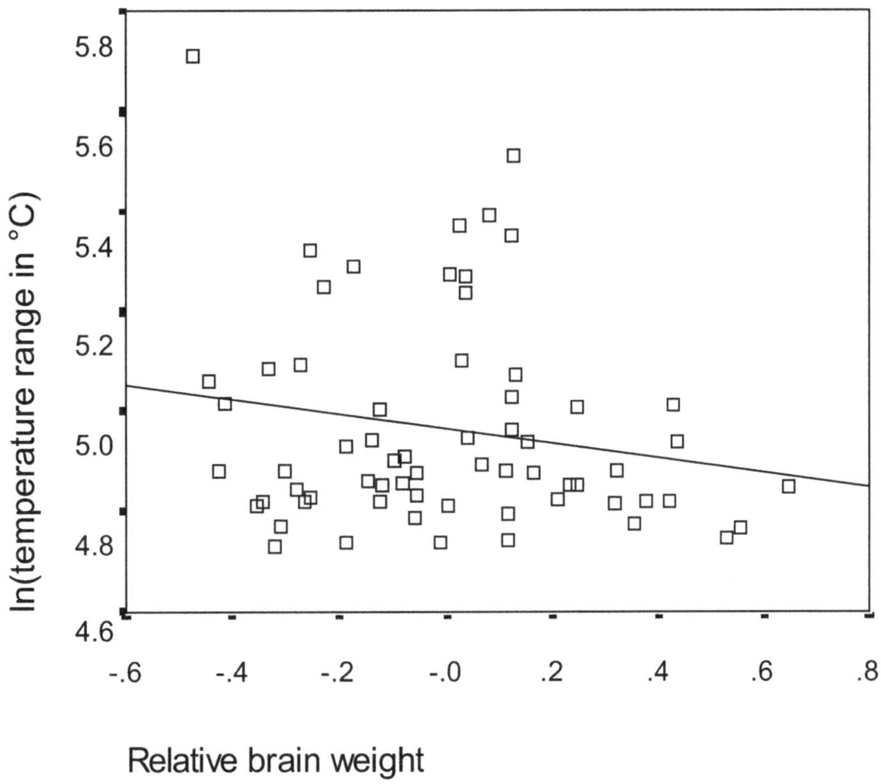

FIGURE 4.34. TEMPERATURE RANGE IN °C (NATURAL LOG TRANSFORMED) AND RELATIVE BRAIN WEIGHT. RELATIVE BRAIN WEIGHTS ARE CALCULATED BY TAKING THE RESIDUALS OF A LOG-LOG PLOT OF BRAIN WEIGHT (G) AND FEMALE BODY MASS (KG). THE RAW DATA, WITH EACH POINT REPRESENTING ONE SPECIES.

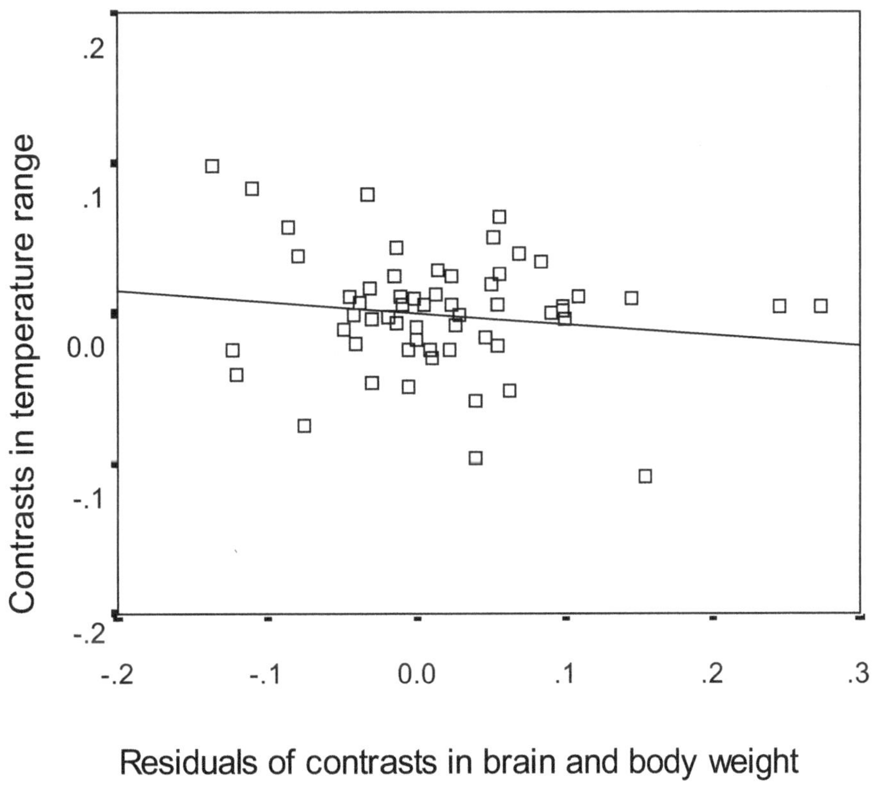

FIGURE 4.35. TEMPERATURE RANGE IN °C (NATURAL LOG TRANSFORMED) AND RELATIVE BRAIN WEIGHT. RELATIVE BRAIN WEIGHTS ARE CALCULATED AS THE RESIDUALS OF A PLOT OF THE INDEPENDENT CONTRASTS OF ABSOLUTE BRAIN WEIGHT (G) AND FEMALE BODY MASS (KG), BOTH VARIABLES NATURAL LOG TRANSFORMED. THE INDEPENDENT CONTRAST DATA.

These differing results may reflect a relationship between the temperature range within a species range and body mass. There was a relatively large value for brain weight for the contrast calculated between *C.galeritus* and *C.torquatus*. Removal of this outlier did not change the significance of the regression. There was no significant relationship between relative brain weight and temperature range using independent contrasts ($r^2 = 0.022$, $p = 0.26$, $n = 58$) (Figure 4.35). There was a violation of the statistical assumptions of CAIC for the dependent variable in these analyses.

In addition there were no significant results or evidence for different patterns in South America and Africa. Primates in Africa may experience somewhat higher ranges of temperatures than in South America (Figure 4.22) but this did not seem to produce different patterns in relation to observation frequencies. This analysis of the effects of species' behavioural flexibility on their tolerance of seasonal variation in climate might be affected by the choice of measurement, given that primate species are more affected by variation in rainfall than temperature. For future analysis it would be interesting to obtain equivalent seasonality data for rainfall.

Interannual variation

There was not a significant relationship between innovation frequency and interannual variation in rainfall, using the raw data or independent contrasts (using independent contrasts, $r^2 = 0.061$, $p = 0.088$, $n = 48$) (Figures 4.36-7). There were two outliers with large values for contrasts in innovation frequency in this analysis (these were the contrasts between *P. troglodytes* and *P.paniscus*, and the contrast calculated at the node between *P.anubis* and *P.papio*, and *P.cynocephalus* and *P.ursinus*). Reanalysis without these contrasts did not change the significance of the results, and did improve the meeting of assumptions.

There was no significant relationship between natural log transformed interannual variation in rainfall and social learning frequency (using independent contrasts, $r^2 = 0.030$, $p = 0.235$, $n = 48$) or tool use frequency (using independent contrasts, $r^2 = 0.055$, $p = 0.105$, $n = 48$). There was one outlier with a large value for contrast in social learning frequency (the contrast between *P.troglodytes* and *P.paniscus*) (Figure 4.39). Removal of this outlier did not change the significance of this result. There was no significant relationship with social learning frequency controlling for innovation frequency.

The results for analysis using Brunch confirm these conclusions. The results were $t = 0.038$, $p = 0.970$, $n = 24$, Kolmogorov Smirnov $Z = 0.526$, $p = 0.945$ for innovation frequency, $t = -0.355$, $p = 0.726$, $n = 24$, Kolmogorov Smirnov $Z = 0.440$, $p = 0.990$ for tool use frequency, and $t = 0.501$, $p = 0.621$, $n = 24$, Kolmogorov Smirnov $Z = 0.482$, $p = 0.975$ for social learning frequency. These results confirm the null hypothesis that there is no relationship between behavioural flexibility and the interannual variation in rainfall within the primate species geographic range.

There were no significant relationships between the alternative brain size measurements and interannual variation in rainfall, using raw data and independent contrasts. The results using independent contrasts for absolute brain weight were $r^2 = 0.004$, $p = 0.645$, $n = 58$, and for brain weight corrected for body weight were $r^2 = 0.027$, $p = 0.21$, $n = 58$ (Figure 4.41). There was an assumption violation for the dependent variable in both of these analyses. None of the contrasts had a residual greater than ± 3, but there was one contrast with a relatively high value for brain weight. Removal of this outlier (the contrast between *C.galeritus* and *C.torquatus*) did not change the significance of the regression. The results with neocortex ratio were $r^2 = 0.002$, $p = 0.843$, $n = 23$. All assumptions were met for this analysis. The results obtained using the brain size data support the lack of significant results given by the observation frequency dataset.

I also carried out these analyses using continental datasets. There was a significant positive relationship between innovation frequency and interannual variability within the range for African but not South American primates, using independent contrasts ($r^2 = 0.163$, $p = 0.046$, $n = 24$ for Africa, $r^2 = 0.026$, $p = 0.453$, $n = 23$ for South America). This pattern was not present in the raw data. The significant result for innovation frequency in Africa remained when interannual variability was natural log transformed to meet the assumptions of CAIC. While the patterns of interannual variability are similar in the two continents (Figure 4.23), South America has a wider area with relatively high mean rainfall. This may mean that the degree of temporal variation in rainfall is less important as a selective pressure in South America.

Life history and range size

Hypothesis: primate species with relatively fast reproductive rate and life history can tolerate environmental change and disperse more successfully than other species, and will therefore have a wider geographic range.

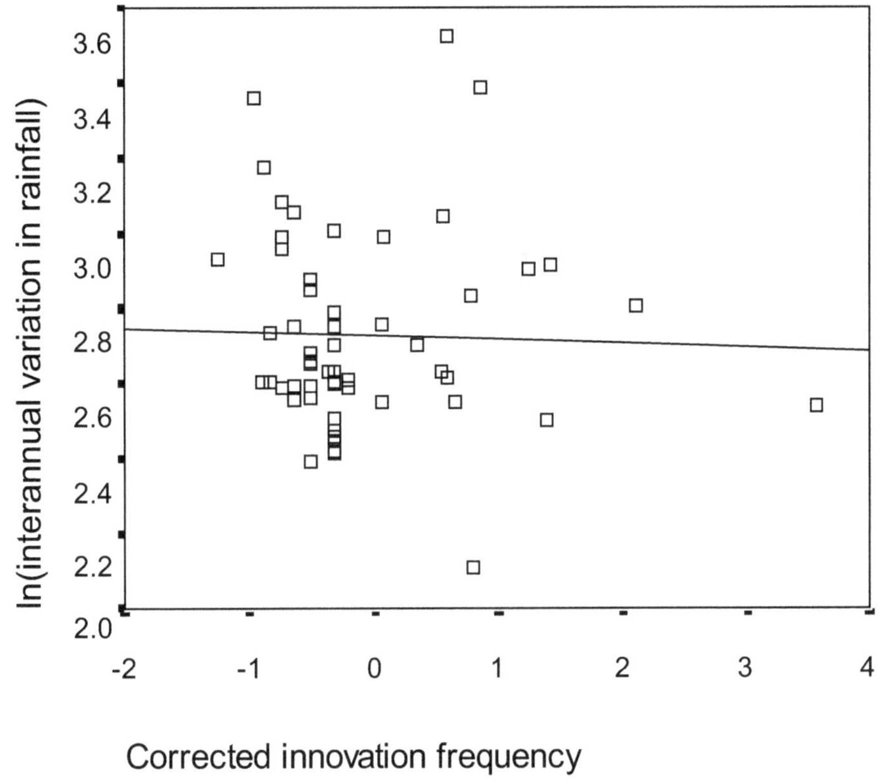

Figure 4.36

INTERANNUAL VARIATION IN RAINFALL IN % (NATURAL LOG TRANSFORMED) AND CORRECTED INNOVATION FREQUENCY. INTERANNUAL VARIATION IS CALCULATED AS THE COEFFICIENT OF VARIATION IN ANNUAL RAINFALL. THE VALUE FOR EACH SPECIES IS THE MEAN FOR THE GEOGRAPHIC RANGE. THE RAW DATA, WITH EACH POINT REPRESENTING ONE SPECIES.

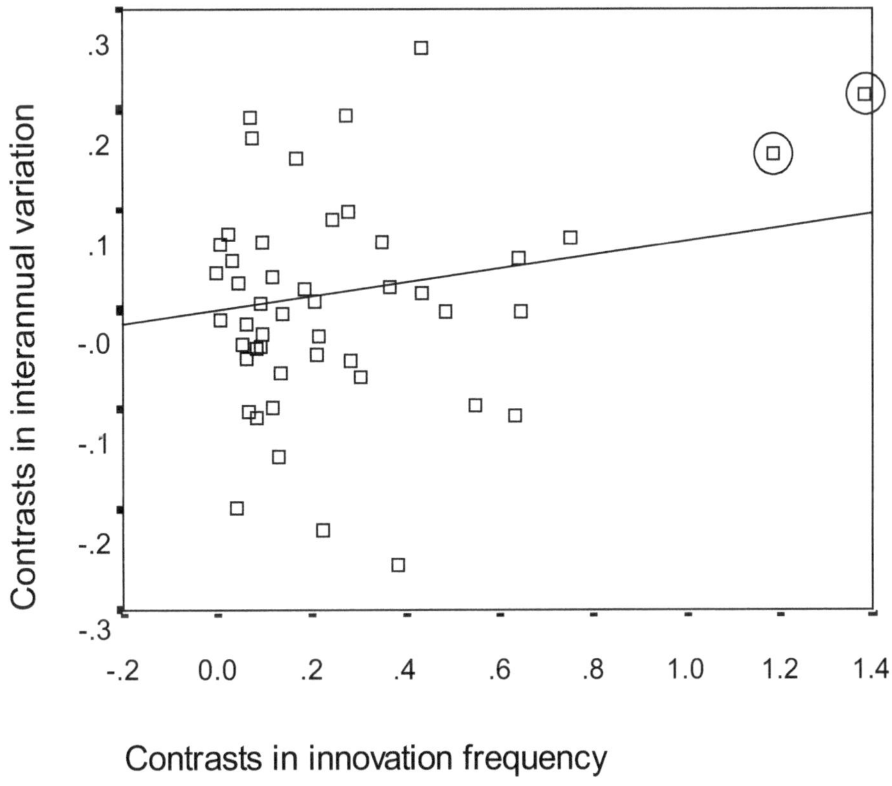

Figure 4.37

INTERANNUAL VARIATION IN RAINFALL IN % (NATURAL LOG TRANSFORMED) AND CORRECTED INNOVATION FREQUENCY. THE INDEPENDENT CONTRAST DATA. OUTLIERS CIRCLED

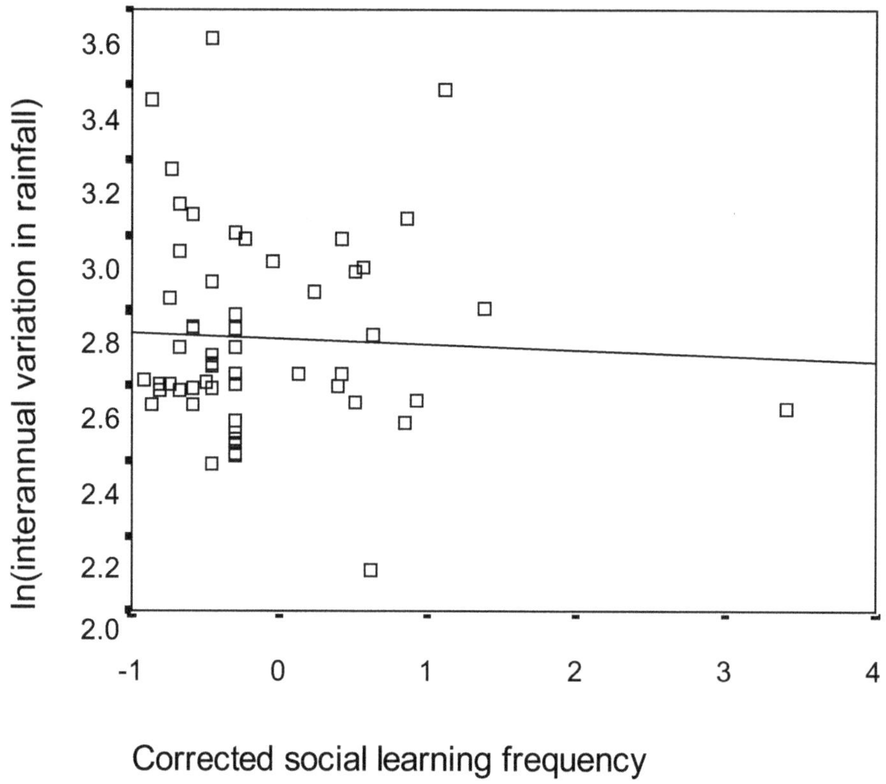

Figure 4.38

Interannual variation in rainfall in % (natural log transformed) and corrected social learning frequency. The raw data, with each point representing one species.

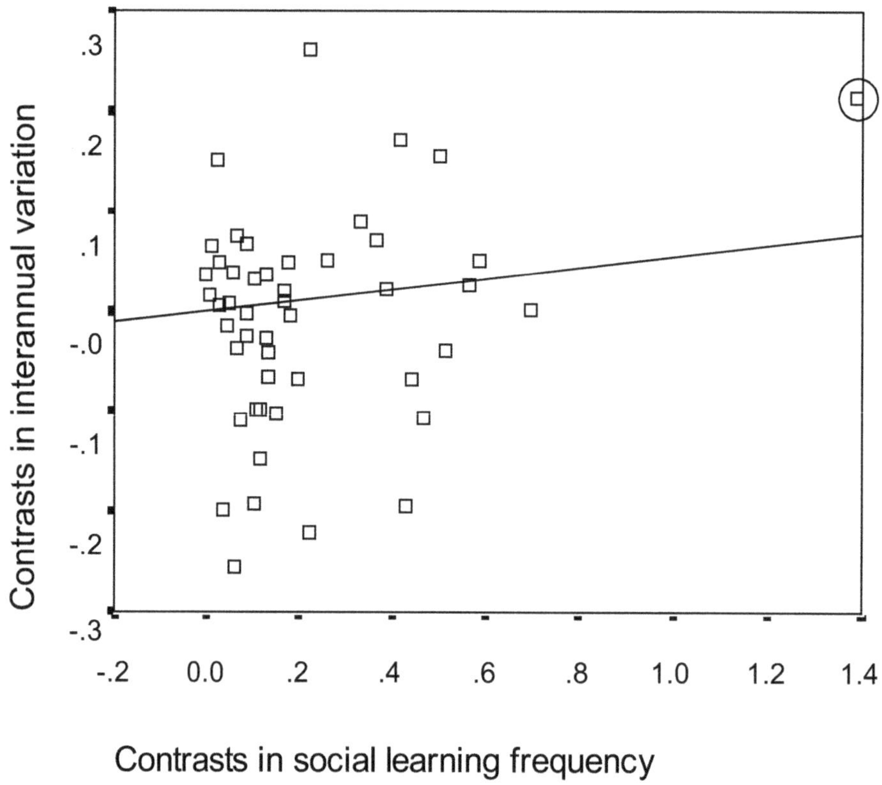

Figure 4.39

Interannual variation in rainfall in % (natural log transformed) and corrected social learning frequency. The independent contrast data. Outlier circled.

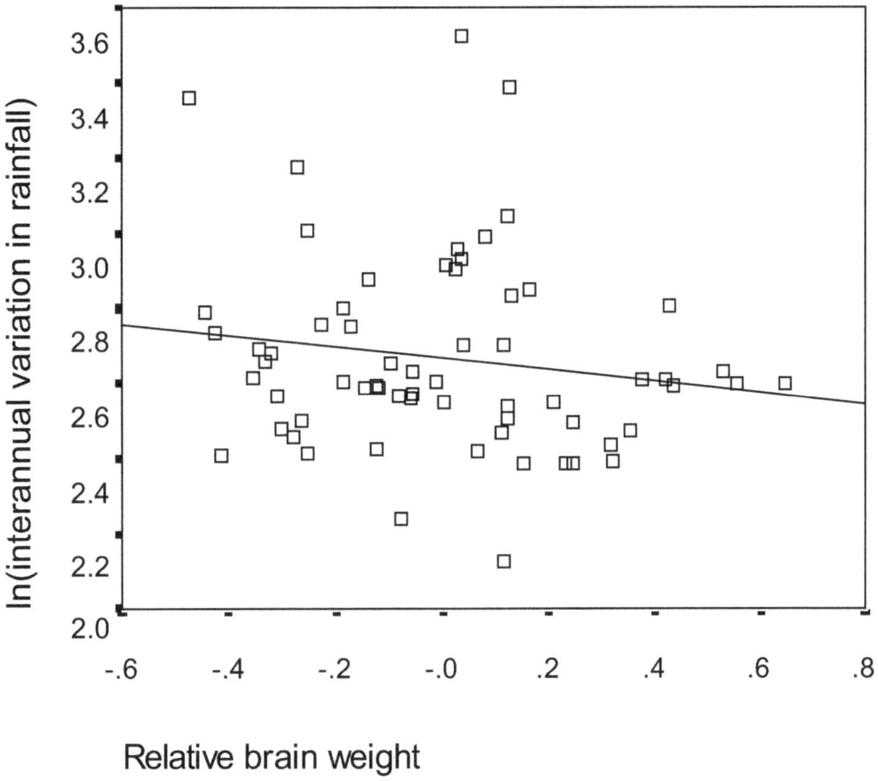

Figure 4.40

INTERANNUAL VARIATION IN RAINFALL IN % (NATURAL LOG TRANSFORMED) AND RELATIVE BRAIN WEIGHT. RELATIVE BRAIN WEIGHTS ARE CALCULATED BY TAKING THE RESIDUALS OF A LOG-LOG PLOT OF BRAIN WEIGHT (G) AND FEMALE BODY MASS (KG). THE RAW DATA, WITH EACH POINT REPRESENTING ONE SPECIES.

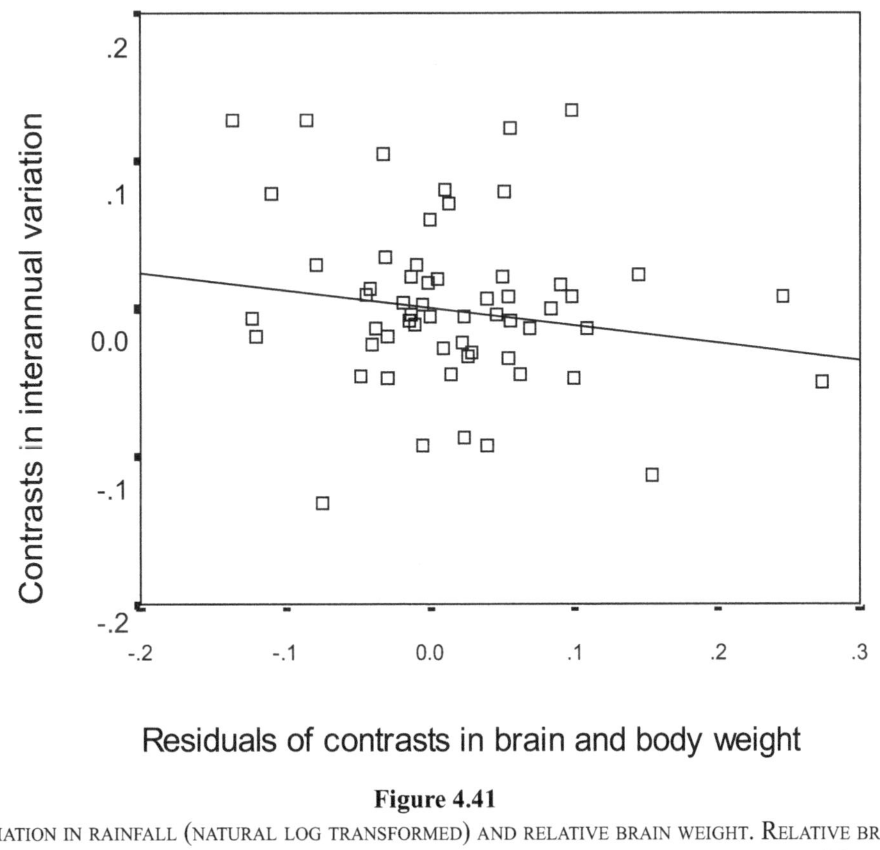

Figure 4.41

INTERANNUAL VARIATION IN RAINFALL (NATURAL LOG TRANSFORMED) AND RELATIVE BRAIN WEIGHT. RELATIVE BRAIN WEIGHTS ARE CALCULATED AS THE RESIDUALS OF A PLOT OF THE INDEPENDENT CONTRASTS OF ABSOLUTE BRAIN WEIGHT (G) AND FEMALE BODY MASS (KG), BOTH VARIABLES NATURAL LOG TRANSFORMED. THE INDEPENDENT CONTRAST DATA.

There was a slight negative trend in geographic range size and gestation length, which would support the hypothesis that species with faster life history schedules have larger ranges (Figure 4.42). However there was a slight positive trend in geographic range size and maximum lifespan (Figure 4.44). It is interesting that despite the strong relationship between life history variables and body mass, the two life history variables present a somewhat different distribution with geographic range size. These patterns were not significant at species level, either independently or when controlling for body mass. It is striking that there was a slight negative change in geographic range size with body mass in the raw data (Figure 4.46); large body mass is associated with large range size in orders other than the primates. This pattern could be influenced by human interference: large bodied primates are particularly vulnerable to hunting pressure.

Using independent contrasts there was very little slope in gestation length and body mass, and a slight positive trend for lifespan. The results for gestation length and geographic range size were $r^2 = 0.000$, $p = 0.889$, $n = 51$ (Figure 4.43). There was one assumption violation for the dependent variable for this analysis. Controlling for body mass did not greatly change the significance ($r^2 = 0.002$, $p = 0.770$, $n = 51$). The results for lifespan and geographic range size (using independent contrasts) were $r^2 = 0.009$, $p = 0.576$, $n = 37$ (Figure 4.45) and again were not greatly changed by controlling for body mass ($r^2 = 0.003$, $p = 0.757$, $n = 37$). Finally, the relationship with body mass was not significant (using independent contrasts, $r^2 = 0.000$, $p = 0.878$, $n = 86$) (Figure 4.47). There were violations of the assumptions of CAIC for the dependent variable (geographic range size) in this analysis. There was one outlier with a very large value for contrast in body mass, this was the contrast calculated at the node between the Tarsiers and the Anthropoids. Removal of this contrast did not change the significance of the regression. These results contradict the hypothesis that species with faster reproductive rates are likely to have larger geographic ranges.

Discussion

Based on these results, it seems that species' behavioural flexibility (as measured by observation frequencies) is related to but does not wholly explain variation in niche breadth. This provides evidence that these are largely separate areas of behaviour, and clears up a possible criticism of observation frequencies as a measurement of behavioural flexibility.

The results suggest that there may be a significant relationship between innovation frequency and geographic range size: however this result was not robust. Alternative methods may be necessary to establish the existence of such a relationship for certain. However these results do make it clear that behavioural flexibility does not have a strong effect on species geographic range size. This interpretation is confirmed by results obtained using different evolutionary assumptions. Brain size provided an alternative measure of behavioural flexibility, and the results of analyses using alternative measures of brain size support this conclusion. This is not consistent with the hypothesis that behavioural flexibility is an important factor allowing some primate species to counter environmental change or invade new habitats more successfully than other species.

The results provide strong evidence against the hypothesis that social learning frequency has coevolved with geographic range size. There was no evidence that social learning frequency, independent of innovation frequency, influenced species geographic range size or tolerance of climatic variability. This negative conclusion is confirmed by the results of parallel analyses using alternative evolutionary assumptions.

Based on these results, it seems that local distribution is not more strongly effected by behavioural flexibility than regional distribution. While analysis using species as data points supports the suggestion that home range size may be more affected by behavioural flexibility than geographic range size, this relationship appears to be due to phylogenetic effects. These results confirm the important of controlling for phylogeny. Linear regression analysis does not support the 'necessity hypothesis' that species will innovate more when under environmental stress. It is therefore unlikely that this effect is obscuring a relationship with geographic range size. This result also suggests that there is no greater threat from human activities for innovative species.

These results provide some evidence that behavioural flexibility affects the degree of climatic variability tolerated within a species geographic range. There was a significant positive relationship between spatial variation in climate and innovation and tool use frequency when phylogeny was controlled for. This supports the hypothesis that behavioural flexibility allows species to cope with greater spatial variation in climate. However, this result was not backed up by analyses carried out using alternative evolutionary assumptions, or with brain size data, and is therefore not very robust. In any case, the regression coefficients obtained in these analyses suggest that behavioural flexibility can only explain a small proportion of the variation in tolerance of spatial variability in climate.

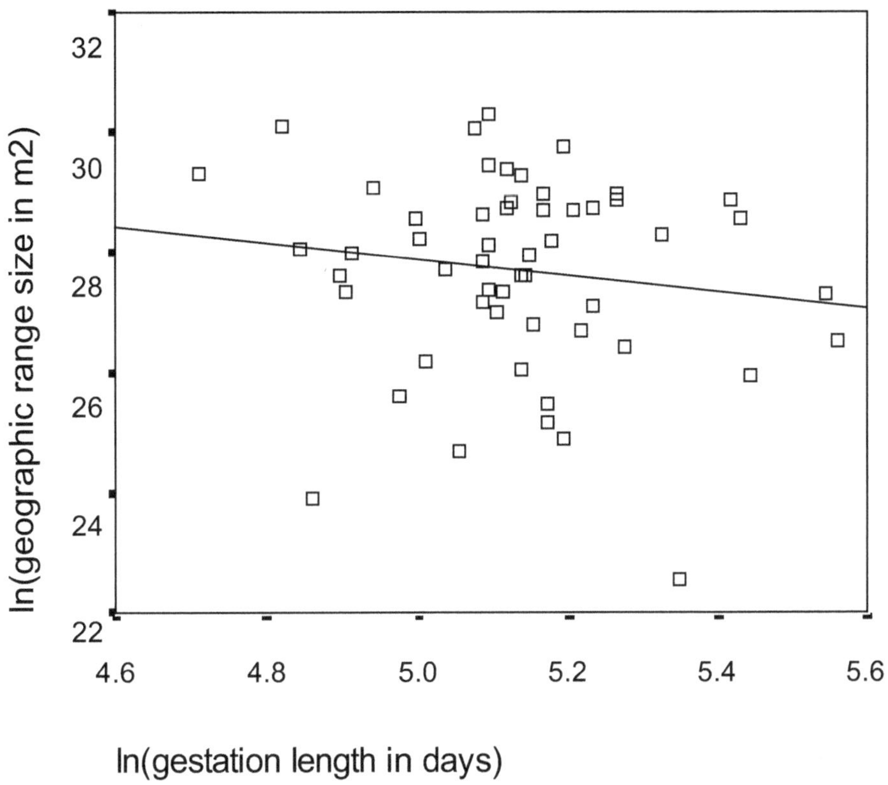

Figure 4.42
GEOGRAPHIC RANGE SIZE IN M^2 AND GESTATION LENGTH IN DAYS, BOTH VARIABLES NATURAL LOG TRANSFORMED. THE RAW DATA, WITH EACH POINT REPRESENTING ONE SPECIES.

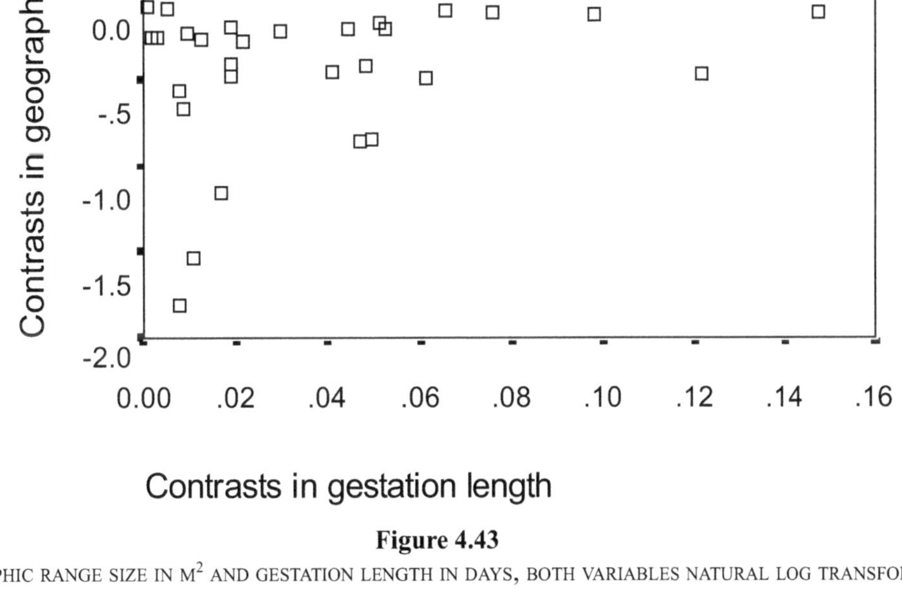

Figure 4.43
GEOGRAPHIC RANGE SIZE IN M^2 AND GESTATION LENGTH IN DAYS, BOTH VARIABLES NATURAL LOG TRANSFORMED. THE INDEPENDENT CONTRAST DATA.

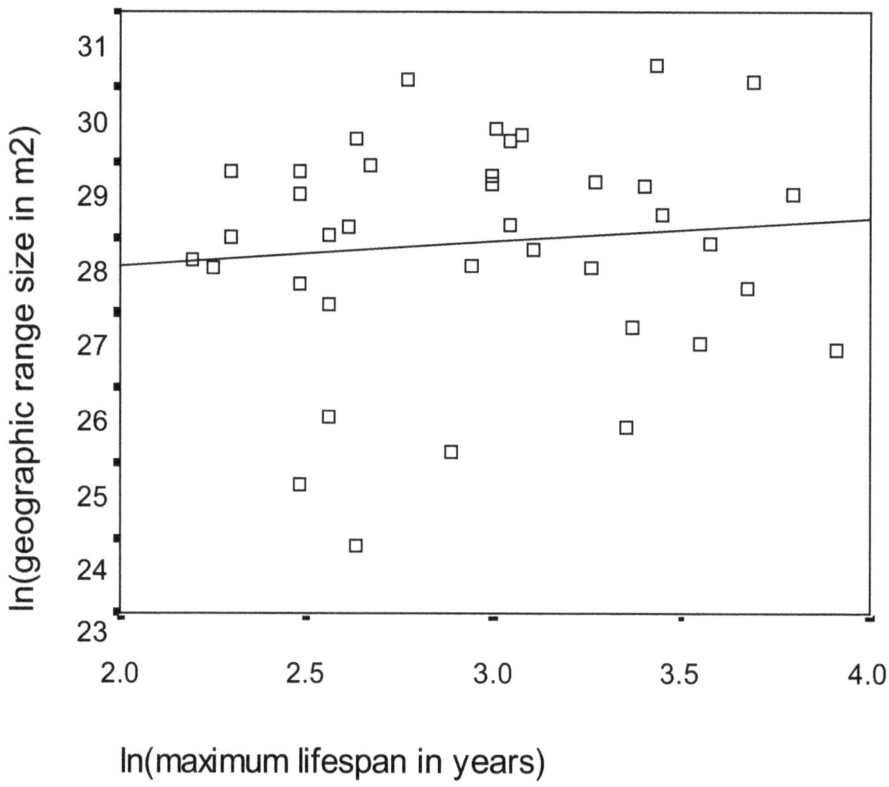

Figure 4.44

Geographic range size in m^2 and maximum lifespan in years (both variables natural log transformed). The raw data, with each point representing one species.

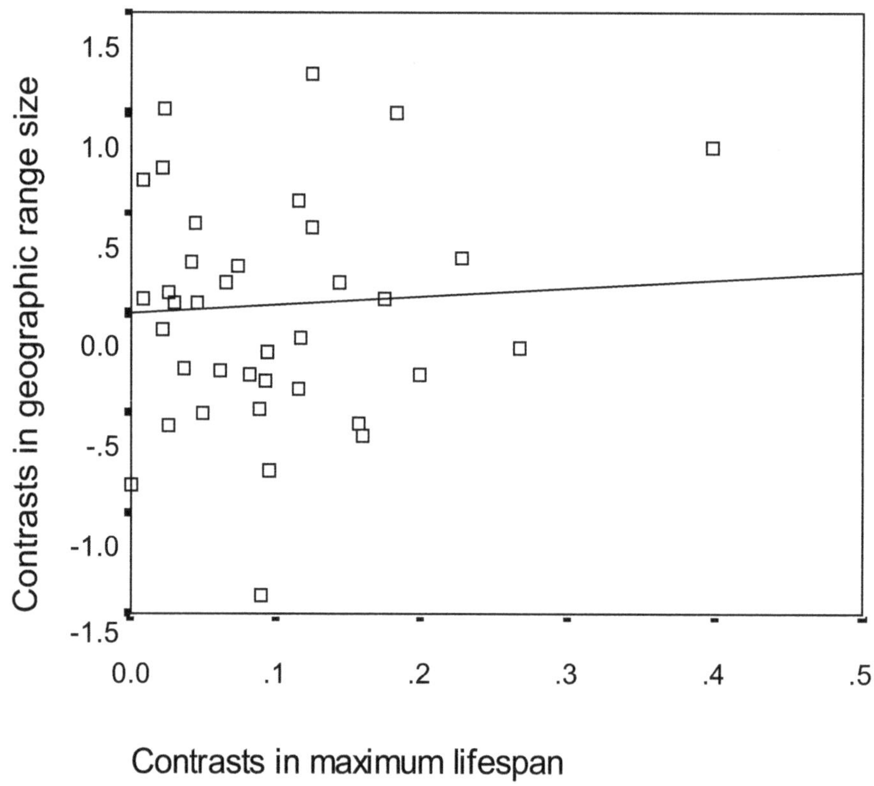

Figure 4.45

Geographic range size in m^2 and maximum lifespan in years (both variables natural log transformed). The independent contrast data.

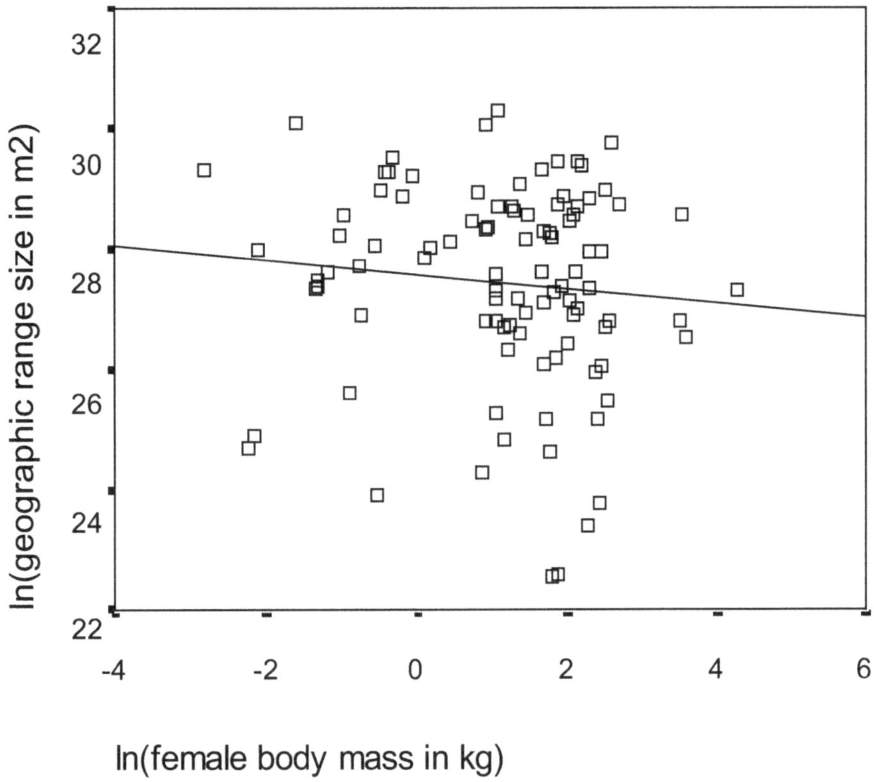

Figure 4.46

GEOGRAPHIC RANGE SIZE IN M^2 AND BODY MASS IN KG (BOTH VARIABLES NATURAL LOG TRANSFORMED). THE RAW DATA, WITH EACH POINT REPRESENTING ONE SPECIES.

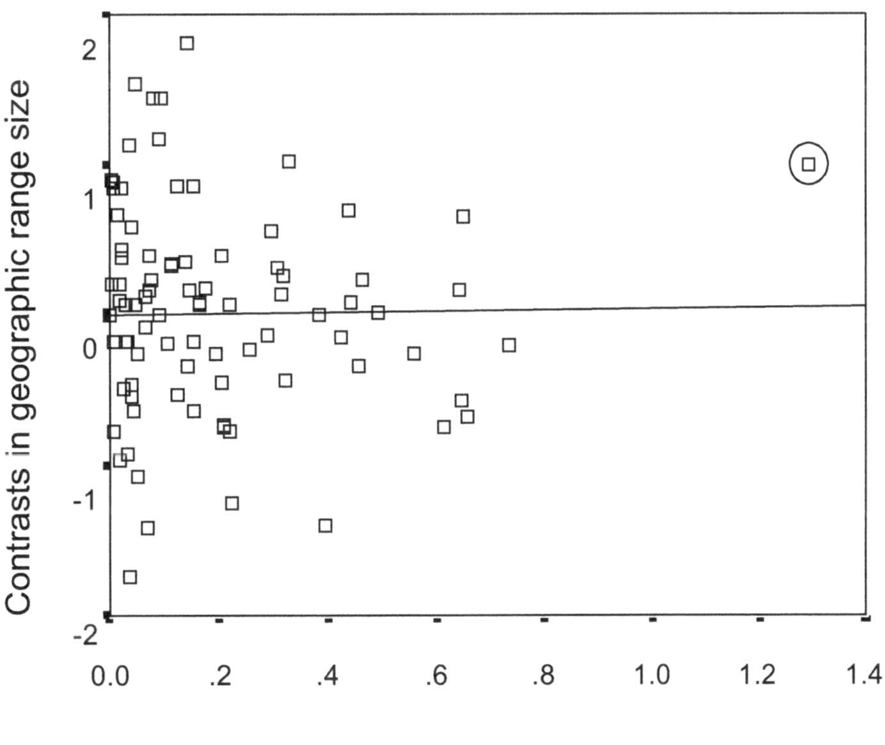

Figure 4.47

GEOGRAPHIC RANGE SIZE IN M^2 AND BODY MASS IN KG (BOTH VARIABLES NATURAL LOG TRANSFORMED). THE INDEPENDENT CONTRAST DATA. OUTLIER CIRCLED.

Behavioural flexibility was not a good predictor of the degree of climatic variability over time tolerated within a species' geographic range. This was true of measurements using both the seasonal and interannual scales. These results contradict the hypotheses that more behaviourally flexible species can tolerate greater climatic variability; that behavioural flexibility is a response to climatic change; or that climatic variability has selected for a propensity to behavioural flexibility or social learning. It is also possible that other time scales or aspects of the environment have a stronger effect on primates.

These results provide some evidence for different patterns in the different continents. Innovation and tool use frequencies were significantly related to geographic range size in South America, but not in Africa and Asia. However these results were not robust. A difference in the continents was also apparent in the brain size raw data: Africa and Asia showed a negative trend, while the trend in South America was positive. These patterns might suggest that behavioural flexibility has been more important in the evolution of geographic ranges in South America than Africa. None of these relationships are significant after controlling for phylogeny and data transformation or scrutiny of outliers, and they do not provide strong support for treating continental datasets separately.

By contrast, there were significant positive relationships between spatial and interannual variability in rainfall and innovation frequency for African primates but not South American primates. This suggests that climatic variability has been more important in the adaptation of African primates. Primates seem to be more strongly affected by patterns in rainfall than other climatic variables (Cowlishaw and Hacker, 1997), and a wider area of South America has relatively high mean rainfall. This may mean that variability in rainfall is a less important selective pressure in South America than Africa, and would explain the apparent conflict in results for the continents for geographic range size and climatic variability. These differences between the continents are consistent with ones described by Harcourt (2000).

The results of the analysis of life history variables provide no evidence that life history is an important predictor of geographic range size in primates. These results are not consistent with the hypothesis that a combination of dietary niche breadth and relatively rapid life history parameters is important in tolerance of environmental change or range expansion.

Finding a suitable model for calculating independent contrasts proved problematic, particularly for the observation frequency and geographic range size databases. It has been stressed that phylogenetic methods always perform much better than cross-species analyses (Purvis *et al.*, 1994), and the regression model is quite robust with regard to minor violations of the assumptions (Shennan, 1997, p.159).

In general the results are very consistent through analysis using the raw data and independent contrasts. In addition there is consistency in the results produced using alternative measurements, such as observation frequencies and brain size measurements. Such consistency suggests that these conclusions are robust. Careful examination of outliers and their effects on the regression decreases the likelihood of accepting a significant relationship when there is none. The use of an alternative algorithm in CAIC (called 'BRUNCH') to generate contrasts, which only makes the assumption that evolution in different branches is independent, adds confidence to my conclusions in difficult cases.

Problems of interpretation are raised when the result obtained using independent contrasts is not the same as that using other methods or measurements. For instance, the significant relationship between spatial variability in climate and observation frequency differs in the raw data and independent contrasts. It may be that alternative techniques, such as an experimental or case study approach, would be necessary to get any further in characterising patterns in geographic range size and behavioural flexibility. For instance, it would be interesting to look at case studies of innovation, tool use and social learning in a cross-section of primate species in particular environments. Behavioural flexibility may affect range size only in cases of species with particular characteristics and environmental problems. For instance it may be especially useful in primates that co-exist with humans. In addition, CAIC is strongly affected by data quality, as discussed in Purvis and Webster (1999). It may be that the extent of occurrence measurement of geographic range size produces too much error, and that a more detailed measurement of geographic range size would be necessary to improve cross-species analyses.

Conclusion

The general message of these analyses seems to be that hominins were unlikely primate species to have large ranges! There is little indication from the primate data that their increasing brain size and associated intelligence would have allowed them to extend their geographic ranges to an area beyond those of modern apes. There is no support for the hypothesis that increased social learning is the key to primate range expansion. Nor does a relatively rapid life history influence primate geographic ranges. This suggests that other variables may be more important. Niche breadth (Eeley and Foley, 1999) and climatic variability at the centre of the range (Cowlishaw and Hacker, 1997) are better predictors than the variables analysed here. The primate analyses suggest that both models 1 and 2 are flawed as an explanation of the evolution of hominin geographic ranges. It may be necessary to look outside the primates for an explanation of hominin range expansion.

Chapter 5

DIETARY ADAPTATION AND DISTRIBUTION IN AFRICAN MAMMALS

Abstract

In the previous Chapter, I argued that a primate-based model of behavioural flexibility and social learning or life history parameters is not able to explain hominin range expansion. A third model in Chapter 3 focussed on dietary niche. Dietary quality, and meat eating, can be linked to spatial patterns at a range of geographical scales from local to regional and of units from the individual organism to the species or higher taxonomic group. In this Chapter I will describe the characteristics of carnivore ecology in greater detail. According to model 3, there are dietary constraints on the distribution of large-brained primates, and a change in dietary niche was critical in overcoming that constraint in *H.erectus*. In this Chapter I will discuss current evidence for trends in hominin diets.

An investigation of large-scale spatial trends in African mammal distribution was carried out according to large taxonomic groups that are broadly related to diet. GIS (Geographical Information Systems) software was used to produce maps showing spatial patterns in the distribution of body mass, species richness, biomass, and home range size for each group. Visual and statistical comparison of spatial patterns in the different orders and in the distribution of environmental variables provides the basis for identification of trends and interpretation of determining factors. The aim of this analysis is to evaluate the effects of dietary niche on distribution.

Analysis of data for modern mammals indicates that there are some strong differences in the spatial distribution of ecological variables. Geographic range sizes differ in scale, but have a similar frequency distribution in the carnivores, ungulates and primates. Primates show a different spatial distribution pattern from carnivores and ungulates in diversity, species range boundaries, body mass, and biomass. This is not the case for primate and carnivore individual home range size; however primates differ from carnivores in the distribution of home range size relative to body mass. Patterns in carnivore distribution seem to be strongly related to ungulate diversity and biomass, while primate distribution reflects primary productivity. This suggests that dietary niche may be an important factor in differentiating variation between higher taxonomic groups.

Introduction

Carnivore ecology

> 'Patterns of variation among extant species provide a useful framework for evaluating hominid evolutionary trends, particularly between 2.5 and 1.5 my ago. During this time, changes in three key variables – body size, ecosystem productivity and diet – all would have contributed to substantial increases in hominid home range size.' (Leonard and Robertson 2000, p.645)

As discussed in Chapter 1, orders of species differ in the scale of their geographic ranges and in their adherence to ecological rules such as Rapoport's rule. These differences between taxons are determined by evolutionary history and ecological characteristics. Thus the characteristic dietary niches of the carnivores, primates and ungulates may be the basis for differences in the scale and nature of their geographical distribution.

A suite of ecological variables is strongly affected by trophic level: these include body mass, home range size and population density. Initially, all energy on earth comes from sunlight. But transforming this into the tissues of an organism costs energy. As far as mammals are concerned, the cost of energy transformation also usually means that the number of individuals at each trophic level decreases markedly from that on the step below it (Shipman and Walker 1989). Thus carnivore population density is lower than that of herbivores. The reason for this difference is obvious: if predators are too numerous, they will consume their prey faster than the prey can reproduce and will quickly starve (ibid.). Population density is inversely dependent on body size in carnivores and herbivores (Damuth 1981; Shipman and Walker 1989), but in carnivores population density controlling for body mass is lower (Shipman and Walker 1989).

A related fact is that carnivores with a similar body size tend to have larger home ranges than herbivores. Home range size increases with body mass in both herbivores and carnivores. Since there is less energy available at higher trophic levels, large carnivores have to range more widely to find enough food.

Large home range size is likely to be associated with a larger geographic range size. Carnivore and primate species characterised by low population densities are more likely to survive if they also have large geographic ranges (Purvis, Gittleman *et al.* 2000). Increase in home and day range might promote dispersal through increased mobility and an increase in knowledge of the environment. Meat has a number of advantages as a dietary resource in range expansion: it is a reliable resource in seasonal environments, and can be eaten safely when moving into new habitats.

Carnivorous species are not necessarily larger than herbivores. However, carnivores with a large body mass are subject to a number of ecological constraints. There are energetic constraints on prey size for carnivores, such that carnivores over a certain body mass will hunt larger prey (Carbone, Mace *et al.* 1999). In addition, as discussed above, larger bodied mammals tend to have a lower population density, larger home range and larger geographic range, and these trends are exacerbated for species at higher trophic levels.

The diet of modern humans is strikingly different from that of other primates. It can therefore be assumed that at some point in human evolution, a shift in dietary niche occurred, involving a substantial increase in the amount of meat consumed. This would have important implications for the ecology of the species in which this shift occurred.

As described above, the density rule based on trophic level predicts that almost any species moving from the primary to secondary consumer trophic level is liable to have too high a population density. Shipman and Walker (1989) suggest that this could be adjusted by a number of means: a reduction in total numbers; a reduction in body size; or an expansion in home range size, causing a net expansion in geographic range size.

This discussion has identified the importance of diet in determining a species' ecological characteristics. It has been suggested that there was a major dietary shift in the course of human evolution. The evidence for hominin diets will be discussed below. It has been suggested that geographic range size is one factor that may have been influenced by dietary niche in hominin evolution (Shipman and Walker 1989). The rest of this Chapter will investigate the way in which distribution and ecology interact with dietary niche.

Hominin diet

Change over time

The evidence for hominin diets comes from a range of sources, including skeletal morphology, isotopic analysis, archaeological evidence and comparison with living species. Some of the results from these different methods suggest that changes in dietary niche, whether an increase in diet breadth or a dietary shift involving increased meat eating, may have been important in human evolution.

Craniofacial and dental evidence provides information about changes in hominin diets. The gracile australopithecines had thick enamelled, flattened molars, which would not have been well suited for slicing through tough fruits, leaves or meat (Teaford and Ungar 2000). The morphological characteristics of these australopithecines would have allowed them to eat both hard foods and soft foods that were not particularly tough, and abrasive and non-abrasive foods. This would have left these early hominins well suited for life in a variety of habitats (Teaford and Ungar 2000).

More extreme morphological trends are noticeable in a number of species of early hominin that are characterised by the development of cranial structures that allowed great force to be applied between large upper and lower cheek teeth. These include *A.aethiopicus*, *P.boisei* and *P.robustus*. Microwear studies of australopithecine teeth have suggested that *P.robustus* often consumed harder, more fibrous food items while *A.africanus* subsisted more on leaves and fleshy fruit (Grine 1986; Kay and Grine 1988).

A reduction in tooth size and less robust jawbones are evident in early African *H.erectus*, as well as some of the early *Homo* specimens (Klein 1999), suggesting that less force was required in chewing the foods eaten by these species. According to Wood and Collard (1999), the teeth and jaws of early African *H.erectus* are more similar to *H.sapiens* than to the australopithecines, suggesting that the diet of this species had mechanical properties similar to those of the modern human diet. By contrast, the teeth and jaws of *H.habilis*, while small in absolute terms, are closer to those of the australopithecines when related to a proxy of body size (Wood and Collard 1999). These studies of fossil morphology suggest a contrast in diet between the australopithecines and *H.erectus*.

Dietary breadth

Stable carbon isotope analysis of *A.africanus* from Makapansgat demonstrates that this early hominin ate not only fruits and leaves but also large quantities of $\delta^{13}C$ enriched foods such as grasses and sedges or animals that ate these plants, or both (Sponheimer and Lee-Thorp 1999). This result suggests that *A.africanus* regularly exploited relatively open environments such as woodlands or grasslands for food (Sponheimer and Lee-Thorp 1999). This contrasts with chimpanzee habitat use: isotopic studies of two populations of 'savannah' chimpanzees indicated little use of open grassland areas (Schoeninger, Moore *et al.* 1999). This comparison suggests that a relatively broad dietary niche characterised the hominins from an early period.

A similarly mixed rather than strictly vegetarian diet for *P.robustus* has been indicated by analyses of Sr/Ca and $\delta^{13}C$ isotope ratios (Sillen 1992; Lee-Thorp, Vandermerwe *et al.* 1994). In addition, the results of analysis of three *Homo* cf. *ergaster* individuals from Swartkrans were indistinguishable from those for *P.robustus* from the same site, showing that proportions of C_3 and C_4 based foods in their diets did not differ (Lee-Thorp, Thackeray *et al.* 2000). Both show a reliance on C_3-based foods coupled with a small but significant proportion of C_4 dietary carbon (Lee-Thorp, Thackeray *et al.* 2000). This complicates interpretation of the dietary distinction between these species suggested by contrasts in craniofacial and dental morphology.

However this does not mean that they were eating exactly the same foods. The range of C_3 foods could have included items such as fruits, nuts, edible leaves, gum, roots, geophytes, and vertebrates and invertebrates which live on C_3 plants (Lee-Thorp, Thackeray *et al.* 2000). Available C_4 plants are more limited, including vertebrates and invertebrates feeding on grasses, and possible grass seeds or C_3 sedges: animal foods are by far the more likely source (Lee-Thorp, Thackeray *et al.* 2000). The dietary differences between the two species are most likely to be found in the C_3 category. Sillen *et al.* (1995) suggest that *Homo* was more active in gathering underground plant resources, on the basis of elevated Sr/Ca in one of two *Homo* specimens. This view has found support in a model proposed by O'Connell *et al* (1999).

Evidence for insectivory in the hominin diet comes from bone tools from Swartkrans Members 1-3 which were probably used to dig into termite mounds (Backwell and d'Errico 2001). This suggests that part of the C_4 signature in some of the Swartkrans hominins is due to insect eating. Most of the remaining archaeological evidence for hominin diets relates to meat eating.

Meat eating

There is evidence that one of the functions of the earliest stone tools was to process meat: bones with cutmarks and percussion marks on them have been recovered from Bouri, Middle Awash dated to 2.5 my ago (de Heinzelin, Clark *et al.* 1999). These are contemporary with the earliest stone tool sites, situated nearby at Gona (Semaw, Renne *et al.* 1997; Semaw 2000). This early evidence is not entirely surprising, given that there is increasing documentation of regular meat eating by extant primates in the wild. Meat eating has been observed with relative frequency in chimpanzees, baboons and capuchin monkeys (Rhine, Norton *et al.* 1986; Boesch and Boesch 1989; Fedigan 1990; Perry and Rose 1994; Stanford, Wallis *et al.* 1994; Rose 1997; Uehara 1997). Chimpanzee groups that live in savannah ecosystems eat more meat than groups that live in dense tropical forests (Boesch and Boesch 1989; Stanford, Wallis *et al.* 1994; Uehara 1997). It might therefore be expected that early hominins would incorporate some meat into their diet, because of their close resemblance with chimpanzees in physiology, and because of being adapted to more open areas (Dominguez-Rodrigo 2002). The high level of C_4 in *Australopithecus* may confirm this expectation (Sillen 1992; Sponheimer and Lee-Thorp 1999; Lee-Thorp, Thackeray *et al.* 2000).

Relative brain size increases throughout early human evolution. Brain tissue is metabolically expensive. Thus species with a large brain relative to their body size will have higher metabolic requirements. According to the expensive tissue hypothesis, the high metabolic requirements of a larger brain were balanced in hominins by a corresponding reduction in gut size, made possible by an increasing emphasis on nutritious, easily digestible meat (Aiello and Wheeler 1995). Reduction in gut size in *H.erectus*, if true, could also be explained by increased preparation of food by cooking (O'Connell, Hawkes *et al.* 1999). Another study suggests that increased dietary quality is associated with higher energy expenditure in both primates and humans (Leonard and Robertson 1996). This contradicts the predictions of co-evolution between brain size and gut size. However, Leonard and Robertson (1996) confirm the association between the increased energy demands of large brain size and a higher quality diet in extant primates. Again, this would suggest that major increases in brain size are likely to involve a major change in diet. However, evolution of the human hunting and gathering economy may have both necessitated and allowed for a higher-quality diet (Leonard and Robertson 1996).

Hominin meat eating becomes visible in the archaeological record at an early date. The best evidence for the nature of carnivorous behaviour in early hominins comes from early Pleistocene sites such as those at Olduvai Gorge and Koobi Fora, where bone and stone are associated in large numbers. In total, excavations at FLK Zinj, Olduvai Gorge, produced about 2,500 artefacts and 60,000 bones, including microfauna. Sometimes these sites involve close associations of stone tools with very large mammals: for instance the elephant butchery site in FLK N Level 6 at Olduvai Gorge (Leakey 1971), and the association of *Elephas recki* with 400 stone artefacts at site 15, Olorgesailie (Potts 1994). Hominins and carnivores were both involved in the formation of some Plio-Pleistocene sites, as suggested by the presence of both cut-marks and tooth marks on archaeological faunal assemblages (Bunn 1981; Potts and Shipman 1981). The production of stone tools has often been seen as a means by which hominin species could dramatically increase the amount of meat in their diet. However confirming this hypothesis requires us to account for all of the processes by which animal bones become part of the archaeological record.

Researchers have analysed archaeological sites with stone and bone assemblages to ascertain hominin involvement with animal carcasses by examining skeletal part profiles and the quantity and placement of cutmarks from stone tools. One example of this approach is Bunn and Kroll's (1986) seminal analysis of the FLK Zinj site in Olduvai Bed I. However many early analyses suffered from methodological and theoretical problems. Skeletal part analysis has tended to ignore the fact that patterns of bone transport and accumulation vary in modern humans, even within one particular group of people (Dominguez-Rodrigo 2002). Secondly, researchers have focussed on transport and discard, which are only the first part of the processes through which bone accumulations at archaeological sites are formed (Dominguez-Rodrigo 2002).

Cut mark patterns may be useful to indicate whether hominins had primary or secondary access to the carcass (Dominguez-Rodrigo 2002), thus illuminating questions of how much meat was eaten and the method of procurement. Dominguez-Rodrigo has carried out analyses of bone surface marks at the FLK Zinj site (Bed I, Olduvai Gorge) and the FxJj50 site (Koobi Fora). The FLK Zinj site is more than 1.75 my old (Leakey 1971), while FxJj50 is over 1.5 my old (Dominguez-Rodrigo 2002). The results suggest that hominins were processing carcasses that had substantial amounts of meat on them, and therefore had early access to the carcasses (ibid.). Upper limb bones and shafts from carcasses at both sites are the most often cut-marked appendicular sections, indicating early access by hominins to fleshed carcasses, because they are utterly devoid of flesh at carnivore kills (ibid.). Similar analyses of bone surfaces at the ST Site Complex Oldowan sites at Peninj, Lake Natron, dated to about 1.5 my ago, indicated again that hominins had primary access to fully fleshed carcasses, and that carnivore activity was limited to post-depositional ravaging (Dominguez-Rodrigo, de Luque *et al.* 2002). Thus the earliest evidence for access to relatively large amounts of meat comes from *c.* 1.8-1.5 my ago, when such behaviour could be attributed to early African *H.erectus*, early *Homo* or the robust australopithecines.

The 'hunting hypothesis' suggested that the inception of big game hunting in hominin behaviour would be associated with a suite of human-like behaviours including paternal provisioning, extended periods of juvenile dependence, central place foraging, a sexual division of labour, and the nuclear family (Isaac 1978). This argument assumes that associations of large mammal fossils and stone tools reflected a pattern of behaviour involving butchery and transport of large mammals from kill sites to base camps for further processing. However evidence for increased meat eating in the archaeological record, as discussed above, does not necessarily imply other aspects of this model. There is evidence that such sites could have been the result of near-kill locations, repeatedly visited by hominins (Dominguez-Rodrigo, de Luque *et al.* 2002). In addition this evidence does not tell us whether meat was a major part of the hominin diet.

Several lines of evidence, including the spatial patterning of sites (Jablonski, Whitfort *et al.* 2000) and community analyses (Turner 1992) suggest that hominins and carnivores shared similar niches by at least middle and later Pleistocene times. Successful colonization of Eurasia by African hominins may have hinged on their ability to obtain flesh, since access to plant foods is intensely seasonal throughout much of temperate and periglacial Eurasia (Stiner 2002). Thus the expansion of the hominin range into higher latitudes may have had important consequences for hominin dietary requirements and for community ecology.

General trends

There is strong evidence for dietary breadth as a part of the hominin adaptation. While dental morphology in the robust australopithecines suggested dietary specialism, isotopic analysis has revealed a mixed diet. However there is no clear pattern of increasing hominin niche breadth that might explain range expansion: as discussed above, it seems likely that meat was a part of australopithecine diets and was not a new factor in the diet of early *Homo*.

So what is the evidence for a dietary shift in human evolution? Increased documentation of primate meat eating, combined with the isotopic evidence, makes it seem probable that early hominin species ate some meat. Archaeological evidence shows that one use of the earliest stone tools was butchery. However the earliest evidence for access to relatively large amounts of meat comes from *c.* 1.8-1.5 my ago, when such behaviour could be attributed to early African *H.erectus*, early *Homo* or the robust australopithecines. Dental morphology suggests that the diet of *H.erectus* had different mechanical requirements from that of earlier species, which could reflect increased meat eating. However an alternative explanation stresses the processing of vegetable foods. Analysis of a range of characteristics suggests that early African *H.erectus* occupied a new adaptive niche (Wood and Collard 1999). However isotopic analysis of *H.erectus* in southern Africa shows no great difference in exploitation of C_4 and C_3 foods by *H.erectus* and *P.robustus*. Finally, *H.erectus* outside Africa may have been more closely associated with the carnivore niche. Thus the evidence for a shift in dietary niche involving eating large amounts of meat in early hominins is equivocal.

Method

Discussion

The aim of the research described in this Chapter is to evaluate the effects of dietary niche on distribution, in order to test the predictions of model 3. As in the analyses described in Chapter 4, I used data from large numbers of species to identify general trends. However in other ways the approaches are quite different. Instead of comparing variation in parameters across species, I compared trends at higher taxonomic levels. As discussed in Chapter 2, taxa with consistently differing characteristics are likely to have a different scale of geographic range size variation, and may also differ in their adherence to biogeographical rules and frequency distribution. One characteristic that might be expected to have such an effect is dietary niche or trophic level. Comparison of taxa with consistently differing dietary niche will provide a test of this hypothesis.

Three taxonomic groups were compared: the carnivores, ungulates and primates. These groups were chosen because they broadly reflected dietary niche and trophic level. Because of their adaptive niche the carnivores provide the best analogy for the change in diet attributed to *H.erectus*. Some primates do eat meat: however meat rarely makes up a large portion of primate diets (Richard 1985, p.136). Not all species in the order 'Carnivore' are dedicated meat eaters: some carnivores are classified as omnivorous, insectivorous, piscivorous or vegetarian on the basis of the food type that takes up at least 60% of their diet (Gittleman 1989). However most carnivores are highly dependent on animal products, and all African carnivores incorporate some animal products in their diet (Kingdon 1997). Most ungulates are herbivorous: however the pigs are opportunists and eat a wide range of food types (ibid.). Primates are unusual in their lack of anatomical specialisation for eating animal matter or leaves and grasses, and can employ quite flexible feeding strategies as a result (Chivers 1991). Primate diets are quite diverse, including gum, insects, herbs, leaves, fruit and meat. Food preference in primates is strongly related to body size (Chivers 1991). The advantage of subsuming a certain amount of variation and using orders as a whole is that it allows identification of very general trends in a comprehensive database.

The study area is the African continent. This area was selected because data on species geographic distribution was available for all African mammals, with a few exceptions (IEA 1998). In addition this area is particularly relevant to early human origins. This does mean that the distribution of a few species will not include their entire geographic range.

Maps of large-scale patterns in the distribution of ecological variables were produced, and compared between different orders of mammals. Interpretation was based on visual examination of maps and charts. Biogeographical and conservation studies use maps showing values for variables such as species diversity and mean species range size in order to investigate spatial patterns (Hacker, Cowlishaw *et al.* 1998; Eeley and Foley 1999; Eeley and Lawes 1999). Eeley and Foley (1999) and Eeley and Lawes (1999) also map species habitat and dietary niche breadth and use these maps to describe spatial trends. This approach has the advantage that it is a less labour intensive way of identifying trends across several analytical groups, each containing large numbers of species, than running cross-species statistical analyses like those described in Chapter 4.

This approach makes it harder to investigate the strength of correlation between variables and suggest causation between factors. Any regression analysis of trends in two maps would be greatly complicated by spatial auto-correlation. In addition, this method has less potential for building up a very specific predictive model. However, this method makes a comparison on a large spatial scale and with a large number of species viable.

Maps were produced showing spatial patterns in the distribution of species richness, body mass, biomass, and home range size. The distribution of these variables was plotted because, as discussed above, there are strong relationship between dietary niche or trophic level and a number of ecological variables. In addition this allowed further investigation of the predictions of model 3.

For instance, I argued that the main constraint on the distribution of large brained primates into more open habitats was dietary. I would therefore expect a difference in the distribution of body mass in relation to habitats and resources in the primates and carnivores.

Extent of occurrence maps were used as a source of data on species distribution (IEA 1998). As dicussed in Chapter 4, there are a number of possible ways of mapping species distribution. The extent of occurrence is a simplification of the actual distribution of a species, in that variation in population density is not represented, nor are relatively small areas of absence. This has certain implications for the maps of large-scale patterns in the distribution of ecological variables. It is probable that the values for variables such as species diversity and biomass will be over-estimated in some areas. In addition, mean values for each species were taken for ecological variables such as body mass, biomass and home range size. These variables tend to vary between populations as well as between species. Thus the maps are likely to simplify some of the variation present between areas. However, the maps will represent general spatial trends well, while subsuming some spatial variation at smaller scales.

Data sources

Species geographic range maps (Extent of Occurrence) are from the African Mammals Databank, instituted by the IEA (Institute of Applied Ecology) (1998). Distribution maps for the African elephant were obtained from the African Elephant Database (Barnes, Craig *et al.* 1998). The two rhinoceros species were excluded from the ungulate database: due to their protected status, up-to-date distribution maps are not available for these species.

Body mass data is from Silva and Dowling (1995), Gittleman (1985), and Smith and Jungers (1997). For those species for which exact field data was unavailable, estimates are from Kingdon (1997). Best estimates were selected according to sample size. Mean adult body mass is the mean of male and female weights.

Home range data is from a number of compilations: primate data from Nunn and Barton (2000) and Leonard and Robertson (2000), carnivore data from Gittleman (1982), and Grant *et al.* (1992). A home range is usually defined as the area used by an individual (in solitary species) or a group during its normal day to day activities (Grant, Chapman *et al.* 1992). Home range size is related to group size, and this variable is particularly relevant to orders with a great deal of variation in group size, such as the primates and carnivores. I followed the approach advocated by Milton and May (1976) and divided home range for each species by group size to determine the home range per individual. The mean of home range values was taken where there was more than one value available.

Data on home range was less readily available than that for other variables: I found data for 31 primate species, and for only 21 carnivores. I decided not to interpolate this data: the distribution of absence of data is biased phylogenetically and spatially, with less data available for rare species with small geographic ranges, which tend to live in particular habitats, such as rainforest for primates. Maps were produced showing the distribution of species for which data was absent, to give some idea of the way in which the home range maps might be biased.

Rainfall and temperature data is from New *et al.* (1999), vegetation from Olson (1983).

Calculating biomass

Estimates of population density can be obtained using regression equations describing a general relationship between density and mean adult body mass (Thackeray 1995). Damuth (1981; 1987) provides a number of equations for different groups of species including all mammals, primary and secondary consumers, and further subsets such as terrestrial secondary consumers. For the carnivore data set, I used Damuth's (1987) equation for secondary consumers:

$$\mathrm{Log}_{10}D = -0.95\ \log_{10}MAB + 4.68 \qquad r = -0.92 \qquad (1)$$

Where D is density (km^{-1}) and MAB is mean adult body mass (g). I obtained biomass values for each area by multiplying the derived population density value by mean adult body mass. Biomass values were converted to kg/hectare for comparison with Thackeray's (1995) ungulate analysis.

Damuth's (1987) equation for primary consumers covers all those species that depend primarily on plant foods. This equation was used to estimate population densities for primate and ungulate species.

$$\mathrm{Log}_{10}D = -0.73\log_{10}MAB + 4.15 \qquad r = -0.84 \qquad (2)$$

This is altered from the equation used by Thackeray (1995), which is from an earlier paper (Damuth 1981).

GIS

Geographic range maps were converted to GRID (raster) format for calculations. New maps were produced showing spatial patterns in the distribution of ecological variables using the map calculator facilities of ArcView GIS 3.2 (1999) and ERDAS Imagine 8.4 (1997). The raster maps were simplified by giving values of 1 for presence and 0 for absence. The details of production of particular maps are given below.

Diversity: Individual species geographic range maps were added together using the Map Calculator in ArcView, to produce maps showing total species richness.

Boundaries: ArcView was used to create a simple edge detecting neighbourhed filter, which was then used to process each species' range map. This produced new maps reclassified with values of 1 for cells at the edge of the range and 0 elsewhere. The new maps were added together using the Map Calculator facility in ArcView.

Body mass: Geographic range maps were reclassified with body mass values for each species, using ArcView's 'Reclassify' option. Equations were written in Imagine's Modeller facility to derive the maximum and range of values for each cell based on the reclassified maps.

Biomass: Geographic range maps were reclassified according to the biomass estimate as above. The resulting maps were added together using the Map Calculator facility to obtain an estimate of the total carnivore biomass present in an area. The raster map was converted to contours for comparison with Thackeray (1995).

Home range: Geographic range maps were reclassified with home range values as above. The sum of home range values, as well as the number of species for which home range values were available, were calculated using ArcView's Map Calculator. The mean value was obtained by dividing the sum of the home range values map by the number of species map in ArcView. Maps showing the distribution of species for which home range data was absent were also produced using Map Calculator.

ArcView's Analysis facility was used to analyse pairs of maps and generate histograms showing the frequency distribution of one variable corresponding to particular ranges of values in another variable.

Analysis

African physical geography

Annual rainfall decreases west-east across the continent and at higher latitudes (Figure 5.1). There is an overall gradient in rainfall and primary productivity from the west-central region outwards in Africa (Figure 5.2).

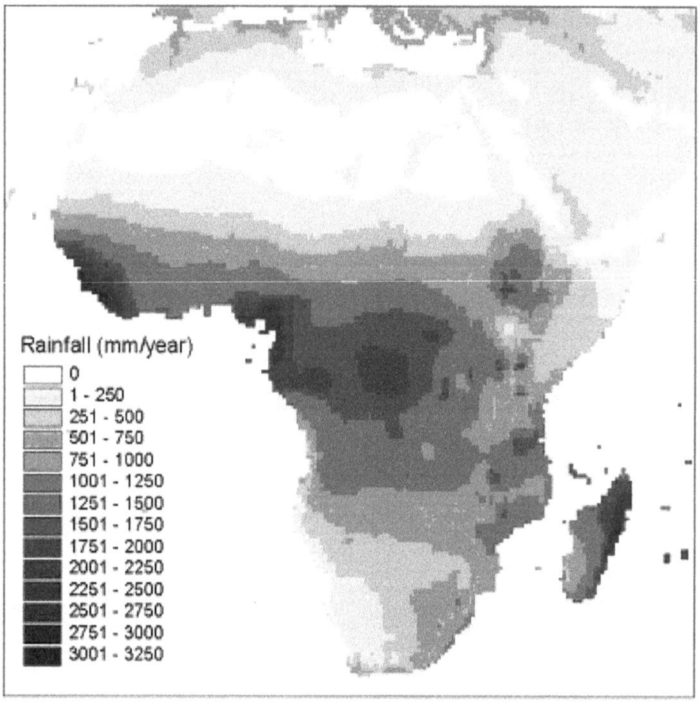

FIGURE 5.1
MEAN ANNUAL RAINFALL IN AFRICA IN MM/YEAR (FROM NEW *ET AL* 1999)

The west-central region is dominated by tropical and sub-tropical broad-leaved forest and woodland. Temperatures are generally higher at low latitudes, but also decrease to the east of the continent (Figure 5.3). There is a general increase in topographic height from the north-west to the south-east of the continent (Figure 5.4). In sub-Saharan Africa, the areas with lower primary productivity have more open habitats including savannah, grassland and scrubland as well as woodlands. Woodlands are particularly prevalent in southern Africa. There are substantial deserts in the north and south (the Sahara and Kalahari respectively). On the northern and southern coasts the climate is more temperate, and there are habitats similar to those found in mediterranean Europe.

Distribution and diversity

The frequency distribution of geographic range sizes is severely right skewed for primates, carnivores and ungulates (Figures 5.5-7). As discussed in Chapter 2, this is a common pattern in geographic range size across an order (Brown, Stevens *et al.* 1996). However mean geographic range sizes are larger for carnivores than for primates (for carnivores, mean = 3481058 km^2, for primates, mean = 1115380 km^2, figures based on the area of suitable habitat within the range (IEA 1998)). Even small carnivore ranges tend to be larger than those of primates and maximum ranges reach a larger size than in the primates (for carnivores, maximum = 16 million km^2, for primates, maximum = 12 million km^2). In addition, a number of carnivore species also occupy areas outside Africa (for instance, the geographic range of the cheetah includes sub-Saharan Africa, North Africa and Southwest Asia). The frequency distribution of carnivore geographic range sizes also has a second smaller peak in the middle of the range of values. Thus while the largest number of carnivores have relatively small ranges, there is also a trend towards increased representation of species with medium sized ranges.

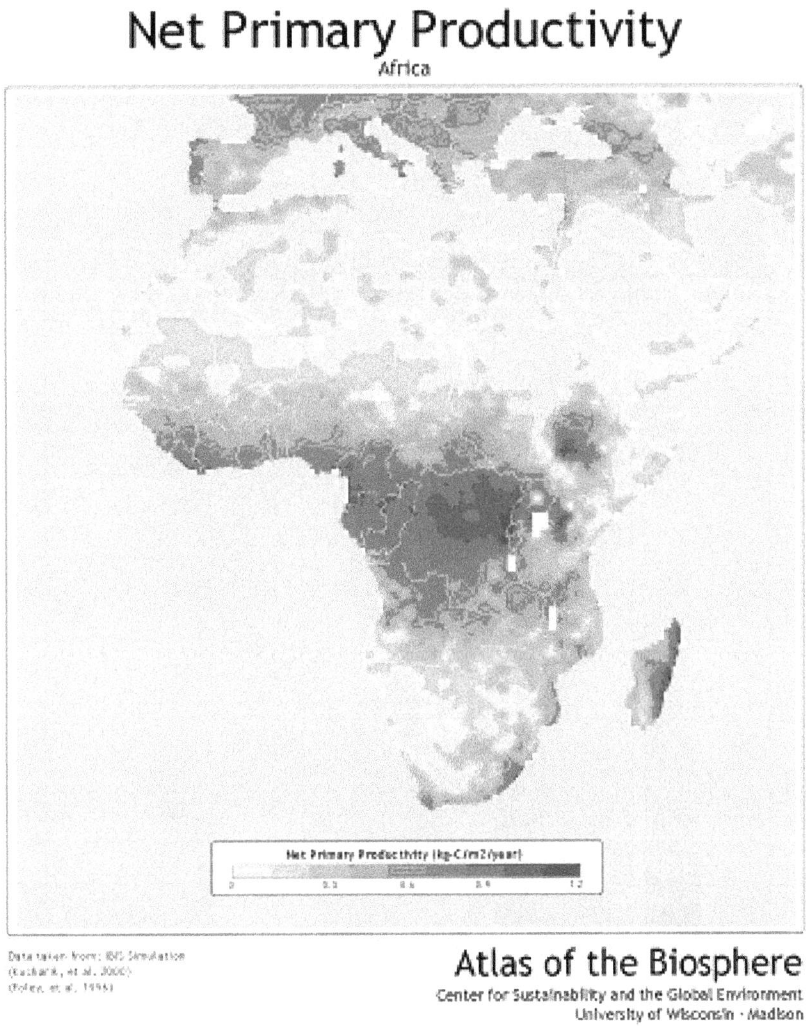

FIGURE 5.2

NET PRIMARY PRODUCTIVITY (FOLEY, 1996, KUCHARIK, 2000)

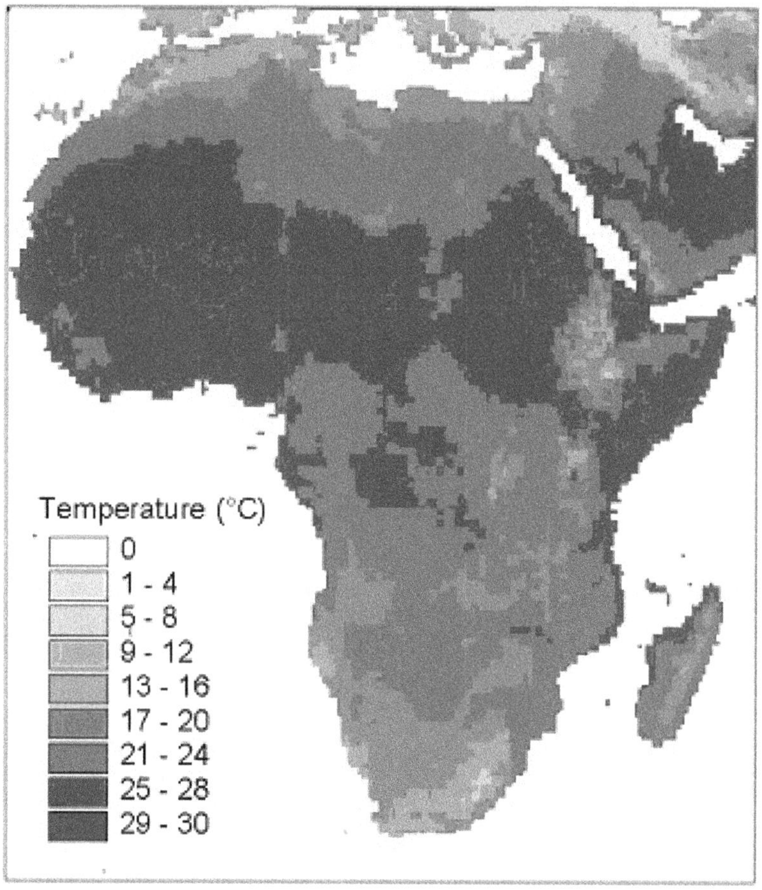

FIGURE 5.3
MEAN DAILY TEMPERATURE (°C). FROM NEW *ET AL* (1999)

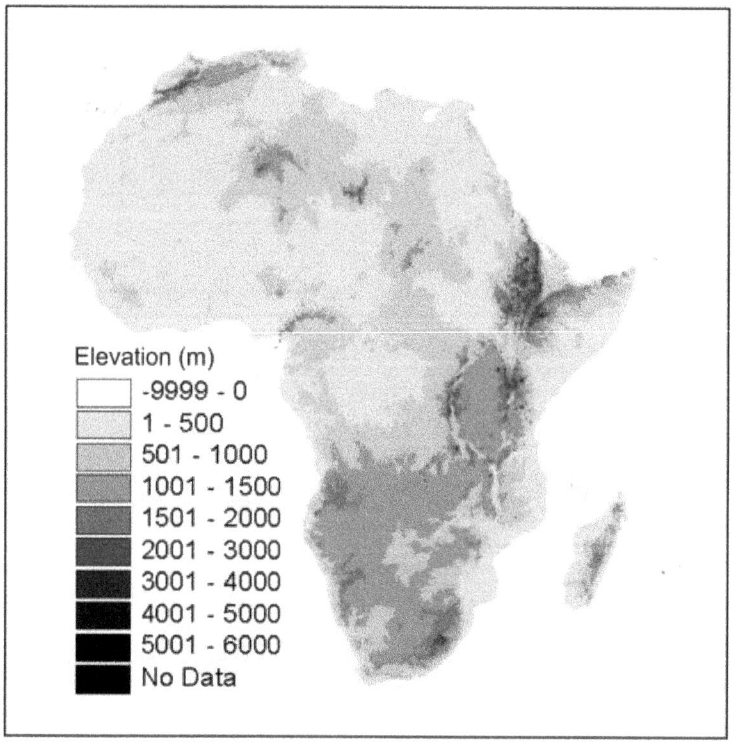

FIGURE 5.4
AFRICAN TOPOGRAPHY (FROM GTOPO30, PROVIDED BY THE USGS-NASA DISTRIBUTED ACTIVE ARCHIVE CENTRE)

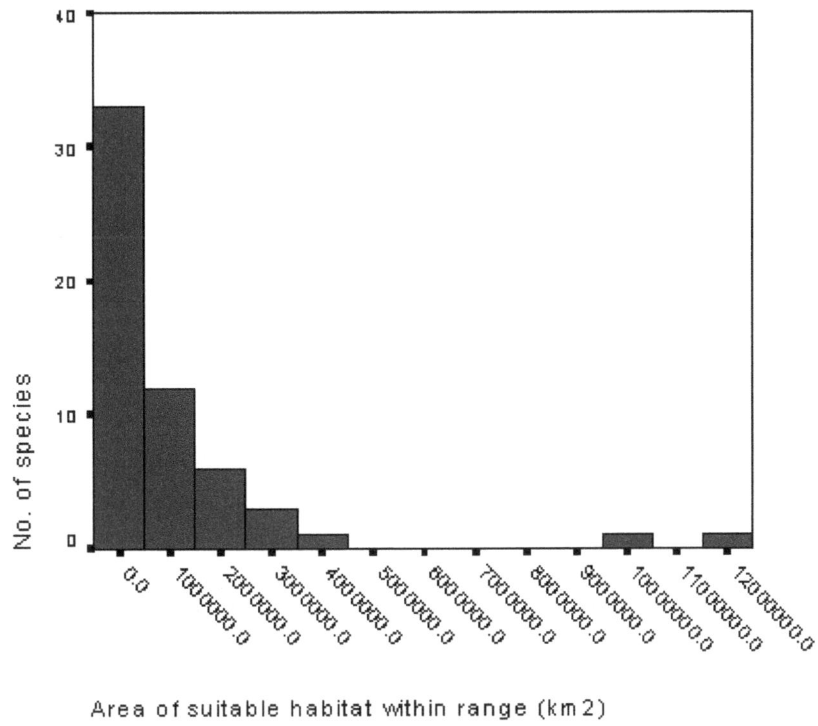

Figure 5.5
Histogram of primate species geographic ranges (km^2), based on AMD assessment of suitable habitats.

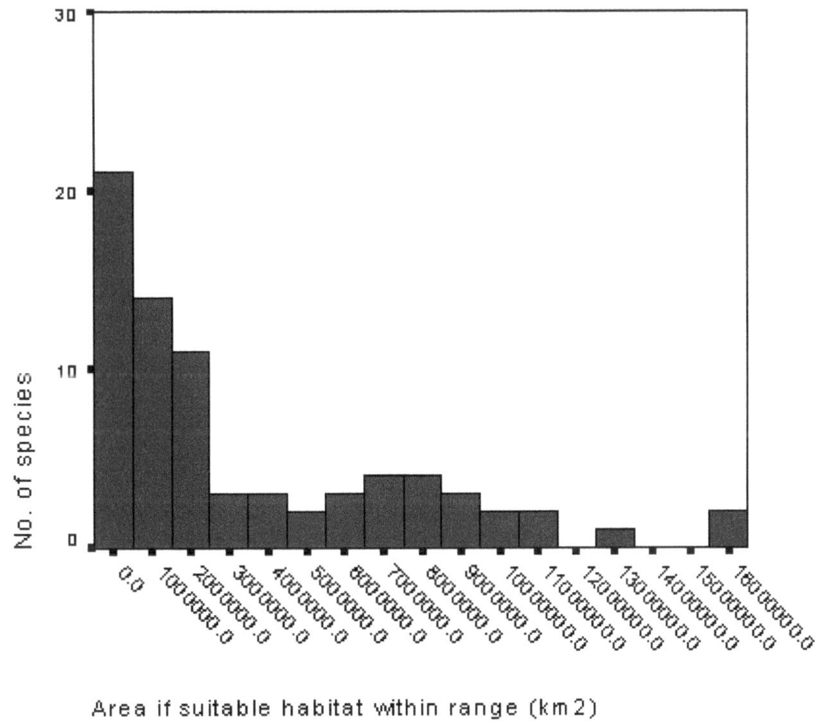

Figure 5.6
Histogram of carnivore species geographic ranges (km^2), based on AMD assessment of suitable habitats

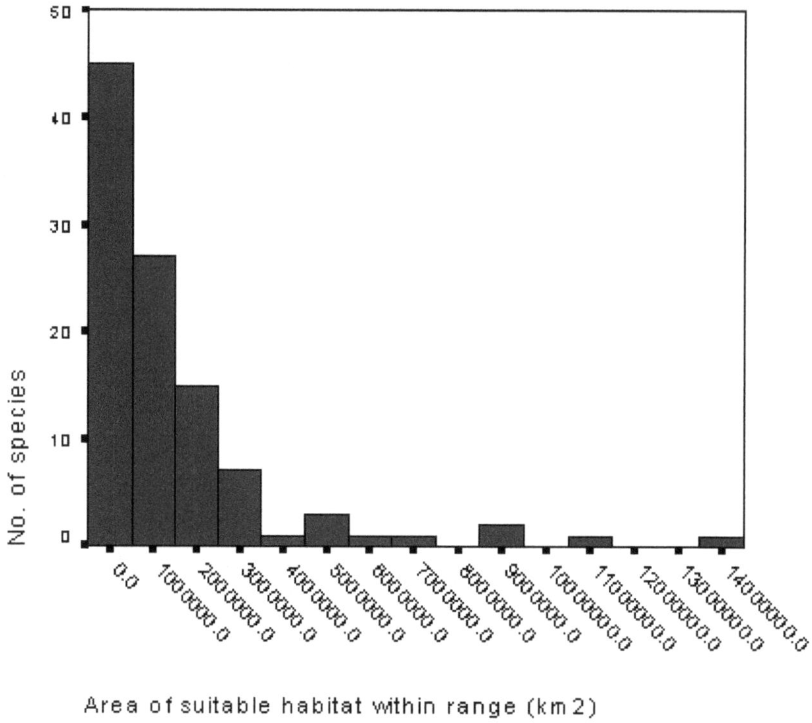

FIGURE 5.7
HISTOGRAM OF UNGULATE SPECIES GEOGRAPHIC RANGES (KM^2), BASED ON AMD ASSESSMENT OF SUITABLE HABITATS

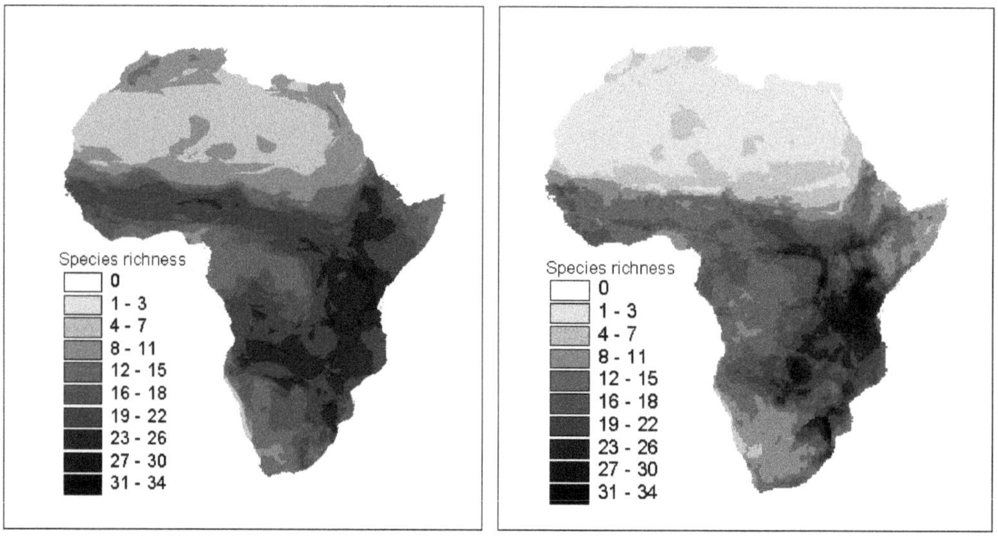

FIGURE 5.8
CARNIVORE (LEFT) AND UNGULATE (RIGHT) SPECIES RICHNESS

The spatial distribution of diversity shows different trends in these groups of species. There is a strong relationship between the distribution of ungulate and carnivore diversity, with hotspots in the same areas (Figure 5.8). Both groups show a trend in which species richness increases to the east of the continent, with lower values in the west-central region and southern Africa. Primate diversity is almost a mirror image, with the highest species richness in the west-central region (Figure 5.9).

These patterns may reflect the continental distribution of climate and topography. There is an overall gradient in rainfall and primary productivity from the west-central region outwards in Africa (Figures 5.1, 5.2). Primate diversity strongly reflects the distribution of rainfall in the continent. As discussed below, primate biomass may be primarily influenced by primary productivity. As would be expected, primate diversity is also clearly related to general trends in habitat. Primate species richness is

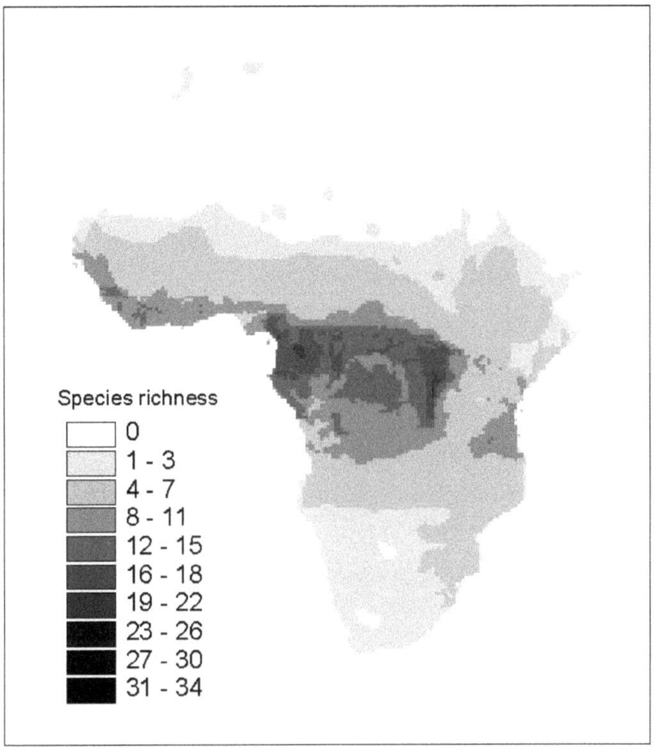

FIGURE 5.9
PRIMATE SPECIES RICHNESS

highest in the west-central region dominated by tropical forest. Primates occur in savannah and woodland habitats, but are considerably less diverse. As discussed in Chapter 2, a number of studies have shown that primate diversity broadly follows general biogeographical trends such as the latitudinal gradient in diversity.

There is evidence that carnivore and ungulate species also follow biogeographical rules. The increase in species richness and turnover in carnivore and ungulate species in the east of the continent reflects the elevational gradient in species richness, as described by Stevens (1992). There is a general increase in topographic height from the north-west to the south-east of the continent (Figure 5.4). Some of the topographic patterns closely parallel those of carnivore and ungulate species richness. Richness is low in the lowland forest of central Africa, while some hotspots are situated in areas of high relief along the Rift Valley. Areas of high relief may have high species turnover, if they act as biogeographical barriers. The east-west trend is much more pronounced than any north-south trend. However there are fewer carnivore and ungulate species in the far south of the continent, as would be expected from Rapoport's rule. This is also true of the north although it is clear from the distribution pattern that this is determined by the extent of the Sahara desert. In addition, the fact that the increasing diversity with altitude trend is weaker in southern Africa may be a latitudinal effect. This is consistent with the suggestion that environmental resistance caused by altitude is stronger in the tropics (Ruggiero, Lawton *et al.* 1998).

Carnivore and ungulate species richness is also related to general patterns of habitat distribution, and is high in savannah and woodland habitats. In addition, as will be discussed below, ungulate biomass is strongly related to particular rainfall and temperature conditions that are optimal in East Africa. Carnivore species richness may reflect the diversity and biomass of ungulates (as potential prey species) present in this area. Ungulate species richness in this area would both support lots of carnivores and provide suitable prey for carnivores with a variety of body sizes and foraging techniques.

A larger area is characterised by relatively high carnivore diversity than high ungulate diversity in Africa. Primate species richness is lower than that of the other groups. This is probably due to the distribution of geographic range size in the orders: there are more wide ranging carnivores with overlapping ranges, thus increasing species richness throughout the continent.

Range boundaries

As discussed in Chapter 2, there are a number of environmental features that might form barriers to

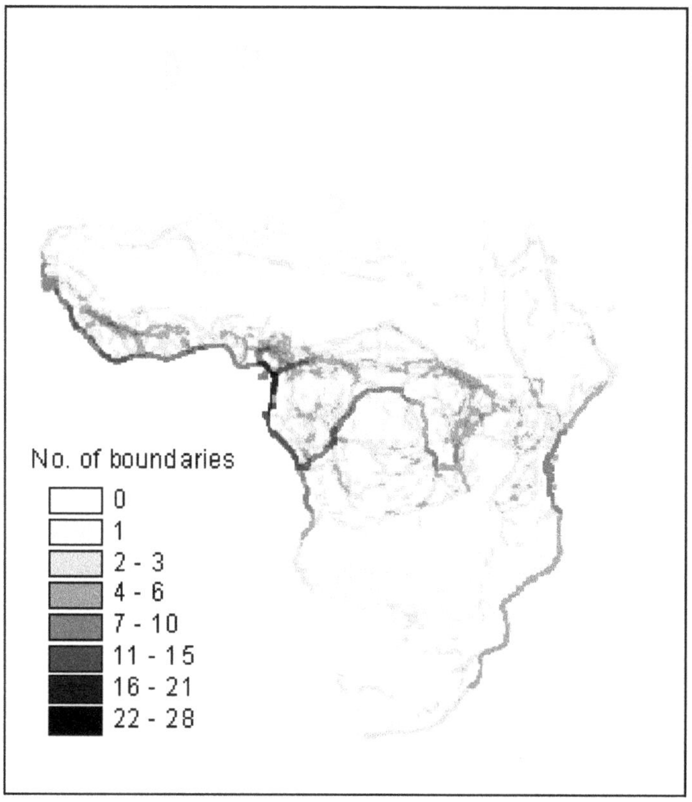

Figure 5.10
Distribution of primate species range boundaries

dispersal and hence to species geographic ranges. These include mountain ranges, rivers, and coastlines and also major changes in habitat distribution and the distribution of other species. Coastlines seem to be particularly important in limiting African mammal distribution, particularly at lower latitudes. Large numbers of carnivore and ungulate species range boundaries occur in areas with high topographical relief in the eastern Rift Valley, suggesting that mountain ranges are significant barriers to distribution for these species (Figures 5.11, 5.12). However these areas are also associated with a number of changes in habitats, providing an alternative explanation (Olson *et al.* 1983).

High species turnover is associated with a number of major changes in the dominant type of habitat. The transition from lowland tropical forest to woodland and savannah habitats is clearly an important range boundary for many primate species (Figure 5.10). As most primates are restricted to forest biomes, the boundary between forest and savannah biomes is particularly important (Eeley and Lawes 1999, p.203). There is also a strong barrier within the tropical forest zone, running parallel to the Zaire river. This and some of the other rivers in the equatorial forest zone are recognised as barriers to primate distribution (Chapman, Gautier-Hion *et al.* 1999; Eeley and Lawes 1999). According to Eeley and Lawes (1999), the role of these rivers in the formation of forest refuges in the past may be an important factor shaping modern distribution.

The transition from moist woodland and woodland mosaics in the south-east to drier and more open habitats in southern Africa (Olson *et al.* 1983) corresponds to large number of range boundaries for carnivores and ungulates and also several primate species. Many species ranges end at the boundaries associated with changes in the dominant habitat and increasing aridity south of the Sahara desert. It is clear that habitat preference and tolerance is important in determining the extent of distribution of many African mammals.

Throughout the continent, areas with a high turnover of species seem to correspond to areas with high species richness. Thus more species' boundaries are located in the south-east of the continent for ungulates and carnivores, and in the west-central region for primates. This does not necessarily imply the influence of competition effects on species dispersal, as the same geographical barriers may affect a wide range of species. Most importantly for this analysis, there are no clear differences in the distribution of range boundaries for primates, ungulates and carnivores that cannot be explained by this association with diversity.

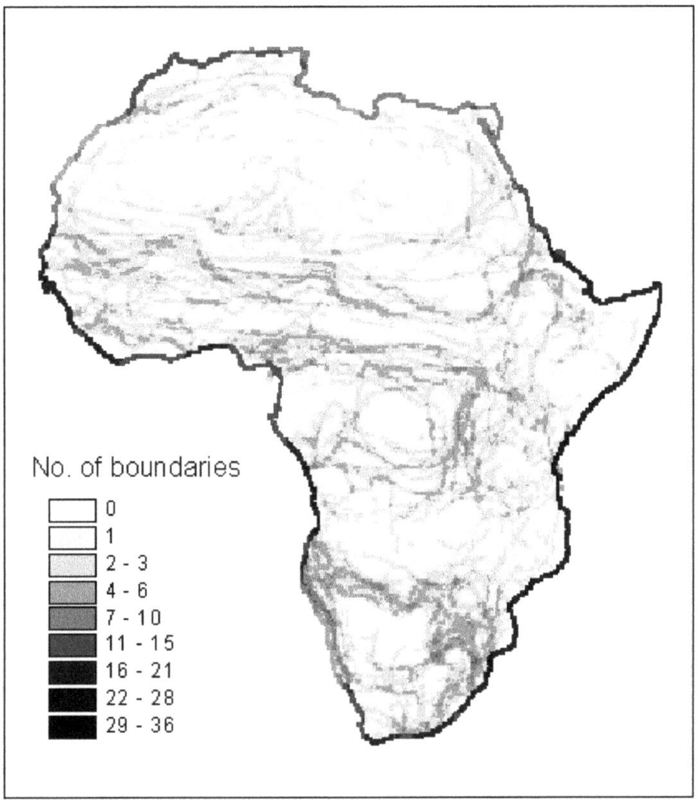

FIGURE 5.11
DISTRIBUTION OF CARNIVORE SPECIES RANGE BOUNDARIES

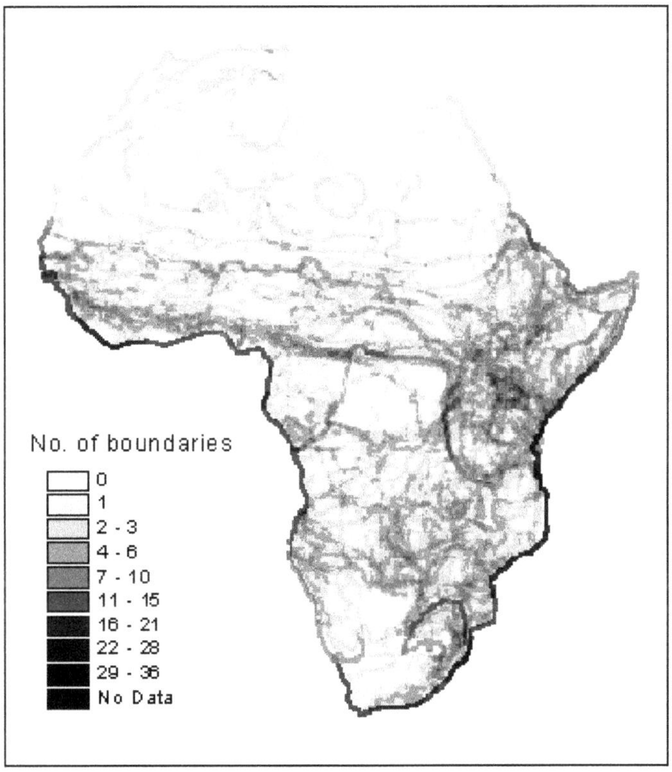

FIGURE 5.12
DISTRIBUTION OF UNGULATE SPECIES RANGE BOUNDARIES

Body mass

Although the maximum body mass is higher for carnivore than primate species, the frequency distribution is very similar and mean carnivore adult body mass (8.9kg) is almost identical to that for primates (8.7kg) (Figures 5.13, 5.14). The key difference as far as spatial distribution is concerned is in the habitat preferences of the species with larger body masses.

Primate species with the largest body mass values are found in west-central Africa, where tropical forest habitats are more abundant, while the species occupying areas to the east and south, characterised by woodland and savannah habitats, tend to have medium body masses (Figure 5.15). Tropical forest environments are also suitable for very small species. Shrub or scrublands south of the Sahara are tolerated only by small-bodied primate species. The energetic requirements of large bodied primate species may only be supported in the highly productive rainforest environments. The diversity of some forest environments offers opportunities for species occupying a range of ecological niches as well as a generally higher carrying capacity based on minimum resources at certain times of year.

Larger bodied carnivores tend to be distributed in the savannah and woodland zone of sub-Saharan Africa; these areas also provide suitable conditions for a wide range of body sizes in carnivore species (Figure 5.16). These areas tend also to be characterised by high ungulate species richness and biomass (Figures 5.8, 5.19). Carnivores that are most tolerant of desert conditions in northern and southern Africa are smaller bodied. The largest carnivores are generally absent from tropical forest where medium to small species are found. This may be because body mass sets a limit on prey size in carnivores (Carbone, Mace *et al.* 1999), and ungulates living in tropical forest tend to be small (Thackeray 1995). Carbone *et al.* (2002) have shown that there is a striking transition from feeding on small prey (less than half of predator mass) to large prey (near predator mass), occurring at predator masses of 21.5-25kg, which is related to energy requirements. This suggests that the distribution of carnivore body mass would be strongly related to the distribution of large or small ungulates.

According to Thackeray (1995), ungulate species that are tolerant of conditions that combine low temperature with low rainfall tend to have a relatively small body mass. The largest ungulates are the African elephant (*Loxodonta africana*), the white rhino (*Ceratotherium simum*), hippopotamus (*Hippopotamus amphibius*), and giraffe (*Giraffa camelopardis*). The hippopotamus has a patchy distribution around rivers. Elephants and giraffes have a wide but patchy distribution, incorporating rainforest, woodland, savannah and shrub in sub-Saharan Africa. Species weighing more than 200kg are distributed throughout sub-Saharan Africa, except in certain parts of southern Africa, with a preference for the eastern side of the continent and savannah or woodland habitats. The smallest species, such as the Okapi (*Okapi johnstoni*), are found in desert, rainforest and also savannah and woodland areas.

Thus, while the range of body masses is similar, the spatial distribution of body mass differs strongly between the primates and the carnivores, and the carnivore pattern again reflects the distribution of ungulates.

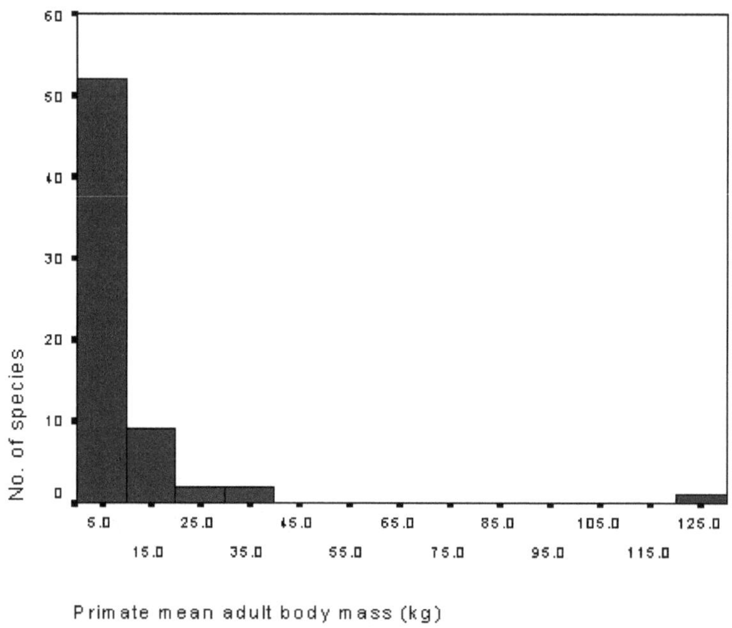

FIGURE 5.13
CHART OF FREQUENCY DISTRIBUTION OF PRIMATE MEAN ADULT BODY MASS (KG)

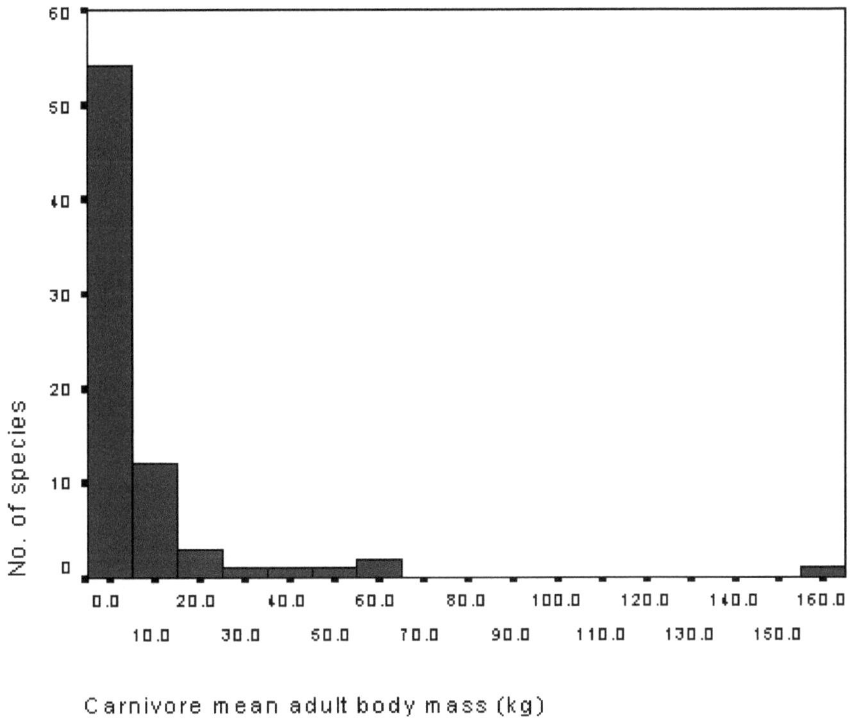

Figure 5.14
Chart of frequency distribution of carnivore mean adult body mass (kg)

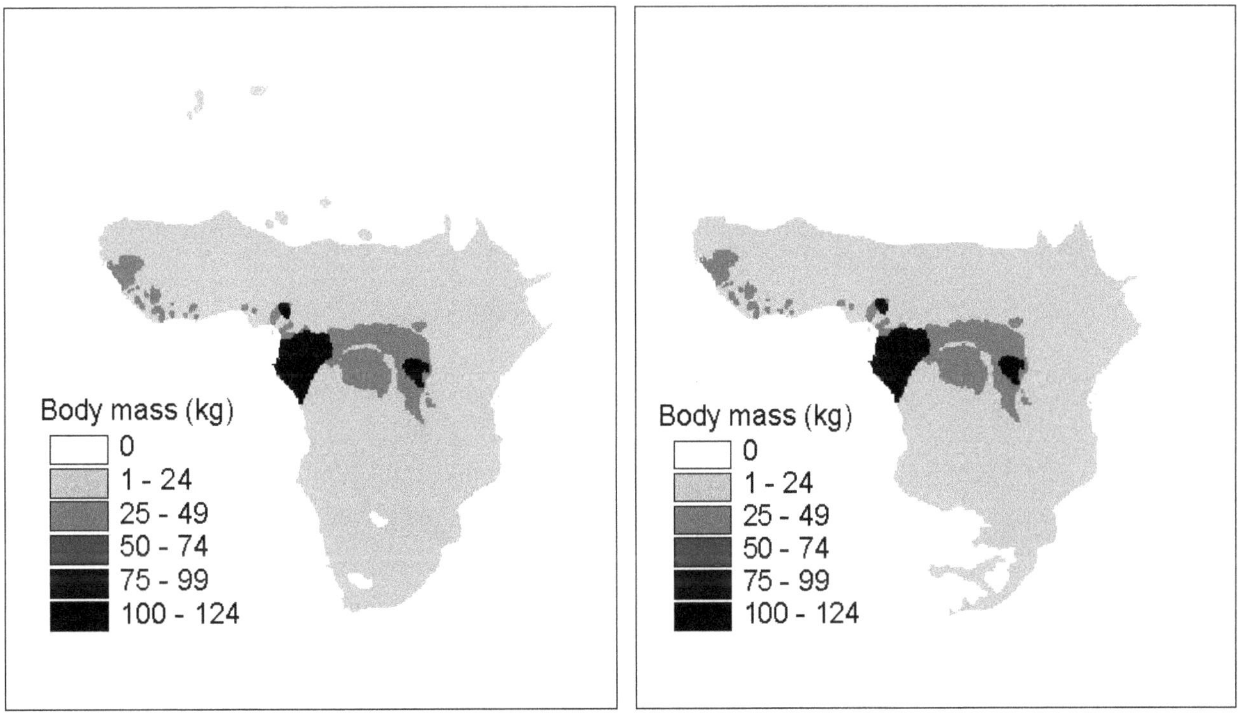

Figure 5.15
Distribution of maximum (left) and range (right) of body mass in primates (kg)

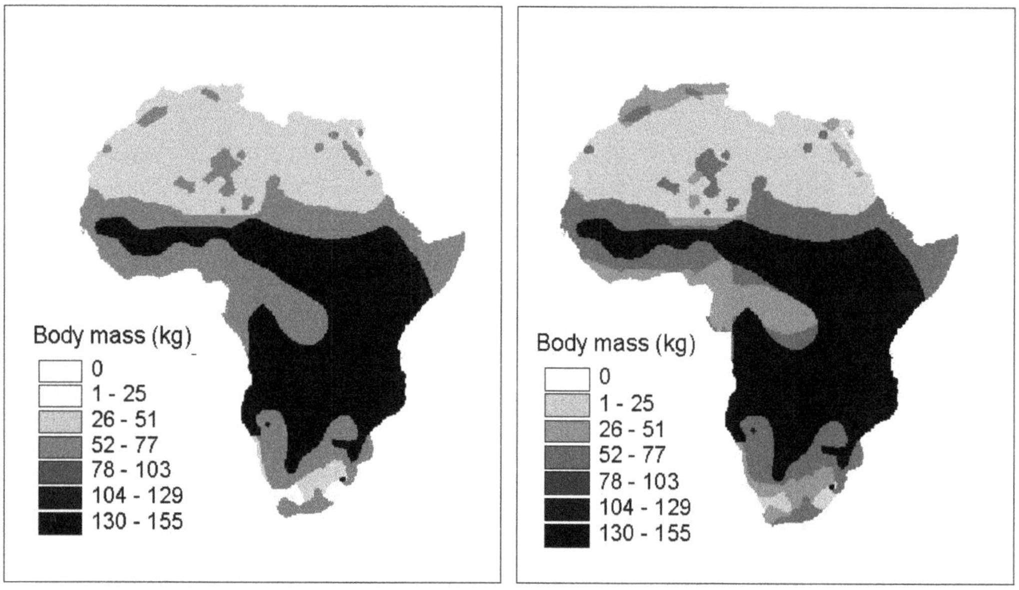

Figure 5.16
Distribution of maximum (left) and range (right) of body mass in carnivores (kg)

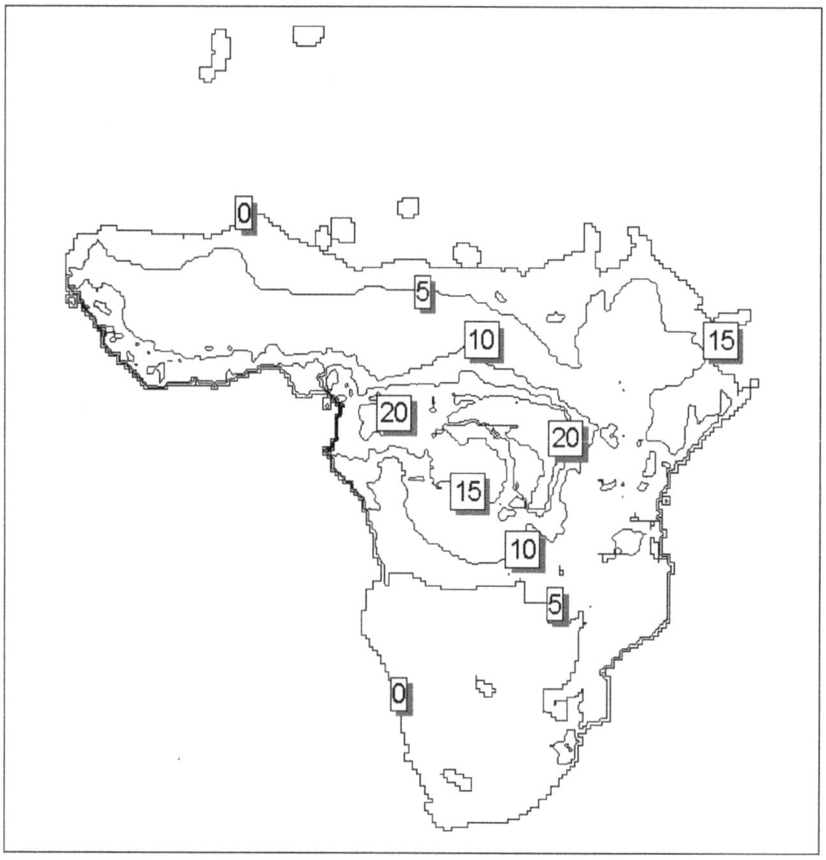

Figure 5.17
Primate biomass in kg/ha

Biomass

How can we explain patterns in diversity and biomass in these groups of species? Biomass is clearly closely related to species diversity. Primate, ungulate and carnivore biomass maps show similar trends to the species richness maps for the same taxonomic group. Thackeray (1995) has calculated a regression equation for the relationship between ungulate species richness and biomass, with an r^2 value of 0.96.

There is also evidence to suggest that environmental and climatic variables influence animal biomass. The climatic variables of mean annual rainfall and temperature, as well as soil quality, have been highlighted as possible predictors of ungulate biomass (Thackeray 1995; Olff, Ritchie *et al.* 2002). Primary productivity has been suggested as a possible determinant of primate biomass (Janson and Chapman 1999).

Alternatively, the distribution of biomass and diversity can be related to species interaction. In particular, carnivore biomass is strongly related to prey biomass (Carbone and Gittleman 2002), which would include density of ungulates. In addition, it is possible that predator relationships may restrict primate ranges.

Particular rainfall and temperature conditions in Africa are associated with particular habitats. Rainfall can be used as a simple surrogate measure for resource productivity (Janson and Chapman 1999). Forested areas tend to occur where rainfall is higher, and areas with high levels of rainfall are generally forested. Savannahs, which range from humid woodlands to dry grasslands, often have less than 1000mm of annual rainfall (Chapman, Gautier-Hion *et al.* 1999). This suggests that climatic variables can be used to investigate the relationship between the distribution of biomass and particular habitats.

However there is a great deal of additional variation. Plant productivity does not increase indefinitely with rainfall, reaching a peak in the New World tropics between 2000-2500mm. Tropical forest can include relatively short-stature deciduous woodlands with only 500mm of rainfall per year and evergreen forests with 60m tall emergent trees in areas of more than 2500mm rainfall (Janson and Chapman 1999). The richness of plant species found in different forested areas can vary greatly; species richness does to an extent vary with rainfall (Chapman, Gautier-Hion *et al.* 1999). Finally, even areas of relatively similar physical and climatic characteristics can produce forests dominated by quite distinct plant families with unique biological traits, depending on their evolutionary and biogeographic history. There is additional variation outside forests. The climatic conditions in grassland and woodland environments are very similar, and differences in tree cover may be largely determined by fires, grazing and trampling by ungulates rather than climate (Richard 1985).

Areas of high ungulate biomass tend to occur where mean annual temperature ranges between 19 and 22°C and rainfall between 750 and 1000mm (Thackeray 1995) (Figure 5.19). In sub-Saharan Africa, these climatic conditions are associated with mixed woodland savannah (ibid.). Environments with intermediate levels of moisture provide both the quantities of vegetation required by large herbivores and the nutritional quality to support small herbivores (Thackeray 1995; Olff, Ritchie *et al.* 2002). More plant available moisture reduces the nutrient content of plants but increases productivity, whereas more plant available nutrients increases both factors (Olff, Ritchie *et al.* 2002). Because larger herbivore species tolerate lower plant nutrient content but require greater plant abundance, the highest potential herbivore density should occur in locations with intermediate moisture and high nutrients (ibid.).

Low ungulate biomass is associated with desert areas with low rainfall as well as forested areas with high rainfall (Thackeray 1995). Desert environments have low productivity, while rainforest environments have higher productivity of less nutritious plants. Ungulates that are most tolerant of conditions that combine low temperature with relatively low rainfall have relatively small body mass and include species that graze only or that graze and browse (Thackeray 1995).

Primate species richness is more directly related to primary productivity than that of carnivores or herbivores. In general, areas with greater rainfall have higher levels of plant productivity (Figures 5.1, 5.2), and the distribution of high annual rainfall suggests a relationship with primate biomass (Figure 5.17), although not all plant productivity is available as food. Food quantity and quality is likely to increase the primate carrying capacity of a forest (Chapman, Gautier-Hion *et al.* 1999). Individual case studies in Africa suggest that the relationship between rainfall and primate biomass may not hold at a less general scale (Chapman, Gautier-Hion *et al.* 1999). Higher primate biomass is significantly related to soil fertility in Amazonian forests (Peres 1999). Other important factors may include habitat heterogeneity; this tends to increase both the species richness and biomass of primate communities (Oates, Whitesides *et al.* 1990). A mosaic of forests of different ages can increase vegetation diversity and the availability of palatable leaves and fleshy fruit and thus lead to high carrying capacities for some species.

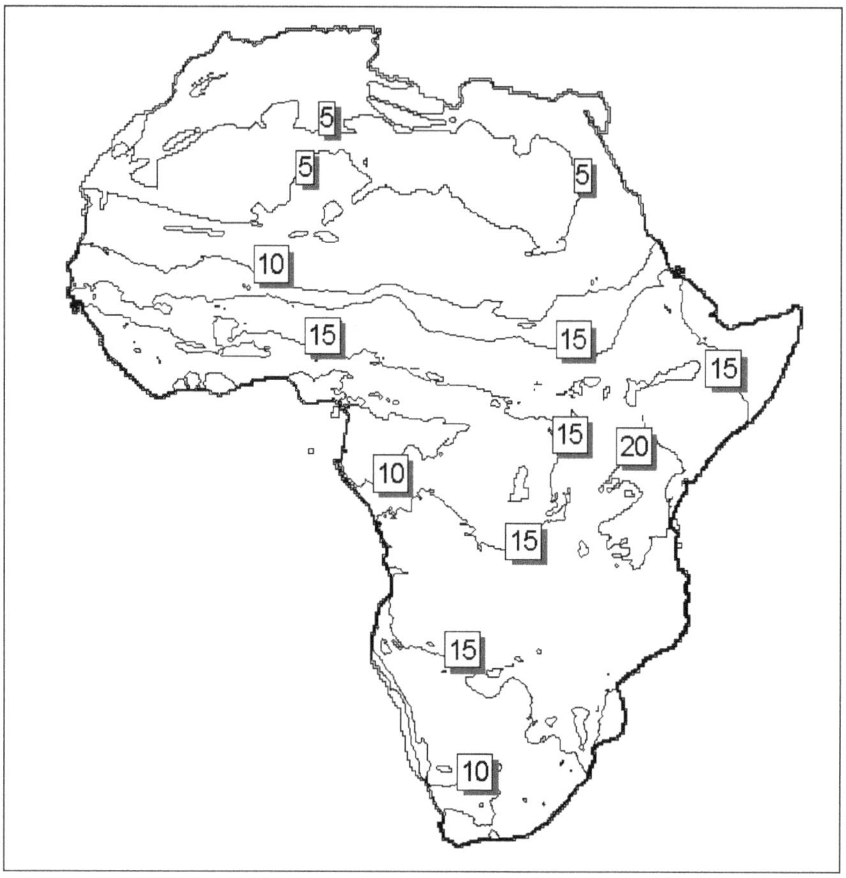

Figure 5.18
Carnivore biomass in kg/ha

Figure 5.19
Ungulate biomass in kg/ha (from Thackeray 1995)

Predation may also be an important limiting factor on population density in primates: examples include hawk-eagle predation on black-and-white colobus populations in Gombe National Park, Tanzania, and leopard predation on vervet monkeys in Amboseli National Park, Kenya (Chapman, Gautier-Hion *et al.* 1999). The spatial distribution of carnivore diversity and biomass (Figures 5.8, 5.18) suggests that increased predator pressure could be one factor keeping primate biomass low outside forest environments.

Examination of the map of primate biomass (Figure 5.17) in relation to climatic variables (Figures 5.1, 5.3) indicates some general trends. High primate biomass tends to occur most often in areas with between 1500 and 1750mm of rain annually (Figure 5.20), and with middle range temperatures of 23-26°C (Figure 5.21). These areas of high rainfall tend to be characterised by tropical or sub-tropical forest habitats and high plant productivity (Figure 5.2). Medium primate biomass tends to occur in areas with annual rainfall between 1000 and 2000mm and temperatures in the twenties (21-28°C), again primarily corresponding to forest habitats. Low primate biomass tends to occur in areas with relatively low annual rainfall (250-1500mm per year) or (a much smaller proportion) relatively high rainfall (2000-3000mm per year). Low primate biomass tends to occur in areas with low (15-20°C) or high (25-28°C) temperature. Areas with lower rainfall primarily correspond to woodland, savannah and mosaic habitats, and at the more arid extreme habitats such as grassland and scrubland. Primate biomass is generally greater in areas with high rainfall than that of either carnivores or ungulates (Figures 5.17-19).

Low carnivore biomass tends to occur in areas with low rainfall (0-750mm per year), although it also occurs in some areas with high rainfall (1750+mm) (Figure 5.22). Low to medium biomass values tend to occur at high temperatures (25-26°C) or to a lesser extent lower temperatures (there is a small peak at 17-18°C) (Figure 5.23). Areas with high rainfall correspond to tropical and sub-tropical forest habitats. Areas with low rainfall and high or low temperatures correspond to more arid habitats in the north and south of the continent, including desert and semi-desert habitats. High carnivore biomass tends to occur in areas with low to medium rainfall (250-1250mm per year) and moderately high temperatures (27-28°C). These levels of rainfall correspond to woodland, savannah and savannah mosaic habitats, and semidesert scrub at the lower end of the rainfall range. Highest biomass tends to be in areas with slightly lower annual rainfall (*c.* 500-1250mm) and temperatures in the lower twenties. These patterns show strong similarities with Thackeray's (1995) description of ungulate biomass distribution.

Carnivore biomass is much lower than ungulate biomass throughout the continent (Figures 5.18, 5.19), as would be expected due to the trophic level effect. Areas of high and low biomass in carnivores and ungulates are notably similar. In both ungulates and carnivores biomass distribution is strongly related to species diversity. However in carnivores there are areas of low biomass relative to the number of species in deserts and rainforest areas: this is probably due to prey size in these environments. The population density of particular carnivore species can be related to the abundance (in terms of mass) of their main prey species (Carbone and Gittleman 2002). At the same time, carnivore prey choice is constrained by body mass (Carbone, Mace *et al.* 1999).

The largest portion of the highest carnivore biomass contour (>20kg/ha) falls in areas with temperatures of 21-22°C (Figure 5.24). This is similar to the distribution of high ungulate biomass (Figure 5.25). This suggests that high levels of carnivore biomass can only be supported by high ungulate biomass. However comparing contour diagrams on a map of Africa or on temperature/rainfall axes indicates the same pattern: carnivore biomass of over 20kg/ha can occur in areas with ungulate biomass of anything from 30-70kg/ha, although it is more common in areas with over 60kg/ha. This is unexpected, given the generally strong connection between carnivore and ungulate distribution. Is this an accurate result? Given that some ungulate species live in tropical forest habitats, which can have 500-2500mm rainfall per year, it seems worth questioning the upper limit to Thackeray's (1995) data.

To investigate the relationship between ungulate and carnivore biomass further, I produced an ungulate biomass map by the same method (Figure 5.26). There is a wider range of temperature values for contours in my maps than indicated in Thackeray (1995), and a much wider range of rainfall values, particularly at the higher end of the range. The histogram of ungulate biomass within contours of carnivore biomass indicates that high carnivore biomass occurs where ungulate biomass is between 24-106kg/ha, but is more common between 48-94kg/ha (Figure 5.27).

There are a number of possible methodological drawbacks to this analysis. Because the analysis is based on general Extent of Occurrence maps, which do not indicate variations in population density and local absence of species, it is probable that biomass will be over-estimated in some areas. This means that the range of values for rainfall and temperature within the higher biomass contours is likely to be higher than the actual values. However Thackeray's (1995) method may also be problematic. Thackeray (1995) calculated values of mean annual rainfall and temperature from metereological data for environments in which particular ungulate taxa are known to have occurred. The use of data from particular environments rather than a map of the climatic variables may better represent the actual climatic preference of species: however it is likely to underestimate the total range of climate tolerated.

While my methodology is likely to exaggerate the extent of areas of high biomass, the fact that my map presents the same broad pattern as that produced by Thackeray suggests that this does not obscure broad trends in the distribution of biomass. The broad distribution patterns suggest a relationship between carnivore and ungulate biomass, while the statistics I have generated indicate that further variation also occurs. As with diversity and body mass, there are strong contrasts between the spatial distribution of primate biomass and carnivore and ungulate biomass.

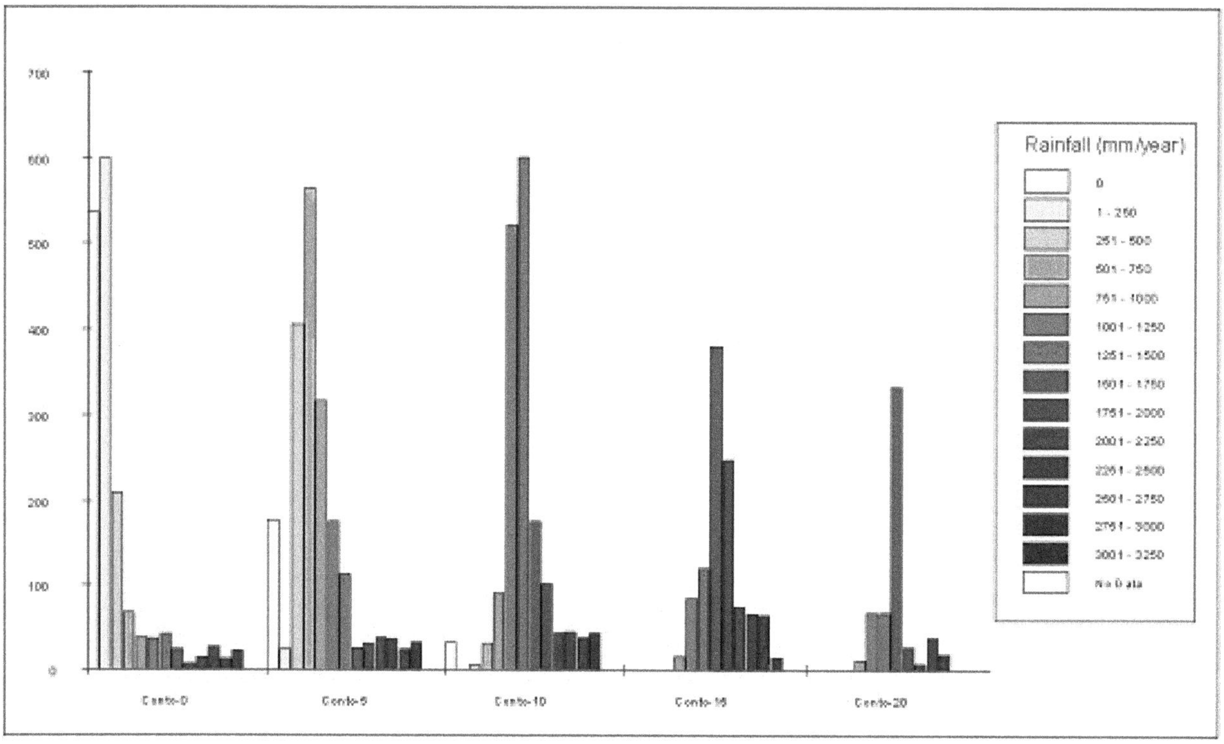

Figure 5.20
Frequency distribution of mean annual rainfall (mm) in each contour of primate biomass

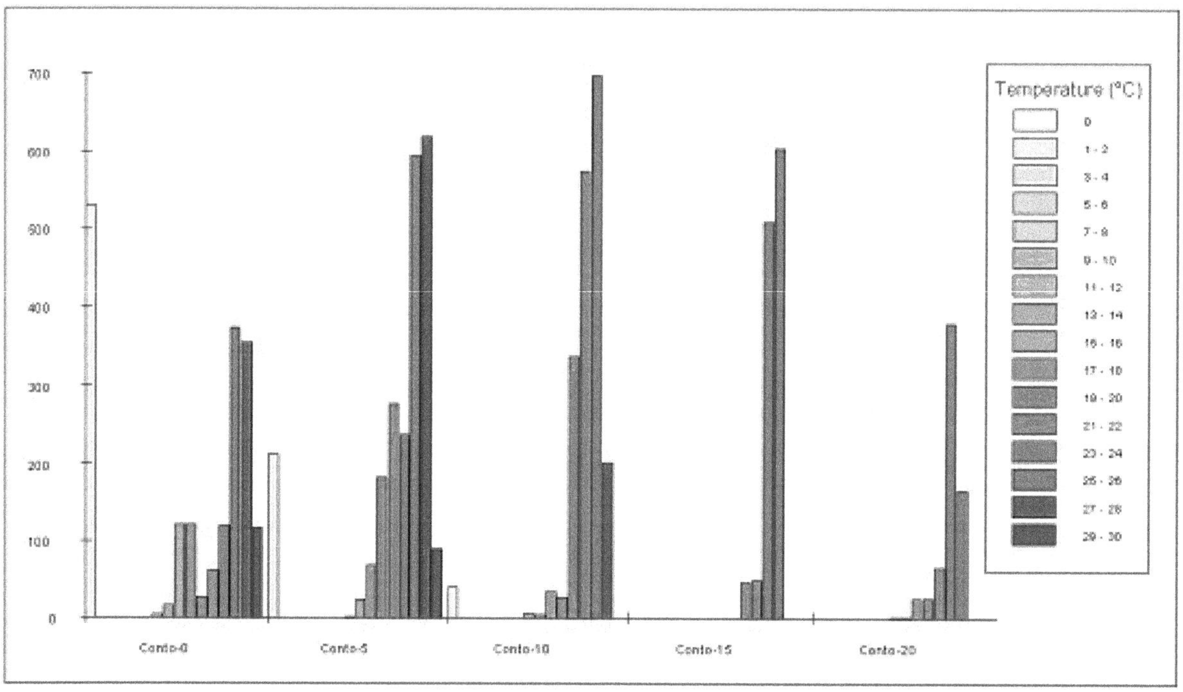

Figure 5.21
Frequency distribution of average daily temperature (°C) in each contour of primate biomass

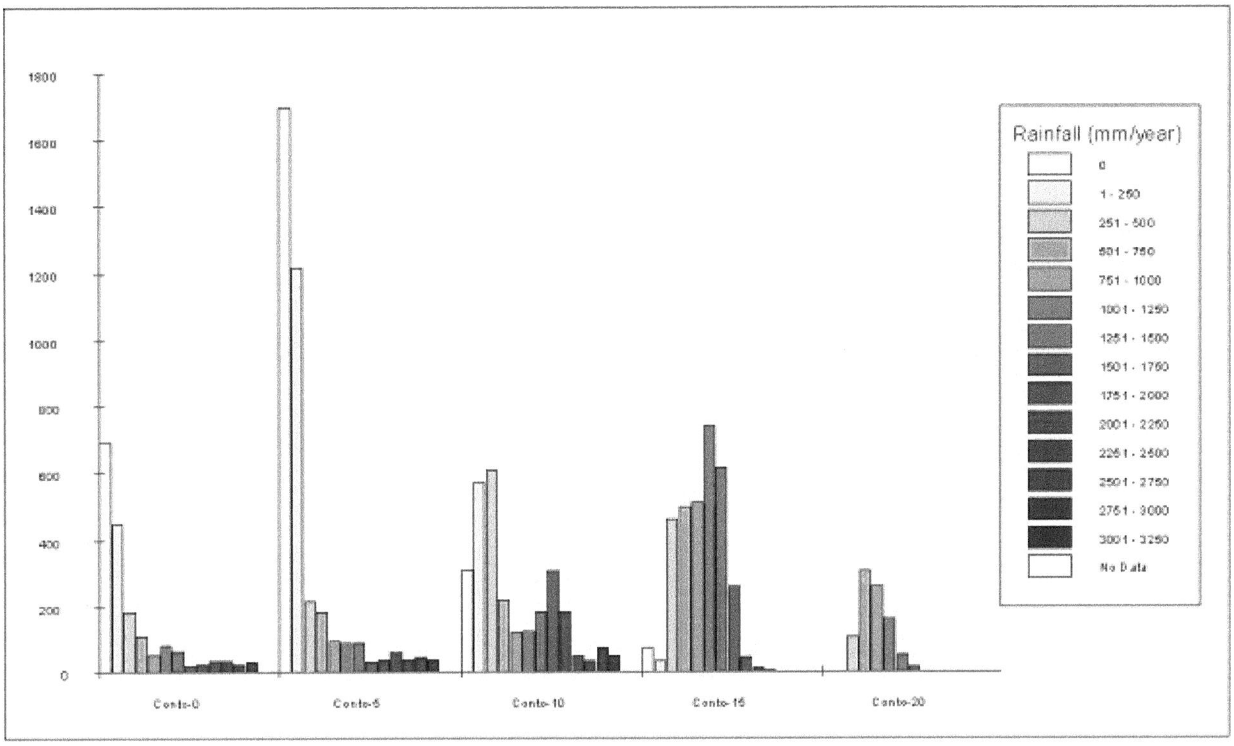

FIGURE 5.22

FREQUENCY DISTRIBUTION OF OF MEAN ANNUAL RAINFALL (MM) IN EACH CONTOUR OF CARNIVORE BIOMAS

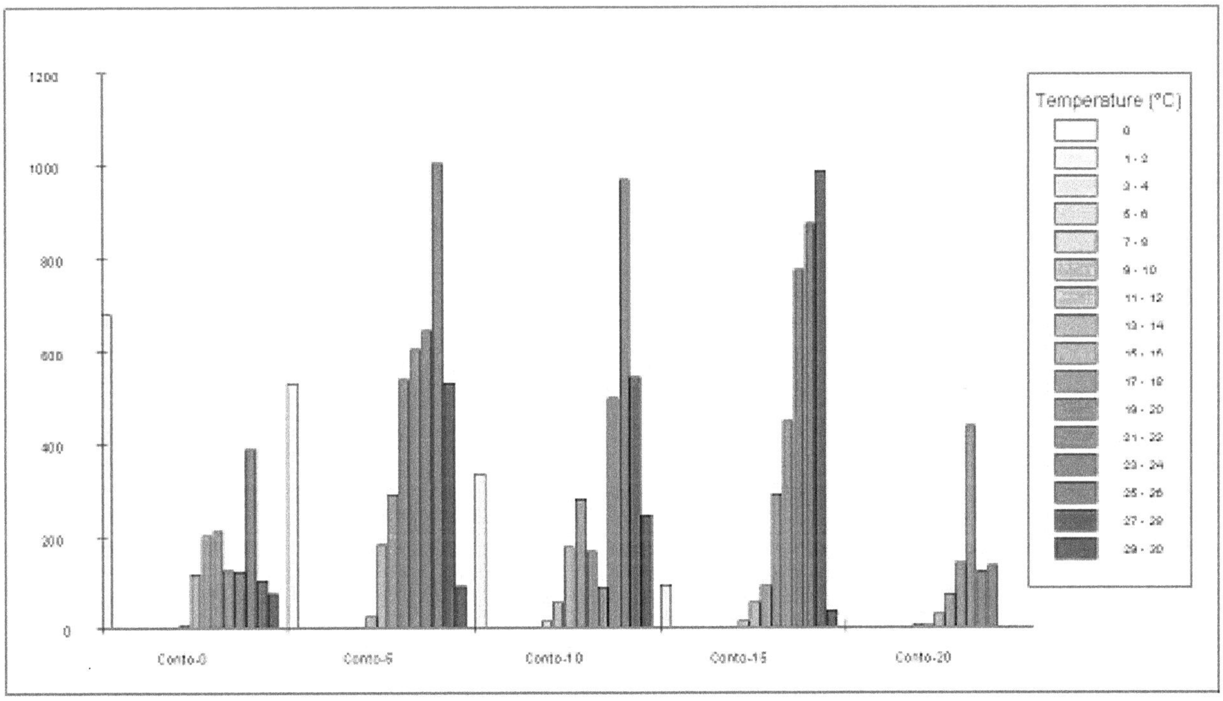

FIGURE 5.23

FREQUENCY DISTRIBUTION OF AVERAGE DAILY TEMPERATURE (°C) IN EACH CONTOUR OF CARNIVORE BIOMASS

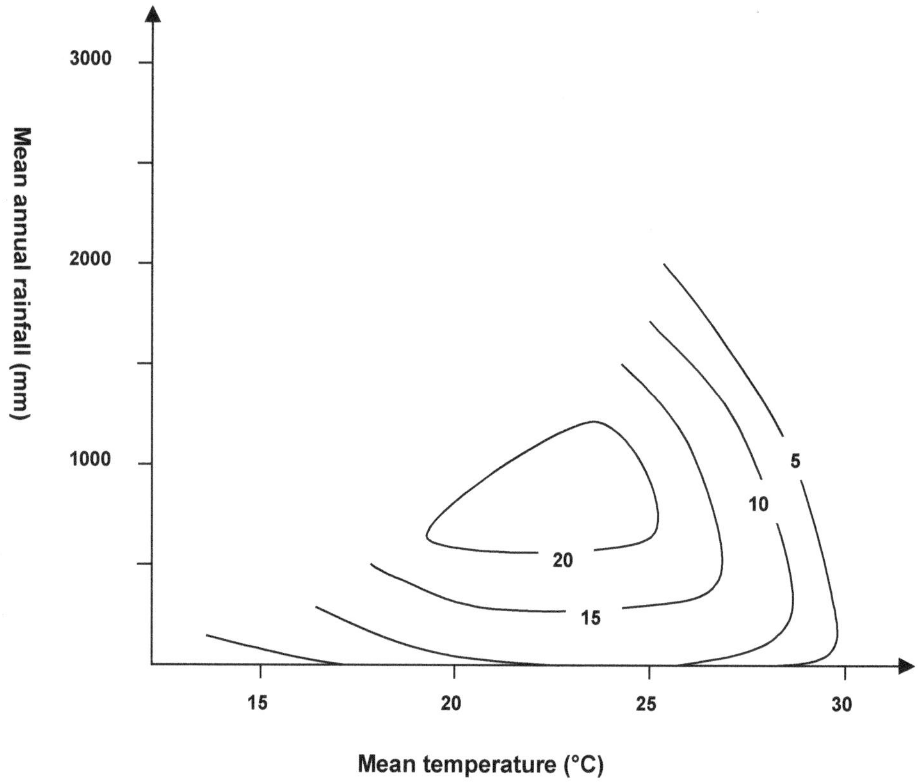

Figure 5.24
Carnivore biomass in relation to mean annual rainfall (mm) and temperature (°C)

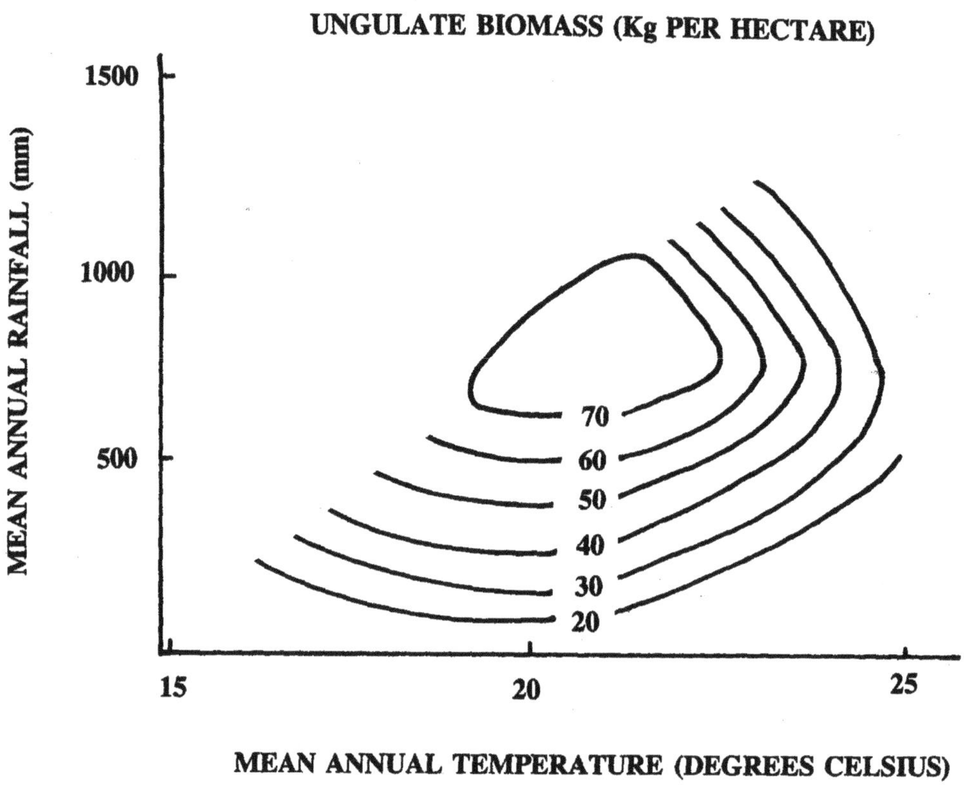

Figure 5.25
Ungulate biomass in relation to mean annual rainfall (mm) and temperature (°C), from (Thackeray 1995)

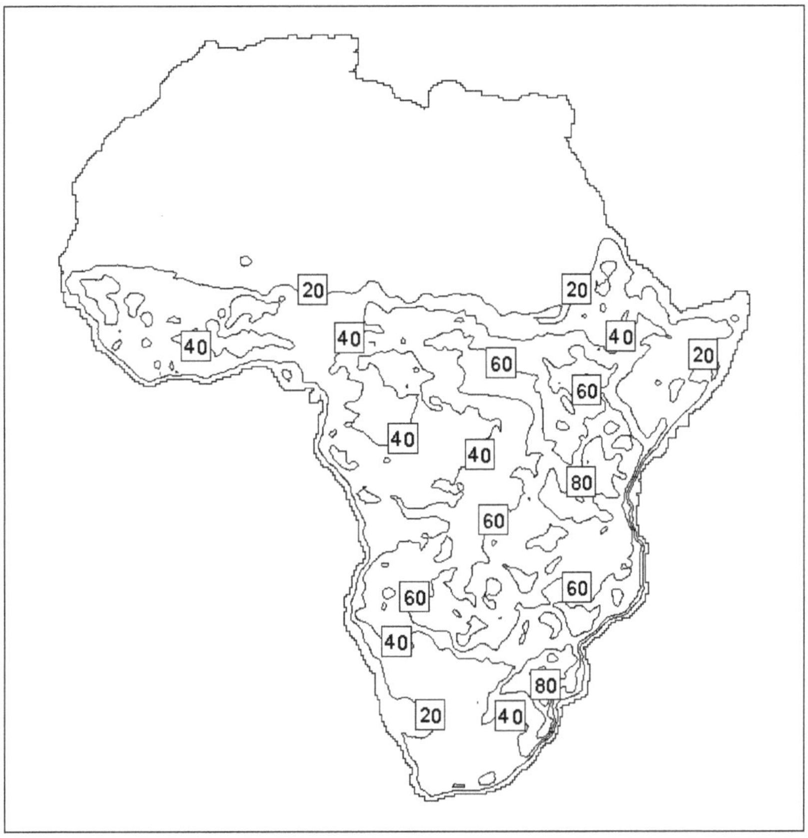

Figure 5.26
Ungulate biomass in kg/ha

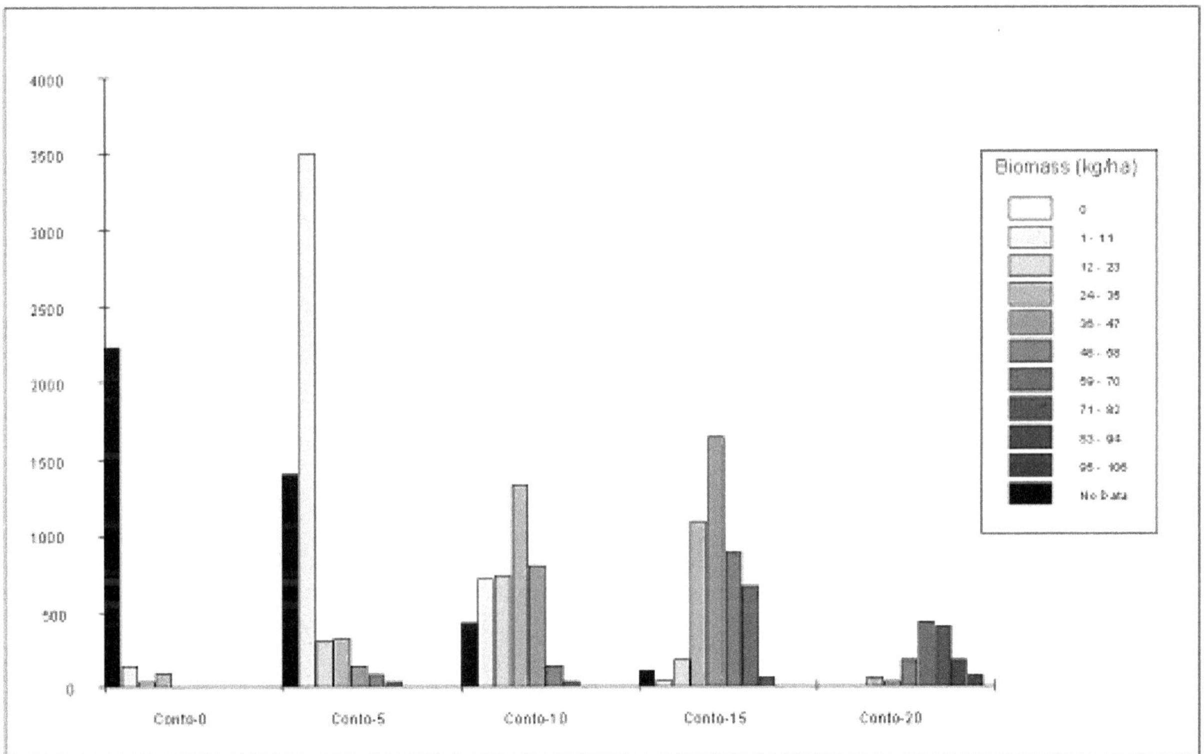

Figure 5.27
Frequency distribution of ungulate biomass (kg/ha) in each contour of carnivore biomass

Home range

A number of factors have been identified as predictors of home range size, and may affect the spatial distribution of large and small home ranges. Home range size increases with metabolic needs, irrespective of taxonomic affinity (Gittleman and Harvey 1982). Home range size does not increase as a simple (linear) function of basal metabolic rate: rather, it seems that relatively large animals have to spend relatively more energy on movement to traverse these larger areas (Leonard and Robertson 2000). In addition, Grant *et al* (1992) have demonstrated that undefended home ranges are larger than defended for ungulates and carnivores.

Diet shows a significant influence on home range size in carnivores (Gittleman and Harvey 1982). Primate home range size is greater for species with higher quality diets (more fruit and animal products) than for those with relatively poor quality diets (mostly folivores) (Leonard and Robertson 2000). Carnivores with a large proportion of flesh in their diets have particularly large home ranges (Gittleman and Harvey 1982). Carnivorous animals tend to have larger home ranges than omnivores or herbivores of a similar size (Harestad and Bunnell 1979). Home range area is related to body weight, and broad correlations also exist with surrogate variables for productivity (rainfall and latitude) (ibid.). Carnivore home range size may be inversely related to prey density: members of a single species may have large home ranges where prey is relatively scarce, in order to meet their metabolic needs, or smaller home ranges where prey is abundant.

The primate species with the largest individual home range sizes (*Gorilla gorilla* and *Pan paniscus*) both live in the west-central rainforest. This is predictable, as these species have large body masses and high metabolic requirements, and rainforests have greater primary productivity. However the largest mean individual home range sizes are distributed north of the rainforest boundaries and in southern Africa (Figure 5.28). Comparison of maps of body mass distribution and home range (Figures 5.15, 5.28) suggests that species north and south of rainforests have large home ranges relative to body mass. This pattern can be attributed to differences in diet and foraging regimes. It has been suggested the primate species that occupy these environments have to range further to meet their energetic requirements. Milton and May (1976) argue that the large home ranges of arid country species (*Erythrocebus patas*, *Papio hamadryas*) are due to their environment, which probably has lower densities of resources per given area. However according to Leonard and Robertson (2000), variation in home range size that is not explained by body mass can be explained by dietary quality. If primate species living outside tropical forest environments have a higher quality diet, this would provide an alternative explanation for their larger home ranges. Although woodland and savannah habitats are characterised by lower values of primary productivity, secondary productivity may also be relevant as a source of high quality food. As discussed above, some primates that live in more open habitats do eat meat as well as other types of food.

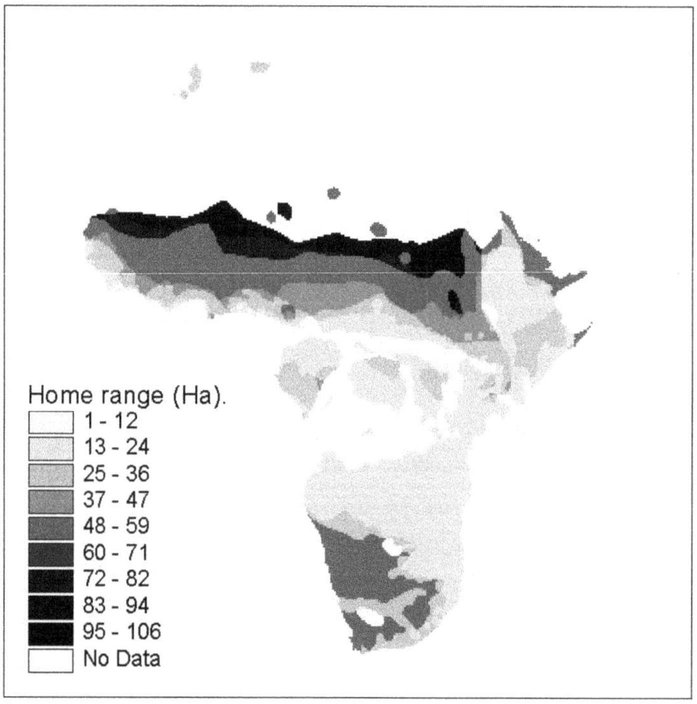

Figure 5.28
Mean primate individual home range (HR_I) in Ha (n = 31)

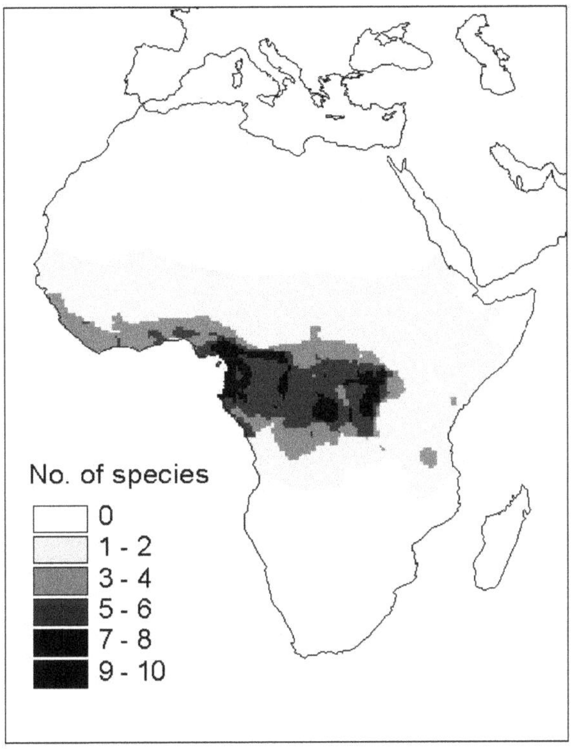

FIGURE 5.29
DISTRIBUTION OF PRIMATE SPECIES FOR WHICH HOME RANGE DATA WAS ABSENT

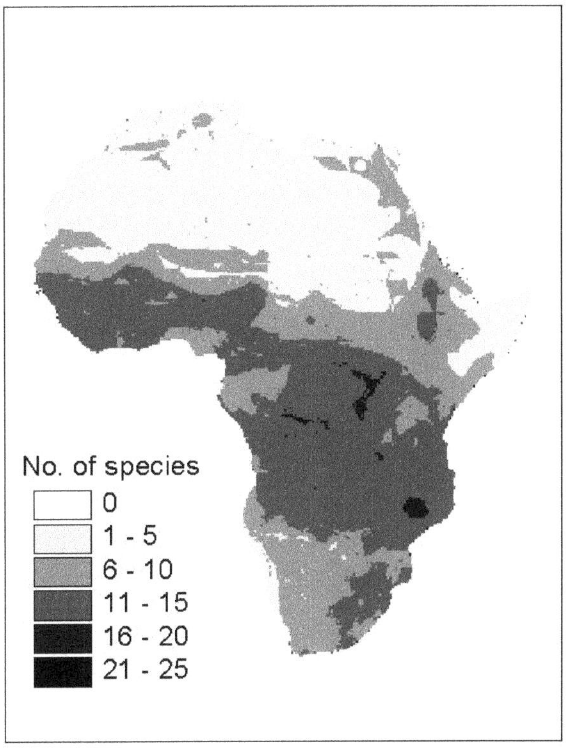

FIGURE 5.31
DISTRIBUTION OF CARNIVORE SPECIES FOR WHICH HOME RANGE DATA WAS ABSENT

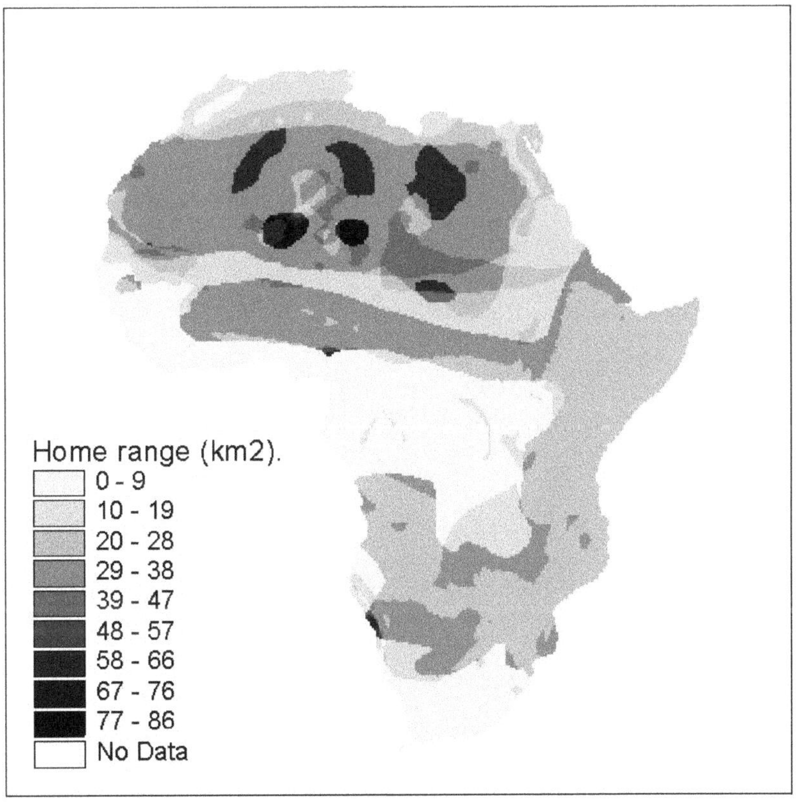

FIGURE 5.30
MEAN CARNIVORE HR_I IN KM^2 (N = 22)

The largest carnivore home ranges tend to belong to the largest bodied carnivores, and are thus distributed in the tropical savannah and woodland areas of sub-Saharan Africa. However the largest mean individual home range values occur in the more arid Sahara desert region, as well as in parts of the savannah, woodland and shrubland habitats south of the Sahara and in southern Africa (Figure 5.30). This suggests that carnivore home ranges may be highest relative to group size when resources are less abundant (and come in smaller packages). There was less data available on carnivore than primate home ranges, and data absence is strongly biased towards forest species (Figure 5.31), which are probably rarer and less widely distributed.

The spatial distribution of individual home range size, unlike that of the other ecological variables, is quite similar for primates and carnivores. This suggests that similar factors determine individual home range size in carnivores and primates. However it also indicates that the effect of body mass on home range size differs in these orders.

Conclusion

Analysis of data for modern mammals indicates that there are some strong differences in the spatial distribution of ecological variables. Geographic range sizes differ in scale, but show a broadly similar frequency distribution between the carnivores, ungulates and primates. Primates show a different spatial distribution pattern from carnivores and ungulates in diversity, body mass, and biomass. Patterns in species range boundaries are associated with those in diversity. In both primates and carnivores, mean individual home range size increases in more open habitats. However primates differ from carnivores in the distribution of home range size relative to body mass.

While other factors will influence the distribution of particular species, there do seem to be broad patterns in distribution that characterise particular orders. Carnivore diversity, biomass and body mass seem to be closely related to ungulate diversity and biomass, suggesting that the availability of prey species is a crucial variable. At the same time, the distribution of primate diversity, biomass and body mass reflects primary productivity. This suggests that dietary niche may be an important factor in differentiating variation between higher taxonomic groups. The maps presented here provide a basis for reinterpretation of spatial patterns in hominin distribution. These results are consistent with the predictions of model 3 as described in Chapter 3. In Chapter 6 I will assess the potential for applying this model to hominin species.

CHAPTER 6

HOMININ DISTRIBUTION IN THE PLIO-PLEISTOCENE

Introduction

In Chapter 4, I demonstrated that primate models based on behavioural flexibility and social learning, or alternatively relatively rapid life history, do not provide an adequate explanation for hominin range expansion. Equivalent analyses by Eeley and Foley (1999) suggest that habitat and dietary niche breadth are better predictors. In Chapter 5, I showed that there are patterns in the distribution of diversity, biomass, body mass and home range size that correspond to dietary niche. This suggests that a model emphasising dietary change may better explain range expansion in early hominin species. However, these models based on cross-species patterns in modern mammals are only useful in interpreting hominin behaviour if we can identify relevant patterns in the archaeological and palaeontological record. The aim of this Chapter is to assess the relevance of these models, by examining the data on hominin niche breadth, biomass, diversity, distribution, and other relevant variables.

I will focus on the period 1.8-0.6 my ago in Africa, and the species *H.erectus* and the robust australopithecines. There are a number of reasons for looking at this period. The occupation of new regions outside Africa is the most dramatic example of range expansion in the early hominin record, although the state of our knowledge of this process is currently limited (Dennell and Roebroeks 2005). Processes of range expansion can also be seen in the occupation of new habitats within Africa. The appearance of early African *H.erectus* marks a major change in hominin adaptive strategy involving morphological and behavioural innovations. This may have involved a shift in dietary niche. Finally, the focus on this period in Africa allows me to compare distribution patterns in more than one hominin species, which is useful for defining the ecological niches of these species.

Context

Fossil context

There were several species present in Africa during the period 1.8-0.6 my ago. These include the robust australopithecines and *H.erectus*, while *H.habilis*/*rudolfensis* was probably present at the beginning of this period. The gracile australopithecines seem to have become extinct before 2 my ago, either through evolution into *Homo* or replacement by *Homo* (Klein, 1999): however it has been argued that the transition to *Homo* came later, and that *H.habilis* and *rudolfensis* can be attributed to the australopithecine genus (Wood and Collard, 1999a). Stone tool assemblages from this period are generally attributed to the Acheulean and Developed Oldowan tradition.

Fossil evidence suggests that the species *H.erectus* emerged about 1.8 my ago. A nearly complete skull (KNM-ER 3733) and a partial skeleton (KNM-ER 1808) from Koobi Fora, are well dated to between 1.8-1.7 my old; a second skull (KNM-ER 3883) is only slightly younger (Feibel *et al.*, 1989, Brown, 1994). A skull and associated skeleton (KNM-WT 15000) from Nariokotome III, West Turkana, are dated to about 1.5 my old (Brown *et al.*, 1985).

The Acheulean industry appears about 1.65 my ago, based on radiocarbon dates from West Turkana in northern Kenya (Roche, 1995). It is also well documented 1.5-1.4 my ago, at Konso in southern Ethiopia, on the Karari escarpment at East Turkana, and at Peninj in northern Tanzania (Asfaw *et al.*, 1992, Isaac and Harris, 1978, Isaac and Curtis, 1974). The Acheulean industry is characterised by the presence of handaxes and other bifacially shaped tools. Stone artifacts including bifaces are stratigraphically associated with possible or probable fossils of *H.erectus* at a number of sites in north, east and south Africa. However there is continuity with the earlier technological tradition: Oldowan tool types are present in assemblages from this period and are sometimes numerically dominant.

Some workers have suggested that the 'early African' component of *H.erectus* is significantly more primitive than is *Homo erectus sensu stricto*, and have proposed that this subset be assigned to a separate species, *H.ergaster* (Tattersall, 1986, Wood, 1991, Wood, 1992). This species is generally recognised as being confined to the African continent, and to a relatively early period. Later African specimens tend to be attributed to *H.erectus*, implying that this species either evolved from *H.ergaster* in Africa or evolved in Asia and migrated to Africa. The best known specimens attributed to *H.ergaster* come from Lake Turkana, and date to between 1.8-1.5 my ago. However the status of *H.ergaster* as a viable species is not universally accepted (Turner and Chamberlain, 1989, Brauer and Mbua, 1992, Rightmire, 1993, Brauer, 1994). Many authorities prefer to regard them as one species, on the basis of the striking derived features shared by *H.ergaster* and *H.erectus* (Brauer and Mbua, 1992, Brauer, 1994, Rightmire, 1990, Walker and Leakey, 1993). Separation depends on the argument that *H.erectus* was also derived relative to *H.ergaster* in some key features (Andrews, 1984, Clarke, 1994b, Groves, 1989, Stringer, 1984, Wood, 1984). These differences are mainly quantitative, and must be assessed on a relatively small number of specimens (Klein, 1999). The novel adaptations of *H.ergaster* and *erectus*, described below, are considerably more striking than the differences.

The lack of consensus on this issue continues, with authors often referring to *H.ergaster* as an alternative designation to 'early African *H.erectus*' (Wood and Collard, 1999a, Wood and Collard, 1999b). The definition of separate species has implications for discussion of species geographic ranges. Given the lack of consensus on this issue, and the failure to identify definitive differences that warrant division of the fossils into two mutually exclusive groups, in this Chapter I will refer to all of the African specimens as *H.erectus*. The effects of changing hominin adaptations on distribution are of interest at the sub-species as well as species level, and I will take chronological trends into account.

H.erectus is differentiated by a number of morphological and behavioural characteristics. It is the first hominin with a large body, fewer differences between the sexes, and a post-cranial skeleton that suggests it was fully upright and committed to bipedalism (Wood and Collard, 1999b, Wood and Brooks, 1999). Relative to its body mass, its jaws and teeth are no larger than those of modern humans and chimpanzees (ibid.). In addition, its teeth, like those of modern humans, grew relatively slowly (ibid.). This suggests that *H.erectus* was the first hominin species to have a pattern of development similar to that of modern humans, with a prolonged growth period for children. However its relative brain size is low compared with later *Homo* (Wood and Collard, 1999b, Wood and Brooks, 1999).

A number of species of early hominin are characterised by the development of cranial structures that allowed great force to be applied between large upper and lower cheek teeth. These include *A.aethiopicus*, *P.boisei* and *P.robustus*. The features that *A.aethiopicus* shares with *P.boisei* and *P.robustus* may reflect a shared dietary adaptation rather than closely shared descent (Klein, 1999). The latter two species are designated '*Paranthropus*' because of their undoubtedly close kinship, and the fact that they represent a highly specialised branch of the hominin family, although a reasonable case can be made for also retaining these species in the Australopithecine genus (Klein, 1999). The number of traits that distinguish the two *Paranthropus* species has shrunk as the sample has grown (Brown *et al.*, 1993, Delson, 1997, Suwa *et al.*, 1997), and arguably the two forms were simply geographic variants of a single widespread species (Klein, 1999).

Two of the robust australopithecines overlap in time with *H.erectus*. *P.boisei* appears in stratigraphic sequences in the Turkana Basin around 2.3 my ago (Omo Shungura Formation Members G-L and West Turkana Nachukui Formation Kaitio Member). This species is present until sometime between 1.2-0.7 my ago. *P.robustus* is present in South Africa from roughly 1.8 my ago until 1 my ago or slightly later. The date of extinction for the robust australopithecines is unclear because the fossiliferous portions of the long stratigraphic sequences in the Lake Turkana basin stop around 1.4-1.2 my ago. At the same time, most sites that probably monitor the immediately succeeding period (Olduvai Gorge, and perhaps Melka Kunturé and the Middle Awash) have not yet provided large fossil samples (Klein, 1999).

Environmental context

The climatic and environmental context for the period under discussion is complex. There is evidence for significant change from earlier periods of human evolution. The early australopithecines may have occupied relatively forested environments (White *et al.*, 1994, Wolde-Gabriel *et al.*, 1994). Pollen sequences suggest that between 3.7-3.2 my ago there were more humid and evergreen species present in the lowlands and highlands in East Africa (Bonnefille, 1995). Between 2.5-2.35 my there was a spell of highland cooling, contemporary with drier conditions in the lowlands (Bonnefille, 1995). At Olduvai, wet, marshy and wooded conditions low in Bed I gave way to aridity and a more open setting at the top of this unit (Potts, 1998b).

Stable isotope studies suggest that a shift to C_4 dominated environments (open, grassy, heat-adapted vegetation) occurred as late as 1.7 my ago, with little evidence of consistently open savannah until after 1.0 my ago (Cerling *et al.*, 1991, Cerling, 1992, Kingston *et al.*, 1994, Sikes *et al.*, 1999). This is evident at some of the sites at which *H.ergaster* fossils have been found. Within the interval 1.8-1.7 my ago, there is isotopic evidence for an expansion of C_4 plant habitats within the Turkana basin (Cerling *et al.*, 1988, Cerling, 1992). Pollen from drier and warmer vegetation than modern is present at Olduvai at 1.77 my ago (Bonnefille, 1995). At this time there was a marked shift in soil $\delta^{13}C$ and $\delta^{18}C$ (Cerling and Hay, 1986). This reflects a change in the percentage of C_4 plants from 20-40% to 60-80%, representing an open, grassy environment (Cerling and Hay, 1986). However, subsequently conditions at both Olduvai and the Turkana basin became somewhat more wooded and humid (Bonnefille, 1995). The percentage of C_4 grasses at Olduvai was slightly reduced from about 1.6-1.4 my ago (Cerling and Hay, 1986). After 1.2 my ago, the carbonate data indicates that Beds III and IV, with greater than 70% C_4 biomass, were warmer and drier than at any time previously (Sikes, 1999).

$\delta^{13}C$ values from the Olduvai sequence further demonstrate the dominance of C_4 vegetation after 1 my ago, and a gradual approach to modern conditions with about 90% C_4 flora in this semiarid area in the midst of the Serengeti plain (Sikes, 1999). By 1 my, at Olorgesailie, in southern Kenya, a C_4 dominated fauna was also present. Only after 1 my ago, when C_3-dominant woodlands and grassy woodlands were generally replaced by C_4-dominant wooded grasslands and grasslands, is greater than 50% C_4 floral biomass consistently present at the majority of the localities analysed to date. Open C_4 grasslands, however, remain the least common vegetation community in the tropical savannah biome (Sikes *et al.*, 1999).

Thus the environmental evidence does not support a sudden, dramatic shift from woody C_3 habitats to open C_4 grasslands at any time (Sikes *et al.*, 1999). The paleosol isotope data records a more gradual approach to open grassland communities with more than 50% biomass (wooded grasslands and grasslands). This overall directional trend was accompanied by environmental fluctuation. According to Potts (1998b), there was one major environmental transition every 33-50 kyr averaged over the Olduvai sequence, which is within the range of the 41 kyr periodicity related to orbital obliquity.

Analysis

Distribution and diversity

H.erectus distribution

Most Plio-Pleistocene fossil and archaeological sites come from the East African Rift Valley or Southern African karstic cave sites. This primarily reflects the distribution of sedimentary traps with good conditions for preserving fossils, and of subsequent tectonic activity making earlier deposits available to prospectors. Discoveries of early hominin fossils in Chad, in Central Africa (Brunet *et al.*, 1997, Brunet *et al.*, 2002), have demonstrated that hominin species ranged widely in Africa from an early period, into areas that may preserve few or no fossil remains. Given these limitations, can we tell anything about the niche of African *H.erectus* from the distribution of fossil and archaeological sites?

H.erectus was present in the Turkana basin at West Turkana and Koobi Fora from a relatively early date (Wood, 1991, Brown *et al.*, 1985, Brown, 1994, Feibel *et al.*, 1989). In addition, possible *H.erectus* fragments from the lower Omo Valley (Shungura Formation) date to 1.4-1.3 my ago (Feibel *et al.*, 1989, Howell, 1978). Remains of *H.erectus* have also been discovered at Konso, Ethiopia, in a time horizon of *c.* 1.4 my (Asfaw *et al.*, 1992). A skull fragment from Gomboré II (Melka Kunturé), Ethiopia, is dated between 1.3 and 0.78 my ago (based on paleomagnetism and radiopotassium dating) (Chavaillon, 1982). *H.erectus* fossils have been found at Olduvai Gorge in Upper Bed II and Beds III/IV: these are not radiometrically dated, but estimates based on paleomagnetism and presumed sedimentation place a skullcap (OH9) near 1.2 my ago and a partial skull (OH12) and various fragments before 1.1 my ago (Leakey and Hay, 1982). A skull from Buia, Eritrea, is probably about 1 my old (based on paleomagnetism and associated fauna) (Abbate *et al.*, 1998). A hominin calvaria and postcranial remains from Bouri, Middle Awash, Ethiopia, dated to *c.* 1 my ago, have been attributed to *H.erectus* (Asfaw *et al.*, 2002). Other fossils of *H.erectus* include cranial fragments from Nyabusosi, Uganda (Senut *et al.*, 1987). All of these sites are situated within the East African Rift Valley.

A partial skull (SK 847) from Swartkrans Member 1 closely resembles the Koobi Fora skulls (Clarke *et al.*, 1970, Clarke, 1994b). Associated fauna at Swartkrans indicates that craniodental fragments of primitive *Homo* date to about 1.8-1.5 my ago, parallel with the appearance of early *H.erectus* in eastern Africa at Koobi Fora and West Turkana (Vrba, 1985a). A crushed cranial fossil from Sterkfontein Member 5 may also be *H.erectus* (Kuman and Clarke, 2000).

Fossils are also found in the north of the continent for this hominin species. Human mandibles and a skull fragment from Ternifine probably date to just after 0.78 my ago, based on fauna and paleomagnetism (Geraads *et al.*, 1986, Klein, 1999). A left femoral shaft attributed to *H.erectus* has also been found at Aïn Maarouf (El Hajeb), Morocco (Geraads *et al.*, 1992). Human fossils have also been found at Sidi Abderrahman and Thomas Quarries, where fauna implies that the sites post-date Ternifine (Geraads, 1980, Geraads *et al.*, 1980). However the latter fossils are in fact too incomplete for species assignment: if they date to after 600 ky ago, they could be attributed to *H.sapiens* (Klein, 1999).

While the focus of this Chapter is on hominin range expansion in Africa, a number of sites outside Africa are relevant to the discussion. Fossils including two complete skulls and several crania and mandibles have been found at Dmanisi (Gabunia *et al.*, 2000a, Vekua *et al.*, 2002, Lordkipanidze *et al.*, 2005). The fossils have been dated to *c.* 1.75 my ago based on their position above a basalt dated to 1.85 my ago, and correlation to the terminal part of the magnetically normal Olduvai subchron, and immediately overlying horizons of the Matuyama chron (Vekua *et al.*, 2002, Gabunia *et al.*, 2001). Faunal remains also support the dating of Dmanisi to the end of the Pliocene or earliest Pleistocene (Vekua *et al.*, 2002). The hominid remains have been classified as a very early type of *H.ergaster* (Gabunia *et al.*, 2000a) and/ or a new taxon, *Homo georgicus* (Gabunia *et al.*, 2002). At 44°N on the southern slope of the Caucasus mountains, with a Eurasian fauna, Dmanisi represents a clear example of range expansion and occupation of new habitats (Gabunia *et al.*, 2000b), although it remains unclear which hominin was involved.

A number of other early dates have been given to hominin fossil sites outside Africa, notably Mojokerto and Sangiran, Java. The Mojokerto cranium now seems to have been found in context (Huffman *et al.*, 2005) and has been given a date of 1.8 my old (Swisher *et al.*, 1994). The specimens from Sangiran have been dated to about 1.6-1.7 my ago (Swisher *et al.*, 1994, Larick *et al.*, 2001). The only other Early Pleistocene hominin fossil evidence consists of three incisors from 'Ubeidiya, Israel (1.4-1.0 my ago) (Belmaker *et al.*, 2002). The principal fossil evidence from Longuppo is probably not hominin (Schwartz and Tattersall, 1996, Wolpoff, 1999). In addition, there are a number of archaeological instances of Early Pleistocene artefacts from Asia, including dates of 1.66 mya from the Nihewan basin in north China (Zhu *et al.*, 2004).

The Acheulean industry first appears shortly after early African *H.erectus*, and is associated with fossils of this species in a number of sites. Stone artifacts are stratigraphically associated with possible or probable fossils of *H.erectus* at Ternifine (Balout *et al.*, 1967), Sidi Abderrahman (Raynal *et al.*, 2001), Thomas Quarries (Raynal *et al.*, 2001), Gomboré II (Melka Kunturé) (Chavaillon, 1979), Konso (Asfaw *et al.*, 1992), Olduvai Gorge Beds II and IV (Leakey, 1975, Leakey, 1977, Leakey and Roe, 1994), and Swartkrans (Clark, 1993). In each case the artefacts include handaxes and other bifacial tools that are the hallmark of the Acheulean stone tool industry (Klein, 1999). Acheulean artefacts are also well known from sites that lack human remains but have been dated to before 600 ky ago, before the earliest *H.sapiens* fossils, and were therefore probably made by *H.erectus*. These include Gadeb 2 and 8 (Clark and Kurashina, 1979) and Melka Kunturé (the Garba II, Garba XII and Simbirro III localities) (Chavaillon, 1979) in Ethiopia, the Karari escarpment at East Turkana (Isaac and Harris, 1978), West Turkana (Roche, 1995), Kariandusi (Gowlett and Crompton, 1994), Kilombe (Gowlett, 1978, Gowlett, 1991) and Olorgesailie (Deino and Potts, 1990, Potts, 1989) in Kenya, and Peninj in northern Tanzania (Isaac and Curtis, 1974) as well as Sterkfontein in southern Africa (Kuman and Clarke, 2000). Acheulean artefacts from the Olpiro Beds at Laetoli may be about the same age as Olduvai Bed II (Harris and Harris, 1981).

The use of stone artefacts to provide information about distribution is complicated. It seems reasonable, based on chronology and association, to suggest that *H.erectus* was the maker of the Acheulean industry. While Oldowan assemblages also occur during this time period, a number of hominin species were present when this technology emerged, and could be the manufacturers. Evidence from later periods provides a caution, indicating that more than one hominin species may produce indistinguishable stone tool assemblages (Shea, 2003).

Dating archaeological sites in Southern Africa from contexts other than karstic cave deposits is difficult. Two series of artefacts have been excavated in river gravel contexts at Three Rivers and Klipplaatdrif in the Southern Gauteng (Kuman, 1998). These alluvial assemblages are in poor context and without fauna but have a technology comparable to that at Sterkfontein, and are considered to be Early Acheulean (ibid.). Sites in southern Mozambique have yielded bifaces similar to those at Sterkfontein, and are currently being studied (Kuman, 1998).

Site distribution suggests a possible limit to hominin ranges at the far South of Africa. There are thirteen australopithecine sites in southern Africa, all of which are limited to an area north of 28°S, the most southerly being Taung. *H.erectus* continues this pattern. This is partly an effect of conditions for preservation and discovery of fossil sites in Southern Africa; however a number of facts suggest that this may represent a real distribution pattern. The prolific faunal site of Langebaanweg in the southwestern Cape Province has yielded no hominins and virtually no primates suggesting a real limit to primate distribution in the early Pliocene (Hendey, 1982). The distribution of Early Acheulean sites suggests an avoidance of the western, more arid interior (Avery, 1995). This area, and the southern, cooler parts of southern Africa, may have been first occupied by later Acheulean people (Klein, 1994).

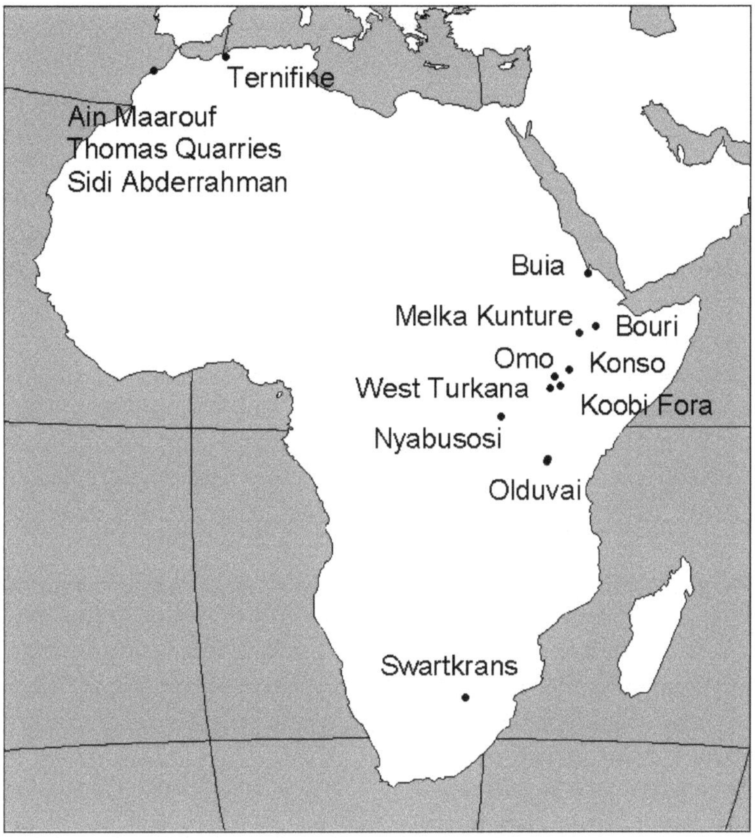

FIGURE 6.1
SITES IN AFRICA WHERE FOSSILS ATTRIBUTED TO *H.ERECTUS* HAVE BEEN FOUND

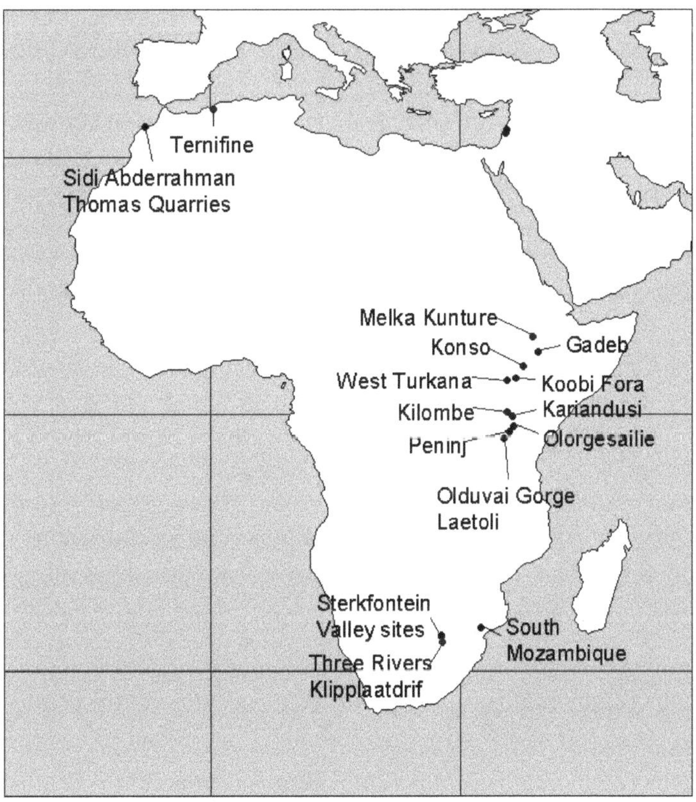

FIGURE 6.2
ACHEULEAN SITES IN AFRICA OLDER THAN 0.6 MY

Beginning approximately 1.5 my ago, there are archaeological traces that indicate the occupation of new habitats in Africa. This includes more intensive occupation of the drier and presumably more open grassy hinterlands of sedimentary basins in the floor of the Rift, as is evident at Karari near Koobi Fora (Harris, 1983). Stone artefacts and fossil evidence suggest that hominins first reached the high plateau of Ethiopia (2600-2400m) at about 1.5 my (Clark and Kurashina, 1979). This is one of the highest altitudes at which assemblages from this time range have been found either in Africa or Eurasia (Clark and Kurashina, 1979). Ethiopian sites such as Gadeb and Melka Kunturé experience extreme fluctuations in daily temperature, with freezing or near freezing conditions at night (Cachel and Harris, 1997).

So how good is the evidence for the distribution of *H.erectus* in Africa? Apart from the discovery of australopithecine fossils at Chad, there is no real evidence for occupation (or the lack of it) of West and Central Africa. Site distribution in southern Africa suggests a range boundary in the south and west of the subcontinent, and this is backed up by evidence of absence, environmental patterns, and trends in distribution over time. The North African sites provide fairly clear evidence of range expansion. As Dennell and Roebroeks (2005) have pointed out, the evidence for an early hominin presence outside Africa is very limited, especially considering the area of land involved. In addition, we have very little evidence for prior absence of hominins, a factor of particular importance given the ambiguous taxonomic attribution of the key fossil specimens (Dennell and Roebroeks 2005). In general, better evidence for expansion into new habitats comes from environmental reconstruction and landscape analysis at particular localities.

H.erectus and mammalian distribution

What are the implications of *H.erectus* distribution, when compared with the distribution of modern mammals? As discussed in Chapter 5, modern primate diversity is highest at lower latitudes and in forest environments. However, modern savannah-mosaic habitats are occupied by baboons (*Papio* spp.), patas monkeys (*Erythrocebus patas*), vervet monkeys (*Cercopithecus aethiops*) and chimpanzees (*Pan troglodytes*) in sub-Saharan Africa (Richard, 1985). Baboons and vervets have particularly extensive distributions within this biome, and baboons also survive at low population densities in more arid semi-desert scrub (Richard, 1985).

Is *H.erectus*' expansion into more open habitats comparable to the distribution of our close relatives the common chimpanzees today? While chimpanzees occupy savannah-mosaic environments, they are mainly found in forest and gallery forest. *H.erectus* appears to have ranged more widely in these environments: their distribution shows more parallels with modern baboons and vervet monkeys. However the larger brain and body size of hominin species puts them firmly in the category of the great apes with regard to dietary quality requirements. *H.erectus* distribution seems to break the primate trend in this respect, and this will be discussed further in the section on body mass below.

The gelada (*Theropithecus gelada*), a specialised herb eater, occupies arid and high altitude environments in the highland plateau of Ethiopia. Hamadryas baboons are also found in this biome, but their density is much lower (Richard, 1985). The gelada is behaviourally and physically adapted to this environment: it has a thick fur coat and often huddles to conserve heat. This suggests that early African *H.erectus*, if not physically adapted, must have possessed behavioural adaptations to this habitat, such as fire (Cachel and Harris, 1997).

H.erectus also occupied higher latitudes. Modern habitats in northern Africa along the Moroccan and Algerian coast are primarily mediterranean forests, woodland and scrub (Olson and Dinerstein, 1998). The only primate species living in North Africa today (*Macaca sylvanus*) has a very limited geographic range. Some modern primates do have ranges extending throughout sub-Saharan Africa, but none occupy habitats in North Africa as well.

However the current distribution of primates is likely to have been affected by climatic and environmental change, particularly the expansion of the Sahara desert, and expansion of human populations. During the Pleistocene *Theropithecus oswaldi* was present in the North African sites of Thomas Quarries and Ternifine as well as in the Turkana basin and at Swartkrans in South Africa (Raynal *et al.*, 2001, Turner *et al.*, 1999). Environmental evidence from North African sites suggests that the climate was rather harsh, open and cold during the Late Pliocene (Geraads *et al.*, 1998) and was still rather open and dry in the Middle Pleistocene (Raynal *et al.*, 2001). This suggests that the distribution of *H.erectus* may have been less unusual than comparison with modern species makes it seem. In addition, it should be noted that carnivores live at a high diversity and abundance in modern savannah and grassland habitats, and some carnivores also have a very wide latitudinal distribution within Africa.

Hominin diversity

The robust australopithecines appear to have been sympatric with early *Homo* at a number of localities. Comparison

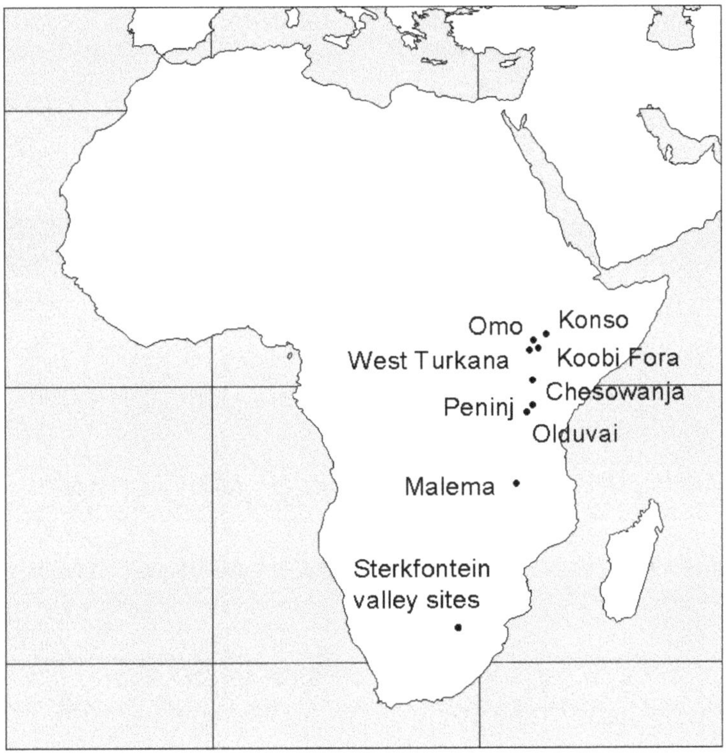

FIGURE 6.3
SITES IN AFRICA WHERE FOSSILS OF *PARANTHROPUS* HAVE BEEN FOUND

of the distribution of fossils of these species may tell us something about the characteristics and differences of their ecological niche: their ability to coexist may also be informative.

P.boisei is known from Konso (KGA 10) (Suwa *et al.*, 1997), Omo (Members G-L) (Suwa *et al.*, 1996), West Turkana (Kaitio Member of the Nachukui Formation) (Klein, 1999), Koobi Fora (Upper Burgi, KBS) (Wood, 1991), Chesowanja (Chemoigut Formation) (Carney *et al.*, 1971), Olduvai (Beds I and II) (Leakey, 1971), Peninj (Humbu Formation) (Klein, 1999), and Malema (Chiwondo Formation) (Kullmer *et al.*, 1999), from roughly 2.3 my ago until sometime between 1.2 and 0.7 my ago. *P.robustus* is known only from the South African sites of Kromdraai B (Member 3) (Broom, 1938), Swartkrans (Members 1-3) (Brain, 1981, Grine, 1993), Drimolen (Keyser *et al.*, 2000), and Sterkfontein (Member 5) (Kuman and Clarke, 2000), from roughly 1.8 my ago until 1 my ago or slightly later.

While it is theoretically possible that the robust australopithecines made and used stone tools (Wood, 1997), the distribution of archaeological sites cannot tell us anything useful about their geographic range. The earliest stone tools come from deposits in East Africa that are 2.6 to 2.3 my old (Kibunjia, 1994, Kimbel *et al.*, 1996, Roche *et al.*, 1999, Semaw *et al.*, 1997). These sites were almost certainly contemporaneous with a number of hominin species, including *H.habilis*, *P.boisei*, and *A.garhi*. Early stone tool technology (Oldowan) continued to be used after the invention of Acheulean technology, and many assemblages contain both.

A temporal and spatial overlap between *H.erectus* and the robust australopithecines is demonstrated by discoveries of fossils in the same stratigraphic levels and localities. In South Africa, Swartkrans members 1 and 2 have provided craniodental fragments from more than 17 individuals of *H.erectus*, in association with much larger quantities of fossils of *P.robustus* (Brain, 1981, Clarke *et al.*, 1970, Clarke, 1994b, Grine *et al.*, 1993, Grine, 1993). Both species may also be present at Sterkfontein Member 5 (Kuman and Clarke, 2000) and Coopers Cave (Berger, pers.com.). Early *Homo* is present with *P.robustus* at Drimolen, in sediments of a similar age to Swartkrans (Keyser *et al.*, 2000).

Fossils of a robust australopithecine have been found in Upper Bed II, Olduvai Gorge, as have *H.erectus* fossils (Leakey, 1971). At Koobi Fora, fossils of *P.boisei* have been found mainly in the KBS member, but also below and above it, spanning a period from just before 2 to 1.4 my ago; fossils representing *H.ergaster* have been found in the KBS member just below and above the Okote tuff, where they probably date from just before 1.7 to 1.3 my ago (Walker and Leakey, 1978, Wood, 1991). *P.boisei* and *H.erectus* also overlap in the Lower Omo River Basin (Shungura Formation), where the former is present from about 2.3-1.2 my ago, and the latter at about 1.4 my ago (Howell, 1978, Klein, 1999, Suwa *et al.*, 1996). At West Turkana, *P.boisei* predates *H.erectus*, appearing in the Kaitio Member (2.3-2.6 my ago) while *H.erectus* is represented in the Natoo member at a level dated to about 1.6 my ago (Brown *et al.*, 1985, Leakey and Walker, 1988, Klein, 1999). At Konso, the assemblage at 1.4 my ago contains nine specimens of *P.boisei* and two fragments of *H.erectus* (Suwa *et al.*, 1997).

The overlap of fossils of *H.erectus* and the robust australopithecines at fossil sites suggests that they may have lived sympatrically and exploited the same habitats. Where two species that are closely related live in the same areas, there is likely to be competition and coevolution. The ability of these species to coexist could be explained in terms of niche differentiation. On the basis of their morphology, the robust australopithecines are characterised as dietary specialists, *Homo* as generalists. However, as discussed in Chapter 5, recent isotopic studies have indicated that the robust australopithecine diet was actually quite varied (Lee-Thorp *et al.*, 2000). Another explanation stresses differences in foraging efficiency (Cachel and Harris, 1997). Comparison with sympatric modern apes, gorillas and chimpanzees, suggests that different habitat ranges and preference patterns may be important (Sikes, 1999).

Competition may be most significant in times when resources are scarce (Foley, 1987b). This is particularly important in highly seasonal environments, such as savannah (ibid.). Foley (1987b) argues that the divergence of *Homo* and the robust australopithecines may reflect different coping strategies during the poor season. The robust australopithecine strategy may have been to forage on low quality plant foods and increase foraging time (Foley, 1987b). For instance, robust australopithecines may have relied on their ability to consume and survive on vegetal foods other than succulent fruits, such as hard cased fruits and seeds or underground storage organs (Sikes, 1999). *Homo* may have focussed on seasonally available high quality foods such as meat, or underground plant storage organs (Foley, 1987b). A further factor in these species' survival in the same areas may have been the distribution of time and activities in each microhabitat (Sikes, 1999). Sikes suggests that the relatively longer day ranges and larger home ranges of one sympatric species may have increased the quantity and diversity of ripe fruit available to each group.

Identification of contrasts in the distribution of these species could highlight differences in habitat tolerance. The distribution of *H.erectus* fossils extends beyond that of *P.boisei* to the north and east, at Buia, Eritrea and Bouri, Middle Awash and Melka Kunturé, Ethiopia. In addition, fossils of gracile australopitheciness have been found at Hadar and the Middle Awash in northern Ethiopia, dating to before 2 my ago. However none of these sites have provided large enough numbers of hominin fossils from the relevant period (2.3-1.4 my ago) to constitute an identifiable range boundary for robust australopithecines.

Clearer evidence of a contrast in distribution, with implications for species habitat tolerance, comes from North Africa. Fossils of *H.erectus* have been found at a number of North African sites, as discussed above. No ancient (Pliocene) pebble tool culture has yet been discovered in Morocco (Raynal *et al.*, 2001) although an Oldowan site with a possibly early Pleistocene date has been described at Ain Hanech in Algeria (Sahnouni and de Heinzelin, 1998). Corroborative evidence comes from the Late Pliocene site of Ahl al Oughlam in Morocco, which has yielded a complete fauna of macro and micromammals including 55 species, but does not include any clearly identifiable australopithecine fossils (Geraads *et al.*, 1998). The consensus of evidence provided by the North African sites suggests that hominin range expansion into that area began with *H.erectus*. This difference can not be attributed to the existence of a major biogeographical barrier: environmental reconstruction by the Pliocene Research, Interpretation and Synoptic Mapping (PRISM) Project indicates that this barrier was less extensive in the late Pliocene and early Pleistocene (1995). This suggests that habitat preference and tolerance for the climatic and environmental characteristics of higher latitudes in Africa may have been an important factor.

Hominin habitat tolerance and preference

A number of sites provide evidence that *H.erectus* occupied relatively open and dry environments. These include several with quite early dates. At Konso, Ethiopia, fossils of *H.erectus* dating to 1.4 my ago were found with fauna indicating a predominantly dry grassland environment (Suwa *et al.*, 1997). The Acheulean sites at Gadeb are situated high on the Ethiopian plateau (Clark and Kurashina, 1979). Environmental evidence from Melka Kunture suggests that highland Ethiopia was probably characterised by riparian forests of montane species and *Acacia* grassland (Clark and Kurashina, 1979, Bonnefille, 1976). At Peninj, stone tools from the Oldowan and Acheulean industry have been found in the Humbu Formation (Isaac and Curtis, 1974). Pollen analysis of the upper Humbu Formation (dated to 1.5-1.35 my ago) has indicated that Peninj may have displayed a limited ecological variation, and that open vegetation environments were predominant (Dominguez-Rodrigo *et al.*, 2001).

At Buia, Eritrea, the furthest north-eastern site for this period, dated to *c.* 1.0 my, the faunal composition is typical of African savannahs (Abbate *et al.*, 1998). At Bouri, fauna from the Daka Member dated to *c.* 1.0 my ago suggests a predominantly savannah environment (Asfaw *et al.*, 2002). The fauna is dominated by bovids, notably Alcelaphines, which are present in a diversity and abundance not recorded at older African sites. The paleoenvironment at Bouri probably involved widepread open grassland nearby and adjacent water margin habitats (ibid.). At Olorgesailie, landscape scale analysis of a 0.99 my old paleosol gave strong evidence for wooded grassland as primary vegetation context (Sikes *et al.*, 1999).

It seems likely that *H.erectus* also tolerated quite varied habitats. The earliest specimens of *H.erectus* come from the Turkana basin, dated to about 1.8 my ago. During the period 1.9-1.5 my ago the Turkana Basin underwent considerable landscape instability. At the beginning of this period, the centre of the Turkana Basin was occupied by a sizeable yet fluctuating lake fed from the north by the proto-Omo River (Brown and Feibel, 1991). Within the period 1.8-1.7 my ago, a major shift in the isotopic composition of paleosols signifies a likely expansion of open C_4 plant habitat within the Turkana basin (Cerling *et al.*, 1988, Cerling, 1992). The lake persisted until about 1.7 my ago, when it was replaced by a very unstable fluvial system (Brown and Feibel, 1991). Environments dominated by an axial river alternated with others consisting of intricate expanses of braided streams (Potts, 1998b). The fact that early hominin toolmakers persisted in the Turkana basin locality through periods of strong environmental change (ibid.) suggests that these species had a strong ability to buffer environmental change with behaviour.

More detailed information is available on a general expansion of habitat breadth in *H.erectus* based on the distribution of stone tools. Raw material transfer patterns at Gadeb suggest that hominins were moving between the plateau, Rift and escarpment edge by 1.5 my ago (Clark and Kurashina, 1979). Landscape analyses in the Turkana basin indicate that toolmakers became less tethered to local rock sources and particular habitats (Rogers *et al.*, 1994). At 2.3 my ago, the availability of stone, water, shade, shelter and presumably food seems to have played a large role in determining where stone tools were discarded. Hominin use of stone appears to have beeen restricted to a few places where all of these resources were available. By 1.6 my ago, many different settings were being used for stone tool activities, and lithic traces were not necessarily tied to sources of raw material. This change coincides with the appearance of *H.erectus*. This increase in the variety of archaeological settings may be in part explained by the environmental changes described above, particularly the increase in more open habitats (Rogers *et al.*, 1994). The evidence from these localities seems to point to increased variability in landscape use soon after the first appearance of *H.erectus*.

P.boisei and *P.robustus*, while classified as different species, were morphologically quite similar. Both species existed for a long time (at least 1 my) with relatively little morphological change (Wood, 1997). During this time there were strong environmental fluctuations (Potts, 1998b). This would suggest that these species also possessed a strong ability to adjust to environmental change by behavioural rather than genetic means.

Based on a survey of fauna from a range of sites, Reed (1997) has concluded that *Paranthropus* occurs in both wooded and more open environs but is consistently associated with well-watered sites. It is not always possible to make a distinction between alternative habitat types on the basis of the data on faunal adaptations collected in Reed's study (de Ruiter, 2000). A preference on the part of *P.boisei* for closed and/or wet habitats has been hypothesised based on mixed locality and horizon tabulations by Shipman and Harris (1988). More detailed data from particular sites is necessary to evaluate these suggestions.

At Sterkfontein, there is evidence for an overall change from tropical to sub-tropical gallery forest, forest fringe and woodland conditions in Member 4 to more open woodland and grassland habitats in the later units (Kuman and Clarke, 2000). However there appears to be a wet localized topography in the *Paranthropus* bearing Oldowan Infill, Member 5 (Kuman and Clarke, 2000). This suggests that *P.robustus* may have had a preference for wetter habitats than *Homo* (fossils attributed to *H.erectus* were found in Member 5 West). A number of studies argue that the faunal evidence throughout the sequence at Swartkrans is consistent with both a preference for open country grasslands, and the presence of a river with its associated habitats nearby (Watson, 1993, Reed, 1997). This interpretation is confirmed by an analysis of the microfauna (Avery, 1995). It seems that *P.robustus* may have tolerated both more open and wooded habitats, while consistently showing a preference for well-watered habitats, as suggested by Reed (1997).

Preference for woody habitats in *P.boisei* and sympatric *H.habilis* has been identified by Sikes (1999) based on oxygen/carbon isotope analysis of Olduvai basal Bed II. However, recent discoveries of *P.boisei* fossils from Konso, Ethiopia have extended the known geographic range and habitat preferences of this species. At this site there is 'clear evidence for the coexistence of *P.boisei* and *H.erectus* within a predominantly dry grassland environment' (Suwa *et al.*, 1997). Fossils of both species were found at locality KGA 10, characterised by a predominantly dry grassland fauna. In contrast, *P.boisei* remains are lacking from the more mesic Konso sites, despite a large combined sample of over 4000 identifiable fossils. This is suggestive of actual association between *P.boisei* and dry grassland fauna (Suwa *et al.*, 1997). This directly contradicts Shipman's (1988) suggestion, indicating that the habitat tolerance and preference of *P.boisei* was quite broad.

The environmental evidence described at the beginning of this Chapter provides a context for species habitat tolerance and preference. The period in which the first appearance date (FAD) for *H.erectus* occurs coincides with an arid climatic pulse. However the increase in wooded grassland and grassland habitats over this period is gradual and these habitats are only in the majority after 1.0 my ago. In addition there is evidence that species present during this time period would have faced frequent environmental change. This has implications for the habitat preference and tolerance and geographical distribution of *H.erectus* and the robust australopithecines. Expansion of wooded grassland and grassland habitats would have relaxed environmental constraints on range expansion in *H.erectus*, given the species' apparent preference for such habitats. In addition, robust australopithecines' preference for more wooded and wet habitats might explain their extinction at some point between 1.2-0.7 my ago. However this explanation is contradicted for *P.boisei* by the preference for more open habitats at Konso.

Body mass

Estimates from McHenry (1994) show that between 4 and 1.5 my ago average adult body weights increased by about 56% from about 37kg in Lucy and her contemporaries (*A.afarensis*), to almost 58kg in early African *H.erectus* (Leonard and Robertson, 2000) (see table 6.1).

Hominin body masses have been estimated on the basis of hindlimb joint size (McHenry, 1994). There are a number of potential errors in this method because of the small sample size, the unique body proportions of hominins, and difficulties of attribution of fossils to male and female categories. Note that the standard errors for each estimate are quite large. However a general trend is clear: the appearance of *H.erectus* in Africa marked a dramatic increase in body size, especially in the female (McHenry, 1994). This may be associated with a change in energy requirements, dietary quality and foraging strategy (Leonard and Robertson, 2000).

Hominin body masses are above the average for both carnivores and primates. Body mass has implications for population density, home range size and geographical distribution. As described in Chapter 5, large primates are more likely to live in forested environments, while large carnivores are distributed in the savannah and woodland zone of sub-Saharan Africa. Body mass also has implications for diet: large mammals have lower energy requirements per unit body weight than small ones, and can afford to process food more slowly, but their total food requirements are great (Richard, 1985).

The largest primates found to be widespread within the savannah biome are the baboons (*c.* 14-32kg for males and 11-15kg for females) (Silva and Dowling, 1995, Smith and Jungers, 1997). The larger bodied chimpanzees (42-59kg for males and 33-45kg for females in *Pan troglodytes*) are found primarily in the rainforest and seasonal forest, but at the limits of their distribution they are found in woodland and savannah mosaic (Richard, 1985). The early hominin species have a body mass close to that of chimpanzees. According to the modern primate

Species	Male body mass (kg)	Female body mass (kg)
A.afarensis	44.6 ± 18.5	29.3 ± 15.7
A.africanus	40.8 ± 17.3	30.2 ± 19.5
P.boisei	48.6 ± 34.6	34.0 ± 13.7
P.robustus	40.2 ± 15.8	31.9 ± 21.5
H.habilis	51.6 ± 22.6	31.5 ± 22.5
African *H.erectus*	62.7	52.3

TABLE 6.1
HOMININ BODY MASS ESTIMATES, FROM MCHENRY (1994)

pattern, we would expect to find hominin species living primarily in forested environments. This prediction is clearly not borne out by evidence for the habitat tolerance and preference of species existing 1.8-0.6 my ago. The increase in body mass in *H.erectus*, accompanied by a preference for relatively open habitats, reverses the primate pattern.

By contrast, as shown in Chapter 5, carnivores with a body mass between 52-77kg could be found almost anywhere in sub-Saharan Africa (excluding the arid parts of southern Africa) and smaller carnivores are distributed throughout the continent. While very efficient dietary generalism might explain the robust australopithecines' ability to live in both wooded and more open environments, the carnivore pattern provides a better model for *H.erectus*.

Biomass

As discussed above, between 2.5 and 1.5 my ago, there was a marked decline in forested areas throughout eastern and southern Africa. There is evidence that the early australopithecines occupied relatively forested environments, and that these were more widespread in East Africa. Stable isotope studies suggest that a shift to C_4 dominated environments (open, grassy, heat-adapted vegetation) occurred as late as 1.7 my ago, with little evidence of consistently open savannah until after 1.0 my ago (Cerling *et al.*, 1991, Cerling, 1992, Kingston *et al.*, 1994, Sikes *et al.*, 1999). This gradual approach to overall grassland communities was accompanied by environmental fluctuation. According to Potts (1998b), over the Olduvai sequence there was one major environmental transition every 33-50 kyr on average.

Transition between woodland and more open wooded grassland and grassland environments would result in changes in both the abundance and distribution of food resources. According to Leonard & Robertson (2000), a key aspect of this change for hominins would have been the associated change in the energetic structure of these ecosystems, with plant productivity declining and animal foods becoming a much more attractive resource. At the same time, particular vegetable foods, notably tubers, may have increased in productivity (O'Connell *et al.*, 1999).

An analysis of ungulate biomass over time in Southern Africa shows relatively low estimates for the period between 3 and 2 my ago, associated with a limited temporal variability in biomass (Thackeray and Reynolds, 1997). Higher biomass estimates were obtained within the period between 2 and 1 my ago, and higher still for the period since 1 my ago, associated with increases in temporal variability. This corresponds to trends in the data from analyses of the floral content over this time. *H.erectus* would have faced an environment in which secondary productivity was increasing, but was subject to high levels of long-term variation.

As discussed in Chapter 5, areas with high ungulate biomass tend to have lower primary productivity than forested areas. This is due to differences in usable forage available to ungulates in forest versus grassland environments (Leonard and Robertson, 2000). Ungulate consumption rates are lower in forests, approximately 5% of primary productivity, compared to 50% in grassland environments (Leonard and Robertson, 2000). A redistribution of forest and grassland resources would thus influence the amount of food resources available to species with different dietary niches. Primate biomass is strongly related to primary productivity. Primates do live in open woodland and savannah environments, but their diversity and biomass is lower throughout such environments. Carnivore biomass is strongly related to ungulate biomass and species richness. Thus an increase in grassland might reduce resources for primates, while providing more resources for carnivores.

Can we test which model fits hominin species by looking at hominin biomass? A number of alternative approaches have been applied to estimating hominin demography; the first of these is based on relative numbers of fossils at particular sites (de Ruiter, 2000). A second approach taken by a number of authors has been to predict biomass from other ecological characteristics such as body mass, brain size, population density and energetic requirements, based on patterns in modern mammals (e.g. estimates of population density by McHenry, 1994). Attempts at estimating hominin population densities from numbers of fossils at particular sites at these time distances tend to require too many assumptions – see for instance Boaz (1979).

Assessing relative numbers of individuals at particular fossil sites can be a complex task, involving detailed assessment of fossil taphonomy (de Ruiter, 2000). For the purposes of this chapter, I will discuss presence/absence and numerical data for fossil specimens. A basic comparison of the number of cranial and mandibular fossils of each species is shown in Table 6.2. I have focussed on these fossil elements because they tend to be relatively well preserved and because a skull is a reliable indicator that one individual was present.

This data suggests a possible pattern. In East Africa, at a majority of those sites where both species are present,

H.erectus	P.boisei	References
North Africa		
Ternifine: 3 mandibles. Sidi Abderrahman: mandibular fragments. Thomas Quarries: a mandible and cranial fragments.		(Arambourg, 1955, Arambourg and Biberson, 1956, Balout *et al.*, 1967, Geraads *et al.*, 1986, Howell, 1970, Howell, 1978, Hublin, 1986, Klein, 1999)
East Africa		
Omo, Shungura Formation, Member K: parietal and temporal fragments L 996-17.	Omo, Shungura Formation, Members G-L: cranium 323-76-896, mandibles L 7a-125, L 74a-21, L427-7.	(Howell, 1978, Klein, 1999, Wolpoff, 1996, Suwa *et al.*, 1996)
East Turkana, KBS Member: skulls KNM-ER 3733, 3883, mandibles KNM-ER 820, 992, 1507 mandible/skull fragments KNM-ER 730, 1466, 1648, 1808, 2592, 731, 1507, 1812.	East Turkana, KBS and Okote Members: crania KNM-ER 405, 406, 407, 732, 13750, 23000. mandibles KNM-ER 403, 404, 725, 727, 729, 801, 805, 810, 818, 1468, 1469, 1483, 1803, 3229, 3230, 3729, 3954, 5429, 5877, 15930.	(Wood, 1991, Brown *et al.*, 1993)
West Turkana, Natoo Member: KNM-WT 15000.	West Turkana, Kaitio Member: cranium KNM-WT 17400, mandible KNM-WT 16841.	(Brown *et al.*, 1985, Walker and Leakey, 1993, Brown *et al.*, 1993, Klein, 1999, Wolpoff, 1996)
Olduvai Gorge, Upper Bed II, Beds III & IV, Lower Masek Beds: crania OH 9, OH 12; mandibles OH 22, 23, 51.	Olduvai Gorge, Bed I, Lower Bed II and Upper Bed II: cranium OH 5.	(Leakey, 1971, Leakey and Roe, 1994)
Konso: mandible KGA 10-1.	Konso: mandible and cranium KGA 10-525, juvenile mandible KGA 10-570.	(Asfaw *et al.*, 1992, Suwa *et al.*, 1997)
Gomboré II: cranial fragment.	Peninj: 1 mandible.	(Chavaillon *et al.*, 1974, Howell, 1978) (Klein, 1999)
Buia, Eritrea: cranium UA 31.	Chemeron: crania KNM-CH 1, 304.	(Abbate *et al.*, 1998) (Carney *et al.*, 1971, Gowlett *et al.*, 1981)
Nyabusosi, Uganda: cranial fragments.	Malema, Chiwondo Formation: partial jaw. HCRP RC 911	(Senut *et al.*, 1987, Klein, 1999) (Kullmer *et al.*, 1999)
Bouri, Middle Awash: calvaria BOU-VP-2/66		(Asfaw *et al.*, 2002)
South Africa		
H.erectus	P.robustus	
Swartkrans, Members 1-3: skulls SK 27, 847 mandibles: 2 (fragmentary).	Swartkrans, Members 1-3: crania SK 12, 13/14, 46, 47, 48, 49, 52, 55, 65, 79, 83, 848, SKW 8, 11, 29, 2581, SKX 265. mandibles SK 6, 12, 23, 34, 1586, SKW 5, SKX 4446, 5013.	(Clarke, 1977, Clarke, 1994b, Clarke *et al.*, 1970, Grine, 1989, Brain, 1981, Grine, 1993, Watson, 1993)
	Kromdraai: cranium and mandible B TM 1517	(Broom, 1938, Klein, 1999)
	Drimolen: skull DNH 7, mandible DNH 8.	(Keyser *et al.*, 2000)

TABLE 6.2

COMPARISON OF *H.ERECTUS* AND ROBUST AUSTRALOPITHECINE DISTRIBUTION IN AFRICA AND NUMBER OF CRANIAL FOSSILS

Approx. my ago	Tuff	MEMBERS	Robust	Non-robust	Indet.
1.88	H?				
	H	G	11	8	0
2.33	G	F	10	3	0
2.35	F	E	3	2	0
	E	D	1	2	0
2.52	D	C	5	4	0
2.6	C4				
2.95	B10	B/C	0	0	1
		B	0	4	1
3.35	B				

P.boisei
H.habilis
A.aethiopicus
A.afarensis

TABLE 6.3

COMPARISON OF NUMBERS OF FOSSILS ATTRIBUTED TO ROBUST AND NON-ROBUST LINEAGES AT OMO SHUNGURA FORMATION, BASED ON FOSSIL DATA FROM SUWA *ET AL* (1996) AND DATES FROM CONROY (1997, P.154)

there seem to be more cranial fossils for *P.boisei.* This is the case at Omo, Koobi Fora, West Turkana and Konso, but not Olduvai Gorge. However, fossils of *P.boisei* are also present in more stratigraphic levels in Omo and Koobi Fora. Examination of the stratigraphy clarifies this situation (Feibel *et al*., 1989). At Omo, the robust australopithecine cranial fossils were all recovered from a level dated to about 2.2 my ago, while the sole *H.erectus* cranial fossil is from 1.5-1.4 my ago (ibid.). At Koobi Fora, the cranial fossils for both species have an age range of just over 1.9 to a little less than 1.5 my ago (ibid.). The evidence for the higher density of robust australopithecine cranial fossils at Koobi Fora is therefore quite convincing. In southern Africa, *P.robustus* fossils are more numerous. The fact that the robust australopithecines are also present at more sites than *H.erectus* does not provide evidence for a wider distribution in southern Africa, as these sites are all situated very close together. *H.erectus* is also present in North Africa. A general pattern could be proposed in which *H.erectus* is more widely distributed, at lower densities than the robust australopithecines. The larger number of cranial fossils could indicate that robust australopithecine population density and hence biomass was higher, although these species were also smaller bodied.

This is a very imperfect comparison. The figures are likely to be biased by a number of factors. For instance, the taphonomy of the southern African cave sites is very different from that in East Africa. However this approach has the advantage that it does not require presumption as to the ecological niche of the hominin species.

Further evidence for the nature of differences in hominin biomass comes from one site with a relatively high sample size of hominin fossils. Hominin remains from Omo represent a major body of evidence for hominin evolution in eastern Africa during the 3-2 my time period (Suwa *et al*., 1996). This site allows us to examine the ecology of the robust australopithecines over the earlier period when they coexisted with early *Homo* and the gracile australopithecine species. Suwa *et al.* (1996) have analysed a subset of the dental material from this site.

The figures presented in Table 6.3 show some interesting trends. There are more fossils of robust australopithecines throughout Members C G. However in Members C-E, the difference is relatively small and there are too few specimens to make a statement about hominin biomass. In these Members, the robust species represented is *A.aethiopicus*, and early *Homo* appears in Member E (Suwa *et al*., 1996). In Member F, there is a relatively large sample and a strong difference in the number of robust and gracile fossil specimens. This Member accumulated within a relatively short time (Conroy, 1997). The robust species in this Member is probably still *A.aethiopicus*, and the non-robust early *Homo* (Suwa et al., 1996).

Early stone tools are found in Members F and G. This might suggest that relatively low biomass in *Homo* in Member

F coincides with increased stone tool use, implying a possible dietary shift. In member G, which was deposited over a longer time period (Conroy, 1997), there are more specimens of both species, and the difference is smaller. The robust species in this Member may be *P.boisei* for the first time, while the non-robust species is still early *Homo* (Suwa *et al.*, 1996).

There is no evidence that the robust australopithecines consistently had a higher biomass than other hominin species. The data from Omo in general suggests that the differences in biomass between the robust australopithecines and *Homo* may have been a relatively late development. It could be argued therefore that a change hominin biomass occurs in *H.erectus*. However further data from like-aged sites would be necessary to confirm this.

Mammalian biomass

It would be very interesting to compare fossil density in carnivores, ungulates, primates and hominins. The distribution of mammalian biomass is likely to have changed over time in response to environmental changes. My examination of modern biomass distribution in Chapter 4 allowed me to identify some connections with environment and the mammal community. Data on biomass from the relevant period would allow more direct interpretation of the hominin part in the pattern, and would also provide a useful control for the discussion of hominin biomass above. Estimates of species abundance are presented in the form of Minimum Numbers of Individuals (MNI) or Quantifiable Skeletal Part (QSP). MNI is claimed to be a poor estimator of abundance, and highly variable in computation (de Ruiter, 2000).

Generating estimates of relative abundance is a complex process, requiring faunal analysis and taphonomic assessment. Owing to different depositional environments in East Africa, few studies of the hominin fossil localities in that region have provided estimates of relative abundance (de Ruiter, 2000). While the southern African hominin sites all occur in caves, East African sites occur in ancient stream or lake deposits that have been exposed by recent gullying. In many places flowing water displaced fossils from their original resting places even before burial, and most were discovered only after they had eroded from their burial sites. In some places, fossils and artefacts occur in near primary position, as at some sites at Olduvai Gorge, Koobi Fora, and the Lokalalei site (West Turkana) (Klein, 1999). These differences have implications for how the fossil sample relates to Plio-Pleistocene communities of animals.

Sites for which estimates of relative abundance are available (from de Ruiter 2002) include Swartkrans (Brain, 1981, Watson, 1993, de Ruiter, 2000), Sterkfontein (Brain, 1981, Pickering, 1999), Kromdraai B (Brain, 1981), Makapansgat (Reed, 1997), Gladysvale (Berger 1994), Olduvai Gorge FLK Zinj site (Bunn and Kroll, 1986), and Koro Toro, Chad Republic (Brunet *et al.*, 1996). Thus there is a reasonable amount of data on abundance at the Southern African sites, but very little from East Africa. This will not provide sufficient information on the geographical distribution of mammalian biomass.

De Ruiter (2000) has re-analysed the faunal remains from Swartkrans in South Africa to provide revised estimates of relative abundance. Reanalysis made significant changes to proportions and interpretation, reducing the number of hominins and increasing numbers of bovids throughout. This suggests that considerable work might be necessary to generate comparable estimates. Estimating the relative abundance of different orders of species at additional sites is beyond the scope of this thesis, but could provide insights into the distribution of hominin biomass and hominin ecology.

Archaeology

In addition to the fossil evidence, the archaeological evidence indicates a major increase in artefact density in Acheulean as opposed to earlier Oldowan assemblages. At Koobi Fora, a diffuse low-density scatter of artefacts is part of the overall archaeological record: however there are also dense occurrences of artefacts and bone that are very different from any archaeological occurrences yet recovered from the Pliocene (Harris and Capaldo, 1993). At 1.6 my ago, a higher density and overall abundance of stone artefacts is evident in the Turkana basin (Rogers *et al.*, 1994). Handaxes were sometimes discarded in vast numbers: for instance at Melka Kunturé, Olorgesailie, Isimila, and Kalambo Falls. However increased tool abundance is likely to reflect differences in hominin use of the landscape (Rogers *et al.*, 1994) and of the tools, rather than hominin biomass.

Implications

Change and variability in climate during this period may have produced an environment in which secondary productivity was increasing, but was subject to high levels of long-term variation. At the same time plant productivity may have declined, and different plants become important. This would have significant implications for species with different dietary niches. Animal resources may have become more attractive; at the same time, other more stable resources may have been crucial to survival.

The method of comparison used here does not produce absolute values of hominin biomass, which could be compared with those for other species. However it highlights a general trend. Numbers of fossils of robust australopithecines and other hominin species from particular localities suggest that the former species may have had a relatively high population density and biomass. By comparison *H.erectus* may have occupied a wider area at lower densities. This could be explained in terms of dietary niche. Carnivorous species tend to live at lower population densities than primary consumers. Thus low population density and range expansion in *H.erectus* could be attributed to a move into the carnivore niche. These patterns of population density also have implications for the way in which *H.erectus* and the robust australopithecines existed in sympatry. Finally, it would be interesting as future research to investigate mammalian abundance data further.

Home range size

The discussion of hominin home range size, and the method used for studying hominin home ranges, has certain theoretical implications. Isaac's (1978, 1981) 'central place foraging' model has been central to archaeological discussion of hominin use of space. He pointed out that only humans (and not apes) repeatedly use a 'home base' or 'base camp' where garbage accumulates (Isaac, 1978, Isaac, 1981). The use of a home base is linked to other distinctively human traits, such as food sharing among group members and division of labour between the sexes (Isaac, 1978, Isaac, 1981). Other studies have built on or suggested alternatives to Isaac's approach with reference to ranging behaviour among other taxa. A number of authors have discussed general trends in home range size throughout the orders of primates (Leonard and Robertson, 2000) and carnivores (Steele, 1996, Gamble and Steele, 1999).

There are a number of possible approaches to reconstructing hominin home ranges. Isaac's model was important for focussing attention on the potential of raw material transfers to provide a measure of the distances involved (Steele, 1996, Gamble and Steele, 1999). Féblot-Augustins (1997a) has conducted a comprehensive review of the raw material data obtainable from Palaeolithic sites, including early stone age sites in Africa. This information can be translated into home range values: alternative methods include using minimum convex polygons (Gamble and Steele, 1999) or calculating the area of a circle with the maximum transfer distance as the radius or the diameter (Foley, 1987a). Alternatively, predictable relationships between home range size and other ecological variables exist in mammalian species. Therefore home range size can be predicted based on other aspects of hominin ecology, such as body mass (McHenry, 1994), body mass and group size (Steele, 1996) or energy requirements (Leonard and Robertson, 2000).

Raw material data has a number of limitations for reconstructing home ranges. The amount of raw material data available for different time periods varies: earlier sites may be preserved more rarely. The distribution of local raw materials affects the distances: for instance, sites with good local raw material sources will rarely give raw material distances that indicate the extent of hominin home ranging. There are large pieces of information missing: hominin movement is only recorded in the archaeological record when it involved transfer of stone. Field surface sampling approaches have mapped the existence of low densities of artefacts distributed between more concentrated occurrences (Harris and Capaldo, 1993); however this broader pattern of landscape use cannot be drawn on in reconstructing home ranges, unless links between sites can be established. It also assumes that raw material procurement was part of daily subsistence activity.

In addition, there is no information about the size of the hominin group from raw material data, and group size has a significant influence on home range size. However this method does have the advantage that it avoids the need to make an assumption as to which ecological niche the hominin species occupied. The data from raw material transfer can provide an archaeological test of the predictions of models based on different dietary niches and trophic levels (Steele, 1996, Gamble and Steele, 1999).

Féblot-Augustins' (1997a) review covered raw material transfer data for the periods 1.9-1.65 my ago, 1.65-1.2 my ago, and 1.2–0.7 my ago. These divisions allow comparison between raw material distances prior to and subsequent to the emergence of *H.erectus*, and cover the period during which the first move out of Africa probably took place. A comparison cannot be made with the earliest archaeological sites (dating from 2.6–2.0 my ago), as at all of the relevant sites the raw materials exploited were exclusively from less than 1km distance (Feblot-Augustins, 1997a). For the periods 1.9-1.65 and 1.65-1.2 my ago, Féblot-Augustins (1997b) found that the closest sources (0-3km and 0-4km) were the most frequently exploited. In addition, a smaller quantity of raw material was procured from further afield, but never more than about 15km away. By contrast, in the period 1.2-0.2 my ago, raw materials were procured from distances as far as 100km. Again closer sources were more frequently exploited (0-11km).

An estimate of home range has been produced based on the assumption that the maximum distance of raw material transport is a good indicator of the extent of distances travelled by hominins under normal circumstances

Period	Average greatest distance (AGD)	Estimated HR HR = πr^2, AGD = D - r
African Lower Palaeolithic:		
1.9 - 1.65 my	7.3 km (n = 12)	41.9 - 167.4 km^2
1.65 - 1.2 my	8.5 km (n = 14)	56.7 - 227.0 km^2
1.2 – 0.7 my	40 km (n = 6)	1256.6 - 5026.5 km^2
0.7 – 0.4 my	11.3 km (n = 3)	100.3 - 401.1 km^2
0.4+ my	14.6 km (n = 11)	167.4 - 669.7 km^2
1.2+ my	21.7 km (n = 20)	369.8 - 1479.3 km^2
European:		
Lower Palaeolithic (1 – 0.3 my)	28.2 km (n = 33)*	624.6 - 2498.3 km^2
Early Middle Palaeolithic (300 – 128 kyr BP)	37.8 km (n = 24)*	1122.2 - 4488.8 km^2
Late Middle Palaeolithic (128 – 40 kyr BP)	57.6 km (n = 110)*	2605.8 - 10423.1 km^2
Early Upper Palaeolithic (40 – 20 kyr BP)	114.7 km (n = 198)*	10332.8 - 41331.1 km^2
Late Upper Palaeolithic (20 – 10 kyr BP)	125.01 km (n = 112)*	12272.8 - 49087.4 km^2

Table 6.4

Average greatest distance (AGD) for the transfer of lithic raw material at different periods. N = number of sites/members in the sample. Data from (Feblot-Augustins, 1997a). Minimum home range area is estimated as the area of a circle for which AGD is the diameter; maximum area is estimated as the area of a circle for which AGD is the radius. *Estimates from (Gamble and Steele, 1999)

(see table 6.4). A minimum home range area estimate is produced based on average greatest distance as the diameter of a circle; maximum area is based on average greatest distance as the radius of a circle. This estimate is assumed to represent total home range, where group size is unknown. It is hoped that this provides a useful, if simplistic, summary of temporal trends in hominin use of space. In future work, it might be preferable to calculate the area within a minimum convex polygon incorporating all the raw material sources utilised around the site (Gamble and Steele, 1999).

There is a gradual increase in the maximum distances of raw material transport at sites over time. This is particularly noticeable in the period 1.2–0.7 my ago in Africa. The high estimate for the period 1.2-0.7 my ago is made less convincing by the lower values for subsequent periods, and it may be that the best estimate is that for the longer period of 1.2+ my ago in the Lower Palaeolithic of Africa. However this does not greatly affect the trend described. This increase could be explained by an increase in hominin group size. Comparison with the robust australopithecines suggests that *H.erectus* had a relatively low biomass and presumably also a low population density. This combination of trends suggests a more thorough restructuring of this species' ecological niche.

The lithic transport data can be compared to home range data for modern primates and carnivores. Primate total home ranges in Africa vary from the tiny (0.009 km^2 for the Demidoff's bush baby, *Galagoides demidoff*), to the maximum values of 45.64 km^2 for the yellow baboon (*Papio hamadryas cynocephalus*), 34.75 km^2 for the chimpanzee (*Pan troglodytes*) and 29.95 km^2 for the patas monkey (*Erythrocebus patas*) (Nunn and Barton, 2000). My estimated minimum home range values for hominin species between 1.9-1.65 my ago fall within these high primate values, and are slightly higher during the next period (1.65-1.2). The maps in Chapter 5 show that average primate individual home range sizes are highest in the more open, dry areas of Africa, although the African apes, which show a preference for wooded habitats, also have large individual home range sizes. The values for 1.2-0.7 my ago are outside the range of primate home range sizes.

Carnivore home range sizes in Africa vary from 0.19 km^2 for the weasel (*Mustela nivalis*), to 1750 km^2 for the African wild dog (*Lycaon pictus*) (Gittleman and Harvey, 1982, Grant *et al.*, 1992). Two hyaenas (*Crocuta crocuta* and *Hyaena brunnea*), the lion (*Panthera leo*) and the cheetah (*Acinonyx jubatus*) also have home ranges larger than 100 km^2. Some of these species also have relatively large group sizes (*Crocuta crocuta* has 32-55 individuals in a group, *Panthera leo* 14.4). My estimates suggest that by the period 1.2-0.7 my ago hominin species had ranges in the hundreds of km^2, a scale most similar to that of the larger and more gregarious carnivores. This is also true of the more conservative estimate for the longer period of 1.2+ my ago in the Lower Palaeolithic of Africa.

According to the primate and carnivore pattern described in Chapter 5, an expansion of more arid landscapes is likely to have led to an increase in hominin home range sizes. Such an increase is evident in the raw material transfer data. Is the scale on which this occurs also representative of a trophic level shift? My estimates of hominin home range size suggests that this is so, possibly in the period 1.65-1.2 my ago and certainly after 1.2 my ago. Modern patterns in species home range size suggest that a change in diet is involved, in terms of increasing omnivory and dietary quality. Increasing body size is also likely to have been influential.

Discussion

H.erectus expanded the hominin range to include the north of the continent as well as the more arid and mountainous regions in the traditional areas of hominin occupation. The overlap of fossils of *H.erectus* and the robust australopithecines suggests that they may have lived sympatrically and exploited the same habitats. This suggests that some form of niche differentiation would have been necessary, whether in dietary niche, back up foods in periods of scarcity, or the distribution of time and activities.

Both *H.erectus* and the robust australopithecines show a tolerance for quite varied habitats: in addition the former species shows a strong preference for the wooded grassland and grassland biome. Regional environmental evidence provides a context for species habitat tolerance and preference. The range expansion of *H.erectus* is associated with a gradual increase in wooded grassland and grassland habitats between 1.7-1.0 my ago. In addition there is evidence that species present during this period would have experienced frequent environmental change. The extinction of the robust australopithecines could also be attributed to this environmental change: however *P.boisei* could certainly tolerate more arid environments.

There is a major increase in body mass between the earliest hominin species and *H.erectus*. This combination of a preference for more open habitats with a larger body mass in *H.erectus* is the opposite of the primate trend described in Chapter 5.

H.erectus may have existed at a lower biomass than the robust australopithecines, and seems to diverge from the earlier hominins in this characteristic. The combination of range expansion with a lower biomass suggests the effects of a shift in dietary niche as described by Shipman and Walker (1989).

Based on the primate and carnivore pattern described in Chapter 5, an expansion of more arid landscapes is likely to have led to an increase in hominin home range sizes. Such an increase is evident in the raw material data. Home range size also increases with body mass. In addition, estimated hominin home range sizes are comparable to those of carnivores rather than primates, possibly in the period 1.65-1.2 my ago, and certainly after 1.2 my ago.

As discussed in Chapter 5, there is strong evidence for dietary breadth as a part of the hominin adaptation. While dental morphology in the robust australopithecines suggested dietary specialism, isotopic analysis has revealed a mixed or omnivorous diet. However there is no clear pattern of increase in niche breadth that might explain range expansion: as discussed above, it seems likely that meat was a part of australopithecine diets and was not a new factor in the diet of early *Homo*. The evidence for a dietary shift in *H.erectus* is also equivocal: dental morphology supports a change, while there is good evidence that hominins had primary access to meat. However isotopic analyses suggest that this species had a mixed diet.

The data on hominin diets and habitat tolerance and preference suggests that both the robust australopithecines and *H.erectus* had broad dietary niches. At the same time, both species are likely to have had relatively high metabolic requirements. The habitat tolerance of the robust australopithecines may be explained by niche breadth according to the primate pattern. However the range expansion, habitat preference, increase in body mass and relatively low biomass in *H.erectus* is more compatible with the predictions of a shift in dietary niche, and this is likely to have involved an increase in meat eating.

Based on these patterns and comparison with African mammals we can perhaps derive some expectations regarding the geographic distribution of these species.

The primate pattern tends to involve somewhat smaller geographic ranges, a strong range boundary between the savannah and desert habitats, and a strong association of biomass and diversity with primary production. By comparison with African carnivores, we might expect a shift in diet to be associated with somewhat larger geographic ranges, a broader habitat tolerance and a strong association with ungulate biomass.

Conclusion

The data surveyed in this Chapter is highly varied, including the distribution of fossil and archaeological sites, landscape analyses, species richness and abundance in faunal assemblages, and raw material transfer data from archaeology. Representation of actual geographic range size, diversity, abundance, and home range size is mediated through factors of preservation, taphonomy, and discovery. While some of the data is very limited (for instance information on geographic range size from site distribution) the combination of data can give us information about hominin distribution and ecology. This is particularly the case when it can be put in context with reference to other species.

I conclude that the range expansion, increase in body mass, and relatively low biomass in African *H.erectus*, contemporary with an increase in home range size, is compatible with the predictions of a shift in dietary niche. This pattern is not fully explicable by breadth of dietary niche according to a primate model, which may be more relevant to understanding the robust australopithecines. Hominin range expansion in Africa 1.8-0.6 my ago can be well explained by a dietary shift including an increase in meat eating.

Chapter 7

DISCUSSION

Introduction

There have been numerous changes in hominin geographic ranges in the course of human evolution. This variability has the potential to give us valuable information about human evolution and hominin behaviour. However interpretation of the archaeological and fossil data for hominin distribution is difficult: it tends to be discovery driven, and suffers from problems of chronological resolution. These limitations have driven a research focus on dating and paleontological exploration. Such limitations also provide a strong incentive to use comparative analysis to set a context for archaeological interpretation of hominin behaviour and distribution.

In this thesis I have set out to create such an interpretative context. My approach has been to carry out comparative analyses of distribution in modern primates and carnivores. An increased understanding of trends in these orders of species allowed me to refine models of the ecology and evolution of hominin geographic ranges: the value of this approach is particularly high where results contradict the assumptions of the models. I was also able to reinterpret hominin distribution in the context of primate and carnivore trends. A wide range of archaeological and fossil data from the period 1.8-0.6 my ago in Africa provided relevant data on key variables for early hominin species. Based on this data, I have started to describe the extent to which early hominin species fitted the primate or the carnivore pattern, and to identify when they diverged from these trends.

Models of hominin range expansion

Model 1

In the discussion of model 1, I argued that species capable of more flexible behaviour should be better equipped to cope with local environmental change, and that this capacity would also be useful in coping with novel environments encountered through range expansion. For relatively long-lived and slow developing species such as the primates, cognitive responses may be a particularly important way of enhancing behavioural flexibility. In particular, I have suggested that an increase in the level of social learning, by increasing the rate of information acquisition by individual organisms, would increase the chances of successful range expansion.

There are a number of feedback mechanisms by which a trend towards increased behavioural flexibility might develop. Environmental fluctuation may select for niche breadth and for species' structures and characteristics that allow organisms to respond flexibly to change. Species that respond to environmental change with behavioural change rather than migration are more likely to experience this form of selection. Range expansion, particularly into higher latitudes, may further increase exposure to environmental fluctuation. In addition, such behavioural flexibility modifies the environment experienced by an organism; thus an animal can influence the selective pressures that affect its reproductive success. For such species, behavioural innovations may have a stronger influence than environmental change on their evolution.

Model 2

According to model 2, a combination of relatively rapid reproductive rate and dietary niche breadth may increase dispersal ability and tolerance of environmental change in primate species. This would contribute to a larger geographic range size over time. Such species would be able to focus on different dietary resources when the environment changed. In addition, populations would recover more rapidly if environmental change caused high levels of mortality. Relatively rapid development tends to be associated with smaller brain size, so this hypothesis contradicts model 1.

Early hominin species appear to have matured slowly, and been relatively long-lived. However modern humans possess a number of unique features, including an increase in post-reproductive lifespan, and a relatively high reproductive rate, related to a decrease in inter-birth interval. It has been suggested that the former characteristic first appeared in *H.erectus* (O'Connell *et al.*, 1999). According to the 'grandmother' hypothesis, an extended post-reproductive life allowed older women to look after and feed children, thus removing some of the reproductive constraints on younger women (ibid.). Such behaviour would also have given that species the potential to speed up (or slow down) reproductive rate in response to circumstances, and thus cope with the environmental constraints and opportunities encountered in range expansion. Based on the primate trend, I have suggested that a combination of reproductive flexibility and dietary niche breadth may have been important in early hominin range expansion.

Model 3

In the discussion of model 3, I suggested that a shift in dietary niche, consisting of increased meat eating, would remove a key environmental constraint on the range expansion of a primate species. A large brain has relatively high metabolic requirements (Aiello and Wheeler, 1995, Leonard and Robertson, 1996). Large brained species have dietary requirements that must be met either by a high quality diet or long time spent foraging, or both. Thus diet tends to put a constraint on the distribution of large-brained species that is absent in other species. The biomass and diversity of plant resources are related to latitude, with tropical habitats the most productive. An alternative source of high quality nourishment such as meat might allow large-brained primate species to extend their range into less productive habitats at higher latitudes.

Like all primates, hominins had relatively large brains, and there is a trend of increasing brain size throughout human evolution. During the Plio-Pleistocene, environmental changes included a reduction of forested areas, and yet hominin geographic ranges increased, notably in the first occupation of Eurasia. There is evidence for a dietary shift with the first appearance of *H.erectus*, and the spatial distribution of *Homo* has been compared to that of fossil carnivores in a number of areas and periods. I argued that this changing distribution indicates the removal of environmental constraints acting on other primates. Thus the range expansion of *H.erectus* may have been dependent on a change in dietary niche and in the environmental factors limiting distribution.

Testing the models

In this thesis, I described three models explaining the evolution of hominin geographic ranges. Each of these models is based on processes that are not exclusive to human evolution, and could therefore be evaluated and refined using comparative data from other mammals. Models 1 and 2 focus on cognitive complexity, social communication and life history parameters. In these characteristics, hominins share similar features with the non-human primates. Therefore modern primates provided a suitable dataset for analysis of these models. Model 3 focuses on aspects of the ecological niche, notably meat eating. Because of the similarity of their adaptive niche to humans in this respect, carnivores made a better source of analogy in order to test this model.

Primate analysis

The biogeography of primate species is of considerable current interest (Cowlishaw and Hacker, 1997, Eeley, 1994, Eeley and Foley, 1999, Eeley and Lawes, 1999, Harcourt, 2000, Ruggiero, 1994). The primates provide data on the distribution of a tropical order, and therefore may provide answers to biogeographical issues. In addition, many primate species are under threat of extinction, and understanding the factors shaping a species' geographical distribution is important for conservation (Cowlishaw and Dunbar, 2000, p.93). My analysis of variation in primate geographic range size contributed to current understanding of primate biogeography in a number of ways. First, use of GIS methods made it possible to generate more accurate data on primate species geographic range area worldwide. I also produced new data on the climatic variability experienced by primate species at a number of different spatial and temporal scales. Finally, my analysis explored the effects on primate distribution of previously unexamined variables, including behavioural flexibility and a capacity for social learning.

Method

The use of GIS methods in this study had a number of advantages. Species geographic range maps were rectified and projected into an equal area format, reducing the distortion to species' area. New maps of climatic variability were created at a number of different scales, based on annual and monthly climate data. GIS methods also made it possible to analyse species range composition in terms of other variables such as climatic variability.

An important methodological consideration in this study was how to measure complex behavioural or environmental variables so that these characteristics could be compared between a large number of species. The area of a species extent of occurrence was taken as a measurement of species geographic range. Other studies have used latitudinal extent (Cowlishaw and Hacker, 1997, Harcourt, 2000, Ruggiero, 1994), however because geographic ranges vary in shape, area may provide a better measure (Gaston *et al.*, 1998a). A new database of primate innovation, social learning and tool use frequencies is based on the primatology and social learning literature, and therefore provides a test fair to all species by measuring responses to social or environmental problems relevant to the animal (Reader and Laland, 2002). Observation frequencies and brain size data were used as alternative measurements of behavioural flexibility in this study. In addition, new measurements of primate tolerance of environmental variability were produced. I focused on climate as a part of environmental variability likely to be relevant to primate species. The scale of variability is likely to be an important factor in its effect on primates. Measurements were based on two time scales of climatic variability relevant to primate life histories (seasonal and interannual variation) and spatial variability.

Comparisons among species are widely used to test hypotheses of how organisms are adapted to their environment. In this study a comparative analysis of a large number of species was used to identify trends in the primates. Regression techniques were used to determine which factors have the greatest predictive value in explaining variation among living primates in the spatial extent of their distribution. Most recent analyses of primate geographic ranges have taken account of phylogeny (Cowlishaw and Hacker, 1997, Eeley, 1994, Eeley and Foley, 1999, Eeley and Lawes, 1999, Harcourt, 2000), using a number of different approaches. I used the CAIC (Comparative Analysis for Independent Contrasts) program to calculate independent contrasts from a phylogeny (Purvis and Rambaut, 1995). This is quite reliable even when evolutionary assumptions are incorrect or there are errors in the phylogeny (Purvis *et al.*, 1994). In my research, I took a number of steps to ensure that no spurious results were accepted. These steps included tests of assumptions, careful scrutiny of any outlying contrasts, and carrying out parallel tests using alternative evolutionary assumptions.

Conclusions for primate biogeography

The results presented in Chapter 4 give some indication that behavioural flexibility influences a small proportion of the variation in primate species geographic range size and tolerance of spatial variability in climate. Alternative methods may be necessary to clarify the minor role of behavioural flexibility in primate distribution indicated by these results.

However, behavioural flexibility is clearly not an important factor in determining which primates have larger geographic ranges. This conclusion is confirmed by results obtained using a number of measurements of behavioural flexibility, and alternative evolutionary assumptions. In addition, more behaviourally flexible primate species do not have a greater ability to tolerate temporal variability in climate. This confirms the previous conclusion, as in Africa primate species geographic range size is strongly related to climatic variability (Cowlishaw and Hacker, 1997, Eeley and Foley, 1999). In addition, primate species with a greater capacity for social learning do not tend to have larger geographic ranges.

These trends have a number of implications for the evolution of primate species geographic ranges. The fact that more flexible primate species do not tend to have larger ranges suggests that this behavioural characteristic does not help primates to counter environmental change or to expand their ranges more successfully. This is confirmed by trends in primate tolerance of spatial and temporal climatic variability. In addition, behavioural flexibility is probably not a response to climatic change. Implications for the evolution of behavioural flexibility in primates are discussed below. In addition, social learning does not seem to be particularly useful in migration in primates, if it is assumed that this would lead to larger geographic range size.

In addition, primates with relatively rapid life histories do not tend to have larger geographic ranges. This conclusion contradicts the predictions of model 2, indicating that a combination of characteristics including wider dietary niche and relatively rapid life history do not form a wide-ranging primate strategy.

There are a number of possible explanations for the results presented in Chapter 4. First, some other species characteristics could account for differences between the ranges of species. For instance, ecological variables identified in earlier studies, such as habitat preference, tolerance of climatic variability, and niche breadth (Cowlishaw and Hacker, 1997, Eeley, 1994, Eeley and Foley, 1999, Eeley and Lawes, 1999) provide a better explanation of variability in primate distribution. The fact that niche breadth is not correlated with innovation frequency indicates that this measurement is not a by-product of foraging innovation. However, these variables still only explain part of the variation in primate geographic range size.

Geographic range size in primates may be a particularly labile trait (Reader and MacDonald, 2003), so evolutionary patterns might be swamped by short-term fluctuations in species range size. Further investigation of the dynamics of the evolution of primate geographic ranges would be necessary to address this possibility.

Range size can be limited by a number of factors, potentially obscuring the role of one factor in predicting range sizes

(Gaston, 1996). The results presented in Chapter 4 indicate that behavioural flexibility does not have a stronger effect on local distribution than on species geographic ranges. In addition, there is no evidence that anthropogenic changes have destroyed a natural relationship between brain size and species geographic range size.

Evolution of behavioural flexibility

The unexpected results presented in Chapter 4 have implications for the evolutionary causes and consequences of innovative capacities and enhanced brain size in primates. These results are inconsistent with the hypotheses that more innovative species can tolerate greater climatic variability, that innovation is a response to climate change, or that climatic variability has selected for a propensity to innovate. This would seem to contradict the suggestion that climatic variability selected for increasing brain size in human evolution (Potts, 1998a, Boyd and Richerson, 2000). Disproving and comparing competing hypotheses is difficult for the comparative student of brain evolution as there are a large number of potential measures of characteristics of interest, several potential alternative analytical methods, and the results can turn on the particular measure or analysis used (Reader and MacDonald, 2003). However there was consistency in the results for measurements based on observation frequencies and brain size. Alternative hypotheses for primate brain evolution, such as those based on the demands of increasing social complexity (Jolly, 1966, Humphrey, 1976, Byrne and Whiten, 1988, Whiten and Byrne, 1997, Dunbar, 1998), may be better supported.

Hominin ranges

A primate-based model of behavioural flexibility and social learning or life history parameters is not able to explain hominin range expansion. This does not necessarily indicate that behavioural flexibility and social learning were not important in hominin range expansion. However, if these factors were important it was probably on a different scale, or incorporated specific behaviours that were particularly useful in encountering new habitats (perhaps a crucial innovation). Alternatively, some other process, such as the ability to exploit a wide range of food types, may have been critical to hominin range expansion.

Future directions

A different method of analysis might be necessary to clarify the role of behavioural flexibility in shaping geographic ranges. An alternative comparative approach could be used to test the relationship between environmental variability and behavioural flexibility or other behavioural characteristics. In this approach, a number of localities with different characteristics of environmental variability could be identified. The behavioural characteristics of a cross-section of the primate community could then be compared between these localities. Such an analysis would provide a check for the results described above.

In the future, a more detailed case study approach would provide further insight into the effect of multiple, species specific factors on geographic range size. A detailed study of particular wide-ranging primate species has the potential to strengthen interpretation of primate distribution.

In addition, a comparison of different populations of particular primate species would provide an alternative way of studying how innovation rates are related to current conditions. For instance, it would be possible to compare different populations of the same species that were under varying degrees of stress (Reader and MacDonald, 2003). Populations on the edge of their species geographic range might be living under the difficult conditions thought to favour innovation according to the necessity (ibid.). Such a comparison would test the 'necessity' hypothesis, according to which primate species innovate more in response to stress from their environments (Reader and Laland, 2002).

African mammal analysis

Contributions to biogeography

The range and pattern of variation in geographic range size is specific to particular taxonomic and functional groups. This is apparent at different levels from closely related species to different taxa (Brown *et al.*, 1996). Different taxa within a region may vary considerably with respect to how they conform to Rapoport's rule (France, 1992, Rhode *et al.*, 1993, Ruggiero, 1994) and this may be determined by a taxon's evolutionary history and ecological characteristics. However, there are few studies that test hypotheses of how a taxon's evolutionary history or ecological characteristics influence specific patterns of geographical distribution. My analysis of African mammal distribution has added to the biogeographical evidence for variation between higher taxonomic groups. There are a number of studies of the distribution of primate and also of ungulate ecological characteristics in Africa and globally (Cowlishaw and Hacker, 1997, Eeley, 1994, Eeley and Foley, 1999, Eeley and Lawes, 1999, Olff *et al.*, 2002, Thackeray, 1995). However this is the first study comparing the distribution of different higher taxa of large mammals in Africa.

In addition, I have investigated the hypothesis that trophic level or dietary niche affects the scale of variation in geographic range size and adherence to ecological rules. Trophic level affects a number of other ecological variables, including body mass, home range size and population density. Both small geographic range size and trophic level are correlated with extinction risk in contemporary species (Purvis *et al.*, 2000). This suggests that there may be geographical constraints on species at higher trophic levels. A comparison of African mammals according to dietary niche has allowed examination of the influence of trophic level on intra-taxa variation in distribution.

Method

An investigation of large-scale spatial patterns in African mammal distribution was carried out according to large taxonomic groups that are broadly related to diet. GIS software was used to produce maps showing spatial patterns in the distribution of body mass, species richness, biomass, and home range size. Visual and statistical comparison of spatial patterns in the different orders and in the distribution of environmental variables provided the basis for my interpretation of determining factors. This analysis aimed to identify differences between the orders that might be due to dietary niche.

Comparison of trends at higher taxonomic levels is less useful for building a specific predictive model than cross-species analysis, but has made comparison possible on a large spatial scale with a large number of species. In addition, this scale of analysis has contributed to questions in biogeography. Taxonomic division into carnivores, primates and ungulates is generally, but not always, parallel to differences in dietary niche. The advantage of subsuming some variation and using orders as a whole was that it allowed identification of general trends in a comprehensive database. Mapping values for ecological variables provided a simple and intuitive way of interpreting trends in distribution.

Conclusions for African mammals

Analysis of the data for modern mammals indicated that there are some strong differences in the spatial distribution of ecological variables. Geographic range sizes differ in scale, but have a broadly similar frequency distribution between the carnivores, ungulates and primates. Primates show a different spatial distribution pattern from carnivores and ungulates in diversity, body mass, and biomass. Patterns in species range boundaries are associated with those in diversity. The distribution of primate and carnivore individual home range size shows similar trends; however primates differ from carnivores in the distribution of home range size relative to body mass. While other factors will influence the distribution of particular species, there do seem to be broad patterns in distribution that characterise particular orders. Carnivore diversity, biomass and body mass distribution seems to be closely related to ungulate diversity and biomass, suggesting the availability of prey species is a crucial variable. At the same time primate diversity, biomass and body mass reflects primary productivity. This suggests that dietary niche may be an important factor in differentiating variation between higher taxonomic groups.

Hominin range expansion

My analyses suggest that dietary niche does influence patterns of geographical distribution. A model of ecological variation based on modern carnivores could potentially explain hominin range expansion: this should be tested using data from the archaeological and fossil record.

There is strong evidence for dietary breadth as a part of the hominin adaptation. While dental morphology in the robust australopithecines suggested dietary specialism, isotopic analysis has revealed a mixed or omnivorous diet. However there is no clear pattern of increase in niche breadth that might explain range expansion. For instance, increased documentation of primate meat eating, combined with the isotopic evidence, makes it seem probable that early hominin species ate some meat.

Archaeological evidence shows that one use of the earliest stone tools was butchery. The earliest evidence for access to relatively large amounts of meat comes from *c.* 1.8-1.5 my ago, when it could be attributed to early African *H.erectus*, early *Homo* or the robust australopithecines. Dental morphology suggests the diet of *H.erectus* had different mechanical requirements from that of earlier species, which could reflect increased meat eating. An alternative explanation stresses the processing of vegetable foods. Analysis of a range of characteristics suggests that early African *H.erectus* occupied a new adaptive niche (Wood and Collard, 1999a). Thus archaeological evidence confirms that meat was a part of the early hominin diet from a relatively early date, and there is some evidence that a dietary shift occurred in *H.erectus*, which may have involved increased meat eating.

Hominin distribution 1.8-0.6 my ago in Africa

Method

The data that could tell us about the shape and size of early hominin geographic ranges is severely limited. Site distribution data is limited in scope and biased by factors of preservation and discovery. As a result, it is only possible to establish the extent of hominin geographic ranges in particular cases. For instance, in North Africa there is faunal data from earlier sites to back up the conclusion that this region was occupied relatively late. However other aspects of species distribution, such as habitat preference and tolerance, can be interpreted from environmental reconstructions and landscape analyses at particular locations. In addition, trends in the distribution of ecological and biological characteristics can be studied in the archaeological record.

In this case study, comparison of the distribution of *H.erectus* and the robust australopithecines illuminated questions of niche diversification and habitat tolerance and preference. Estimates of hominin body mass were available in the literature: other authors have developed consistent ways of calculating body mass from hominin fossils. In order to estimate relative hominin biomass, I compared numbers of cranial fossils attributed to the main hominin species at different sites. Raw material transfer distances may indicate some of the extent of hominin movement in the landscape, and therefore provided an indication of changing home range size. In all, I found that there was sufficient relative data available from the Palaeolithic record for this period to allow me to test some of the predictions of models based on primate niche breadth and the effects of a shift in dietary niche.

Conclusions

Our understanding of the geographic range of *H.erectus* in Africa is limited by the data available: for instance, there is no evidence as to whether this species was present in west and central Africa or not. Nonetheless, there is clear evidence for range expansion into new regions and habitats. This species appears to have been the first hominin to occupy north Africa, by 1 my ago. In addition, evidence from particular sites indicates the occupation of new environments at about 1.5 my ago, in the drier peripheries of the Rift basin and at higher altitudes. This suggests that range expansion began soon after the emergence of early African *H.erectus*.

There is a strong contrast in the spatial distribution and ecological characteristics of *H.erectus* compared to the robust australopithecines. *H.erectus* had a relatively large body mass, existed at a wider range of sites at a lower abundance, occupied a larger home range area, and had a larger geographic range. The combination of increasing body mass with a preference for more open habitats is in direct contrast to the primate pattern. In addition, within a taxonomic or functional group geographic range size tends to increase with abundance, and the archaeological and fossil evidence suggests that *H.erectus* broke this trend. This suite of changes suggests a fundamental shift into a new ecological niche. The values for hominin body mass and home range size are comparable to carnivore characteristics.

There is a range of evidence (from environmental data to morphology) to suggest habitat and dietary niche breadth in both the robust australopithecines and *H.erectus*. There is also evidence that meat was a part of early hominin diets, and that a shift in dietary niche occurred in *H.erectus*. I conclude that robust australopithecine ecology and distribution during this period are consistent with an explanation of the evolution of hominin geographic ranges based on niche breadth. However a shift in dietary niche, probably involving increased meat eating, provides a better explanation for range expansion in *H.erectus*.

Future directions

My examination of modern ecology and distribution in Chapter 5 allowed me to identify patterns and processes that could be used to interpret the hominin data. However the distribution of African mammals is likely to have changed over time in response to environmental change. Data on fauna from the relevant period would allow more direct interpretation of the hominin part in broader patterns. It would also provide a useful control variable in discussions of the hominin data.

For instance, data on faunal diversity 1.8-0.6 my ago would provide a context for patterns in hominin diversity. Data on mammalian species present at key sites is available from Turner *et al.* (1999) and can be supplemented from primary sources. Patterns in fossil diversity are strongly influenced by factors of taphonomy and preservation. For instance, there are strong contrasts in the faunal composition of the southern African cave sites, compared to the largely stream-accumulated East African sites. In addition, diversity at particular sites will reflect the size of the site and fossil abundance. It would therefore be necessary to establish controls for these variables. For instance, diversity figures could be calculated relative to the number of fossil specimens retrieved as an independent measure of sample size. This method would have the advantage of allowing interpretation of hominin diversity with reference to other species from faunal assemblages subject to the same influences.

Data on hominin biomass could also be usefully put in context with other fauna. Generating estimates of the relative abundance of fossil species is complex, requiring faunal analysis and taphonomic assessment. Owing to different depositional environments in East Africa, few studies of hominin fossil localities in that region have provided estimates of fossil abundance. By comparison, data from southern Africa is extensive, but may require reassessment (de Ruiter, 2000). Estimating the abundance of different orders of species at additional sites could provide additional insights into the distribution of hominin biomass and hominin ecology.

Conclusions

Theory in human evolution

In Chapter 1, I argued that evolutionary theory and its subdisciplines including ecology provide the most useful framework for interpreting hominin behaviour from the archaeological and palaeoanthropological record. In this thesis, I integrated biogeography and human evolutionary theory to provide the arguments for my models of the evolution and ecology of hominin geographic ranges. This was a productive approach, bridging the gap between the long time scales of human evolution and local, spatial processes. It also gave me access to a new source of ideas from outside human origins research. The models are particularly convincing where I was able to use examples of trends in the fossil record to back them up.

I argued that the use of comparative data and theory from convergent disciplines could aid interpretation of the evolution and ecology of hominin geographic ranges. I carried out comparative analyses of trends in the geographical distribution of modern primate species and African mammals. This did successfully set limits to interpretation of the archaeological data. Some unexpected results allowed me to reject two hypotheses for the evolution of hominin geographic ranges based on primate trends in cognition and life history. This does not prove that a capacity for innovation or social learning was not important in hominin range expansion. However it does suggest that if this is the case, it must have been on a different scale from the primate trend, or involved novel innovations that were particularly useful in new habitats. The literature on primate biogeography suggests that other variables are better predictors: niche breadth and climatic variability. Further analysis demonstrated that a third hypothesis was more plausible based on patterns in modern African mammals. Thus models based on niche breadth or a shift in dietary niche provide possible explanations for changes in hominin geographic range size.

The aim of this comparative approach is to set the 'degrees of freedom' available in reconstructing hominin behaviour and distribution. Understanding variation beyond these limits may require different tools. For instance, identification of trends in primates and carnivores should allow us to identify when hominin geographic ranges became larger than would be expected for species with their characteristics. At this point, alternative theory and data will be necessary.

The archaeological record provides the only direct evidence of changes in hominin geographic ranges over time, and of hominin behaviour, ecology and biology. I argued that comparative models are most useful when they produce predictions that can be tested in the archaeological record. The utility of such models depends on our ability to identify data in the archaeological record that can be compared with the variables in the model. For the period 1.8-0.6 my ago in Africa, I was able to identify data on hominin distribution, diversity, body mass, biomass and home range size that could be compared with a model based on trends in modern African mammals. I also examined the predictions of trends in primate dietary and habitat niche breadth. I conclude that these models are useful and relevant to Palaeolithic archaeology.

Methods

Comparisons among species are frequently used to test hypotheses of how organisms are adapted to their environment. I found the comparative method an effective means of testing hypotheses about the evolution of the geographic range in primates and in African mammals. The comparative approach worked using different methods (statistical analysis and visual comparison of trends) and at different scales (species and higher taxonomic groups). In some cases, a more detailed study of particular species or groups of species might provide a better impression of the varied factors that shape a particular geographic range.

The use of GIS methods provided a number of advantages in analysis of geographic ranges. I was able to generate more accurate or new data on species distribution, and carry out calculations based on maps of different variables. I was also able to generate new maps of spatial trends in a range of variables for illustration and interpretation. Finally, I could acquire useful GIS data through the Internet.

In order to find out about hominin geographic ranges, I surveyed a variety of data from the primary and secondary literature, including the distribution of fossil and archaeological sites, landscape analyses, environmental reconstruction, species richness and abundance in faunal assemblages, and raw material transfer data from archaeology. The effect of factors of preservation and discovery, and the limited number of sites, means that establishing the extent of a hominin species distribution was only possible in some areas. However it was possible to combine this with other information about the range, such as habitat tolerance and preference. In addition, the nature and distribution of ecological characteristics could be estimated. Further research to contextualise this data would be very useful.

Limitations

In this thesis I have not attempted to provide a definitive account of the data on early hominin distribution. Nor have I interpreted all of the variation in early hominin geographic ranges, in particular early evidence for hominin presence in Eurasia. My aim was to use a case study to assess whether appropriate data was available on hominin distribution and other key variables, to allow comparison with the predictions of the primate or carnivore models.

Implications for discussion of hominin range expansion

My identification of a different trend in ecology and distribution in *H.erectus* compared to the robust australopithecines is particularly interesting in the context of questions raised by a number of recent articles by Wood and Collard (1999a, 1999b). These authors have argued that early African *H.erectus* is the first species that is more closely related to *H.sapiens* than to the australopithecines. They identify a common adaptation, characterised by a larger body mass, a modern human-like physique that would have been adaptive in more open habitats, and dedicated bipedalism. Teeth and jaws were adapted to a diet that had similar mechanical properties to that of *H.sapiens*, and its developmental pattern was modern human-like. On this basis, they argue that the appearance of early African *H.erectus* represents a grade shift in human evolution. The differences in ecology and spatial distribution described in this thesis support the suggestion of a distinct shift in adaptive strategy for this species.

My conclusions are interesting in the light of a larger debate about expansion into Eurasia. Until relatively recently, the prevalent view has been that hominins did not colonise Asia until a million or so years ago (Klein, 1999). There are a number of sites with suggested dates indicating an early Pleistocene phase for this range expansion. Important among these are hominin fossils from Dmanisi, Georgia, with a possible date of 1.75 my ago (Gabunia *et al.*, 2000a, Vekua *et al.*, 2002) and from Mojokerto and Sangiran, Java at 1.6-1.8 my ago (Swisher *et al.*, 1994, Huffman *et al.*, 2005). In addition, in their review of this evidence Dennell and Roebroeks (2005) have pointed out that we currently have little evidence of prior absence in Asia. Thus, there is a great deal of uncertainty as to when, where and which hominins occupied this area.

In this context, the value of my comparative research lies in developing constraints on interpretation. I have developed methods of testing some of the most basic, and often assumed, hypothesis about hominin (and human) colonization. This should set some limits on and challenges for future interpretation. The current research provides the groundwork for formulation of more complex hypotheses, or hypotheses specific to particular ecological contexts and hominin species. It also establishes preliminary expectations for hominin distribution in relation to habitat distribution and ungulate biodiversity, which should be developed further. Theory and comparative evidence from biogeography has considerable potential for improving our understanding of early hominin distribution and evolution. However, as Dennell and Roebroeks point out, increasing the number and quality of fossil assemblages from little known areas and from before and during the earliest dates for hominin presence in the region is essential: this evidence will ultimately provide the basis for new models of hominin distribution.

At the same time, the debate concerning hominin hunting has been revitalised by new evidence from primatology and experimental studies. Faunal remains from Olduvai Gorge, Koobi Fora and Peninj have been re-analysed to produce convincing evidence that early hominins had primary access to carcasses (Dominguez-Rodrigo, 2002, Dominguez-Rodrigo *et al.*, 2002). This strongly contradicts the widely held view of early hominin species as opportunistic scavengers. Reestablishment of hunting as part of hominin behaviour, sets the scene for re-assessing the importance of increased meat eating in hominin range expansion.

Conclusions

Early hominins were unlikely primates to have large ranges. There is little indication from the primate data that their increasing brain size and associated intelligence would have allowed them to extend their geographic ranges to an area beyond those of modern apes. There is no support for the hypothesis that increased social learning is the key to primate range expansion. Nor does a combination of relatively rapid life history parameters and dietary niche breadth form a general wide-ranging primate strategy.

A model of range size based on dietary niche may provide a better explanation for changes in early hominin geographic ranges. Different taxonomic groups of modern African mammals show distinctive patterns of distribution that are related to diet. Archaeological and fossil data from the period 1.8-0.6 my ago indicates a contrast between the ecological characteristics and distribution of *H.erectus* and the robust australopithecines that can best be explained by a shift in dietary niche. Changes in the evolution of the hominin geographic range in *H.erectus* may be part of a carnivore trend. At the same time, the possibility remains that a novel adaptation, innovation or exponential change in some characteristic may have been sufficient to remove the environmental limits on hominin distribution at some stage in the evolution of early hominin geographic ranges.

APPENDICES

List of contents

List of tables

PRIMATE DATA AND RESULTS OF REGRESSION ANALYSIS

Table 1
Primate geographic range size data

Species name	Geographical range (km2)
Allenopithecus nigroviridis	388488
Alouatta belzebul	1471500
Alouatta caraya	2554500
Alouatta fusca	779400
Alouatta palliata	592600
Alouatta seniculus	5421700
Alouatta villosa	333900
Aotus trivirgatus	6618200
Arctocebus calabarensis	979950
Ateles paniscus	6180900
Brachyteles arachnoides	488600
Cacajao calvus	429800
Cacajao melanocephalus	643100
Callicebus moloch	4826300
Callicebus personatus	665300
Callicebus torquatus	1798800
Callimico goeldii	1078900
Callithrix argentata	1223400
Callithrix humeralifer	312500
Callithrix jacchus	2578100
Cebuella pygmaea	1442500
Cebus albifrons	3664200
Cebus apella	11384500
Cebus capucinus	431600
Cebus olivaceus	1986100
Cercocebus albigena	1721578
Cercocebus aterrimus	970411
Cercocebus galeritus	987020
Cercocebus torquatus	1470309
Cercopithecus aethiops	14390917
Cercopithecus ascanius	2897603
Cercopithecus cephus	730981
Cercopithecus diana	355951
Cercopithecus erythrogaster	35263
Cercopithecus erythrotis	93608
Cercopithecus hamlyni	269667
Cercopithecus l'hoesti	403990
Cercopithecus mitis	4275922
Cercopithecus mona	3182177
Cercopithecus neglectus	2926393
Cercopithecus nictitans	1686485

Cercopithecus petaurista	625107
Cercopithecus pogonias	950820
Chiropotes albinasus	587300
Chiropotes satanas	2077700
Colobus angolensis	2329053
Colobus badius	33815
Colobus guereza	2566695
Colobus polykomos	972029
Colobus satanas	294283
Erythrocebus patas	6161038
Euoticus elegantulus	755597
Euoticus inustus	110814
Galago alleni	851845
Galago senegalensis	11878820
Galagoides demidoff	5360213
Gorilla gorilla	728170
Hylobates concolor	609325
Hylobates hoolock	766925
Hylobates klossi	6275
Hylobates lar	1942925
Hylobates pileatus	214825
Hylobates syndactylus	187750
Lagothrix flavicauda	29200
Lagothrix lagothricha	3466500
Leontopithecus rosalia	23900
Loris tardigradus	769800
Macaca arctoides	2885425
Macaca assamensis	1557025
Macaca cyclopis	30200
Macaca fascicularis	2749325
Macaca fuscata	164550
Macaca mulatta	5743150
Macaca nemestrina	3048400
Macaca nigra	83900
Macaca radiata	633450
Macaca silenus	41125
Macaca sinica	60650
Macaca sylvanus	86028
Mandrillus leucophaeus	116947
Mandrillus sphinx	439764
Miopithecus talapoin	1224320
Nasalis larvatus	754425
Nycticebus coucang	3812275
Nycticebus pygmaeus	473925
Otolemur crassicaudatus	4112043
Pan paniscus	435250
Pan troglodytes	2581174

Papio anubis	8454062
Papio cynocephalus	3789567
Papio hamadryas	1359580
Papio papio	389084
Papio ursinus	2983972
Perodicticus potto	3431742
Pithecia monachus	2294800
Pithecia pithecia	1654100
Pongo pygmaeus	325575
Presbytis cristata	1880375
Presbytis entellus	3287175
Presbytis francoisi	101075
Presbytis frontata	447325
Presbytis geei	14650
Presbytis johnii	21200
Presbytis melalophos	1145575
Presbytis obscura	238150
Presbytis phayrei	1210050
Presbytis pileatus	319100
Presbytis potenziani	6300
Presbytis rubicunda	704375
Presbytis vetulus	49975
Procolobus verus	502709
Pygathrix avunculus	107575
Pygathrix nemaeus	524325
Pygathrix roxellana	1385175
Saguinus bicolor	149200
Saguinus fuscicollis	1798200
Saguinus imperator	233500
Saguinus inustus	346000
Saguinus labiatus	264400
Saguinus leucopus	36800
Saguinus midas	1507200
Saguinus mystax	941000
Saguinus nigricollis	487700
Saguinus oedipus	132500
Saimiri sciureus	5220200
Simias concolor	6300
Tarsius bancanus	470975
Tarsius spectrum	53500
Tarsius syrichta	64900
Theropithecus gelada	208426

Table 2. Primate species tolerance of climatic variability

	Spatial variability in mean rainfall (mm/year*10):			Spatial variability in mean temperature (°C*10):			Temperature	Interannual
Species name	SD	mean	CV (%)	SD	mean	CV (%)	range (°C*10)	variability (%)
Allenopithecus nigroviridis	2.698999882	49.33499908	5.470760985	3.243999958	251.0350037	1.292250049	114.612999	8.369999886
Alouatta belzebul	8.001999855	55.70800018	14.36418437	6.006999969	261.5450134	2.29673657	135.2740021	15.64999962
Alouatta caraya	8.293000221	39.81499863	20.82883463	22.64999962	235.2599945	9.627646071	190.0789948	15.78199959
Alouatta fusca	9.326999664	36.70299911	25.41209135	24.03800011	211.8589935	11.34622595	166.720993	21.20400047
Alouatta palliata	27.09000015	64.35700226	42.09332194	22.32600021	241.3049927	9.25219158	131.9589996	15.43700027
Alouatta seniculus	16.22900009	62.93399811	25.78733368	21.38299942	255.1799927	8.379575215	122.8769989	13.63300037
Alouatta villosa	21.92099953	50.86299896	43.09812629	25.6060009	243.0890045	10.53359075	165.2330017	14.52600002
Aotus trivirgatus	20.93099976	55.01900101	38.04322029	37.53499985	246.2429962	15.24307307	140.5249939	14.93599987
Arctocebus calabarensis	7.007999897	48.36800003	14.48891807	9.380999565	246.2250061	3.809929671	123.9680023	11.28499985
Ateles paniscus	19.09600067	59.44499969	32.12381322	26.40800095	251.1909943	10.51311613	134.4689941	14.1590004
Brachyteles arachnoides	7.984000206	37.01800156	21.56788554	19.95999908	216.7489929	9.208808224	164.2749939	19.2689991
Cacajao calvus	6.494999886	65.44400024	9.924515405	3.226000071	262.0249939	1.231180287	118.7099991	11.8579998
Cacajao melanocephalus	9.668999672	78.31800079	12.34582034	9.414999962	258.1470032	3.647146721	113.3180008	10.38000011
Callicebus moloch	14.24499989	54.3769989	26.19673791	9.852999687	257.3569946	3.828533863	149.1109924	13.40499973
Callicebus personatus	7.646999836	34.07600021	22.44101358	17.25200081	221.3150024	7.795224281	164.2480011	21.81500053
Callicebus torquatus	9.437000275	75.07499695	12.57009745	10.94200039	258.0809937	4.239754441	117.1090012	11.84000015
Callimico goeldii	13.74600029	68.80899811	19.97703886	9.906000137	257.5180054	3.846721367	127.9319992	13.29399967
Callithrix argentata	10.46100044	43.89300156	23.83295758	7.596000195	256.3309937	2.963356123	163.3919983	14.15499973
Callithrix humeralifer	3.444999933	58.92699814	5.846216577	5.828999996	263.0809937	2.21566747	131.5690002	12.85400009
Callithrix jacchus	10.1079998	32.32500076	31.26991357	19.75099945	240.0919952	8.226429803	163.0749969	23.93700027
Cebuella pygmaea	10.47000027	69.23000336	15.12350102	5.458000183	260.6350098	2.094116285	122.8600006	12.80500031
Cebus albifrons	15.10400009	62.1590004	24.2989752	26.34700012	254.2019958	10.36459216	127.6360016	13.42000008
Cebus apella	18.17000008	50.02700043	36.32038683	20.5	249.802002	8.206499483	150.4230042	16.50600052
Cebus capucinus	29.26399994	71.41000366	40.98025268	27.47500038	241.3600006	11.3834108	115.3539963	13.88399982
Cebus olivaceus	15.70400047	60.0870018	26.13543695	14.05900002	257.1300049	5.467662174	117.7549973	13.41699982
Cercocebus albigena	5.541999817	44.31399918	12.50620553	13.9630003	240.223999	5.812491822	133.4920044	11.2130003
Cercocebus aterrimus	4.68900013	45.20399857	10.37297646	9.255999565	245.9980011	3.762632023	133.8450012	8.296999931
Cercocebus galeritus	2.54399991	44.92699814	5.662519233	7.513000011	243.9409943	3.079843154	131.798996	10.9289999
Cercocebus torquatus	9.852999687	48.13199997	20.47078803	10.65200043	251.3820038	4.237375893	129.3450012	13.30700016
Cercopithecus aethiops	11.67199993	23.60499954	49.4471517	36.85400009	240.6829987	15.31225732	210.4149933	20.95499992
Cercopithecus ascanius	6.315999985	40.8769989	15.45123212	18.38999939	234.6150055	7.838373062	149.8769989	10.87699986
Cercopithecus cephus	7.09100008	44.48899841	15.93877213	7.568999767	243.602005	3.107117188	124.3789978	12.81700039
Cercopithecus diana	12.72099972	45.47600174	27.97299505	7.169000149	259.196991	2.765850067	131.121994	17.29899979
Cercopithecus erythrogaster	4.102000237	50.59999847	8.106720079	1.934999943	267.2000122	0.724176592	126.0670013	13.53299999
Cercopithecus erythrotis	9.656000137	57.93600082	16.66666667	13.16899967	251.8090057	5.229757225	128.3619995	10.87199974
Cercopithecus hamlyni	4.584000111	38.02500153	12.05522663	24.66399956	220.647995	11.1779849	131.6150055	11.45899963

	Spatial variability in mean rainfall (mm/year*10):			Spatial variability in mean temperature (°C*10):			Temperature	Interannual
Species name	SD	mean	CV (%)	SD	mean	CV (%)	range (°C*10)	variability (%)
Cercopithecus l'hoesti	9.904999733	41.93500137	23.61988651	25.08099937	224.7230072	11.16085072	128.3040009	12.14099979
Cercopithecus mitis	11.07299995	33.06999969	33.48351998	24.62400055	232.2030029	10.60451426	159.8090057	17.03499985
Cercopithecus mona	8.187000275	43.84500122	18.67259675	16.31299973	246.5839996	6.615595395	139.7969971	11.53999996
Cercopithecus neglectus	6.441999912	42.63600159	15.10929654	16.19099998	239.2200012	6.768246761	139.4329987	10.88899994
Cercopithecus nictitans	8.666999817	48.29499817	17.94595744	9.711999893	248.8509979	3.902736969	131.727005	11.76799965
Cercopithecus petaurista	13.78100014	42.34999847	32.54073349	7.918000221	262.57901	3.015473408	139.9889984	17.16200066
Cercopithecus pogonias	6.590000153	47.42599869	13.89533238	7.734000206	245.6829987	3.147959056	123.5289993	11.42099953
Chiropotes albinasus	3.286999941	59.18799973	5.553490498	4.507999897	262.605011	1.716646564	129.4620056	13.5340004
Chiropotes satanas	12.33399963	60.27799988	20.46185948	10.12699986	259.5759888	3.901362334	120.9869995	14.94299984
Colobus angolensis	6.798999786	40.02899933	16.98518549	18.63299942	235.1239929	7.92475459	150.9290009	11.11600018
Colobus badius	4.102000237	50.59999847	8.106720079	1.934999943	267.2000122	0.724176592	126.0670013	13.53299999
Colobus guereza	9.373000145	36.36100006	25.77761923	31.72699928	230.2180023	13.78128511	161.5590057	14.25699997
Colobus polykomos	11.83600044	41.44800186	28.55626305	8.708000183	261.5639954	3.329204454	158.0619965	16.29700089
Colobus satanas	4.925000191	48.29499817	10.19774382	8.467000008	245.6360016	3.446970295	117.9690018	13.01599979
Erythrocebus patas	11.99400043	23.36899948	51.32440713	23.83699989	266.2139893	8.954074863	219.5189972	19.84300041
Euoticus elegantulus	8.038999557	48.39599991	16.61087605	10.46199989	245.3899994	4.263417384	127.1920013	11.68200016
Euoticus inustus	4.506999969	34.35300064	13.11966898	23.48800087	203.7059937	11.53034354	131.3139954	13.4119997
Galago alleni	6.769000053	47.76300049	14.17205784	8.31400013	244.0829926	3.406218534	125.0130005	11.19900036
Galago senegalensis	10.36600018	24.6609993	42.03398269	31.97299957	241.5619965	13.23593944	204.5160065	20.20199966
Galagoides demidoff	9.69299984	40.88399887	23.7085415	19.89999962	244.1759949	8.149859133	152.8249969	12.27700043
Gorilla gorilla	6.892000198	44.51699829	15.48172712	18.06599998	238.6519928	7.570018491	124.1289978	12.20400047
Lagothrix flavicauda	8.126999855	32.36399841	25.11123549	32.07799911	220.727005	14.53288378	147.5449982	15.72700024
Lagothrix lagothricha	15.01599979	64.07800293	23.43393849	21.83799934	256.4129944	8.516728802	128.522995	13.00800037
Leontopithecus rosalia	8.81400013	36.22200012	24.33327839	19.06599998	225.4440002	8.457089106	140.1109924	17.77799988
Macaca sylvanus	2.006999969	12.85000038	15.61867634	16.85099983	149.3249969	11.28478164	301.9500122	28.875
Mandrillus leucophaeus	8.041000366	56.40800095	14.25507061	14.32699966	241.7550049	5.926247389	135.6730042	9.387999535
Mandrillus sphinx	5.551000118	47.19599915	11.76159043	7.960000038	244.0209961	3.262014403	120.0820007	12.94299984
Miopithecus talapoin	8.607999802	40.77299881	21.11201053	17.67900085	236.3300018	7.48064178	139.401001	13.35700035
Otolemur crassicaudatus	7.287000179	25.99600029	28.03123595	20.36400032	220.8789978	9.21952767	200.6490021	19.84900093
Pan paniscus	2.83100009	48.34500122	5.855827941	4.107999802	250.5279999	1.639736797	116.8580017	8.258999825
Pan troglodytes	8.052000046	43.11399841	18.67606889	18.06100082	243.0480042	7.43104264	142.977005	12.66499996
Papio anubis	11.95899963	26.61300087	44.93668223	30.3029995	253.427002	11.95728919	194.7160034	18.40399933
Papio cynocephalus	8.425000191	27.8560009	30.24483026	22.30400085	228.4400024	9.763614348	187.9559937	18.77300072
Papio hamadryas	7.232999802	9.187000275	78.73081077	37.91199875	241.2050018	15.71774982	194.1360016	33.87900162
Papio papio	18.01099968	35.52299881	50.70236265	17.14900017	270.2590027	6.345394602	214.977005	18.20100021
Papio ursinus	7.341000008	14.70199966	49.93198373	27.58099937	198.7160034	13.87960652	247.526001	29.66200066
Perodicticus potto	8.833999634	44.17100143	19.99954574	17.03100014	244.3419952	6.970148591	131.970993	11.91100025

	Spatial variability in mean rainfall (mm/year*10):			Spatial variability in mean temperature (°C*10):			Temperature	Interannual
Species name	SD	mean	CV (%)	SD	mean	CV (%)	range (°C*10)	variability (%)
Pithecia monachus	10.64599991	64.54699707	16.49340851	6.93900013	261.1220093	2.657378499	125.7409973	13.07699966
Pithecia pithecia	12.97200012	63.06200027	20.57023257	13.72200012	256.0700073	5.358690877	114.0739975	13.52600002
Procolobus verus	13.6590004	45.67599869	29.90410892	7.809999943	259.0799866	3.014513026	138.2299957	16.43700027
Saguinus bicolor	2.075999975	55.375	3.748984154	2.214999914	268.4460144	0.825119315	107.9820023	13.91100025
Saguinus fuscicollis	14.80900002	64.12599945	23.0935972	20.51600075	255.6100006	8.026290325	131.2030029	13.85400009
Saguinus imperator	10.5010004	60.87400055	17.25038655	6.751999855	255.1380005	2.6464109	142.6779938	11.72399998
Saguinus inustus	6.201000214	85.09400177	7.287235392	2.673000097	256.7579956	1.041058173	114.2659988	10.94499969
Saguinus labiatus	3.987999916	57.3030014	6.959495695	3.997999907	264.5759888	1.511097029	128.7680054	12.87899971
Saguinus leucopus	10.69400024	68.26699829	15.66496332	26.03199959	261.3330078	9.96123674	119.8669968	14.13300037
Saguinus midas	9.345999718	61.20199966	15.27074241	6.474999905	261.2420044	2.478544719	113.4710007	14.60000038
Saguinus mystax	6.032999992	61.69499969	9.77875034	4.965000153	263.29599	1.885710509	122.1600037	12.81499958
Saguinus nigricollis	10.2869997	78.38800049	13.12318166	7.859000206	258.868988	3.035898686	124.2350006	14.76500034
Saguinus oedipus	41.18500137	72.38800049	56.8947907	22.3010006	254.5099945	8.762328033	114.2669983	13.48900032
Saimiri sciureus	13.3039999	63.94100189	20.80668039	12.52200031	259.072998	4.833386886	124.0159988	13.6079998
Theropithecus gelada	8.420000076	29.18300056	28.85241378	33.38999939	191.0220032	17.47966142	198.6990051	15.66699982

Table 3
Model tests for observation frequency, geographic range size and brain weight

Branch lengths	dependent variable	independent variable (contrasts)	n	r2	slope	p
Equal	square root of expected variance	geographic range size	106	0.008	-3.13E-14	0.36
Equal	estimated nodal values	geographic range size	106	0.582	1.42	0
Equal	square root of expected variance	SQ(geographic range size)	106	0.075	-3.69E-07	0.004
Equal	estimated nodal values	SQ(geographic range size)	106	0.166	8.71E-01	0
Equal	square root of expected variance	ln(geographic range size)	106	0.138	-0.226	0
Equal	estimated nodal values	ln(geographic range size)	106	0.109	-0.676	0.001
Purvis	square root of expected variance	geographic range size	106	0.121	-4.09E-11	0
Purvis	estimated nodal values	geographic range size	106	0.268	4.787	0
Purvis	square root of expected variance	SQ(geographic range size)	106	0.236	-2.02E-04	0
Purvis	estimated nodal values	SQ(geographic range size)	106	0.084	3.221	0.002
Purvis	square root of expected variance	ln(geographic range size)	106	0.32	-119.074	0
Purvis	estimated nodal values	ln(geographic range size)	106	0.057	-3.101	0.013
ln(Purvis)	square root of expected variance	geographic range size	108	0.043	-2.27E-13	0.031
ln(Purvis)	estimated nodal values	geographic range size	108	0.488	1.662	0
ln(Purvis)	square root of expected variance	SQ(geographic range size)	108	0.142	1.57E-06	0
ln(Purvis)	estimated nodal values	SQ(geographic range size)	108	0.141	1.059	0
ln(Purvis)	square root of expected variance	ln(geographic range size)	108	0.102	-0.929	0.001
ln(Purvis)	estimated nodal values	ln(geographic range size)	108	0.2	-0.888	0
Equal	square root of expected variance	corrected innovation frequency	97	0.095	-0.54	0.002
Equal	estimated nodal values	corrected innovation frequency	97	0.341	1.305	0
Purvis'	square root of expected variance	corrected innovation frequency	97	0.093	-1.625	0.002
Purvis'	estimated nodal values	corrected innovation frequency	97	0.297	1.913	0
ln(Purvis)	square root of expected variance	corrected innovation frequency	98	0.043	-2.27E-03	0.031
ln(Purvis)	estimated nodal values	corrected innovation frequency	98	0.488	1.662	0
Equal	square root of expected variance	ln(brain weight)	81	0.001	6.97E-02	0.771
Equal	estimated nodal values	ln(brain weight)	81	0.025	-0.927	0.161
Purvis'	square root of expected variance	ln(brain weight)	81	0.001	-39.288	0.767
Purvis'	estimated nodal values	ln(brain weight)	81	0.008	-6.716	0.426

Table 4
Niche breadth results

dependent variable	independent variable	method	n	r2	slope	p	x vs x@node	x vs SD	y vs y@node	y vs SD
no. of habitats	innovation frequency	species	22	0.234	9.439	0.019				
no. of habitats	social learning frequency	species	22	0.204	9.435	0.031				
no. of habitats	tool use frequency	species	22	0.045	9.512	0.332				
no. of habitats	innovation frequency	CAIC	21	0.33	2.284	0.005			*	
no. of habitats	innovation frequency	CAIC	19	0.258	2.893	0.022			*	
no. of habitats	social learning frequency	CAIC	21	0.3971	2.8248	0.0017	***		*	
no. of habitats	social learning frequency	CAIC	20	0.096	1.554	0.171	*		*	
no. of habitats	tool use frequency	CAIC	21	0.1261	1.6207	0.1049	***		*	
no. of habitats	ln(brain weight)	species	28	0.06	1.34E-00	0.201				
no. of habitats	relative brain weight	species	28	0	0.307	0.927				
no. of habitats	ln(brain weight)	CAIC	27	0	-0.0317	0.9829			*	*
no. of habitats	ln(brain weight)	CAIC	26	0.02	1.398	0.281			*	*
no. of habitats	relative brain weight	CAIC	27	0.006	-1.802	0.699			*	*
no. of habitats	relative brain weight	CAIC	26	0.003	1.625	0.802			*	*
no. of food types	innovation frequency	species	22	0.075	7.82	0.208				
no. of food types	social learning frequency	species	22	0.174	7.758	0.048				
no. of food types	tool use frequency	species	22	0.078	7.767	0.196				
no. of food types	innovation frequency	CAIC	21	0.1067	0.5679	0.1379	*			
no. of food types	innovation frequency	CAIC	18	0.276	1.178	0.021				
no. of food types	social learning frequency	CAIC	21	0.155	0.7715	0.0698	***			
no. of food types	social learning frequency	CAIC	19	0.159	0.791	0.081	**			
no. of food types	tool use frequency	CAIC	21	0.0208	0.2877	0.522	***			
no. of food types	tool use frequency	CAIC	20	0.123	0.53	0.118	***			
no. of food types	ln(brain weight)	species	28	0.207	1.318	0.013				
no. of food types	relative brain weight	species	28	0.079	2.562	0.141				
no. of food types	ln(brain weight)	CAIC	27	0.0394	0.6599	0.311				
no. of food types	relative brain weight	CAIC	27	0	-5.30E-03	0.998				
							***p<0.001, **p<0.01, *p<0.05			

Table 5
Results of regression of observation frequencies and geographic range size (whole dataset)

dependent variable	independent variable	method	n	r2	slope	p	x vs x@node	x vs SD	y vs y@node	y vs SD
ln(geographic range size)	innovation frequency	species	83	0.033	3.48E-01	0.102				
ln(geographic range size)	social learning frequency	species	83	0.012	2.09E-01	0.331				
ln(geographic range size)	tool use frequency	species	83	0.002	7.61E-02	0.682				
ln(geographic range size)	SF controlling for IF	species	83	0.001	-1.10E-01	0.732				
ln(geographic range size)	innovation frequency	CAIC	71	0.0283	0.1491	0.1577	***	**	**	**
ln(geographic range size)	innovation frequency	CAIC	70	0.043	0.395	0.083	***	**		*
ln(geographic range size)	innovation frequency	CAIC	66	0.107	0.829	0.007	*	*	***	*
ln(geographic range size)	social learning frequency	CAIC	71	0.0098	0.0912	0.4087	***	*	**	*
ln(geographic range size)	social learning frequency	CAIC	70	0.017	0.261	0.273	***	*		*
ln(geographic range size)	social learning frequency	CAIC	68	0.006	0.194	0.523	***	*	*	*
ln(geographic range size)	SF controlling for IF	CAIC	71	0	-4.38E-02	0.897				
ln(geographic range size)	tool use frecuency	CAIC	71	0	0.0001	1	***		**	*
ln(geographic range size)	tool use frecuency	CAIC	70	0	3.72E-02	0.875	***			
							***p<0.001, **p<0.01, *p<0.05			

Table 6
Results of regression of observation frequencies and geographic range size (continental datasets)

dependent variable	independent variable	location	method	n	r2	slope	p	assumption violations
geographic range size	innovation frequency	Africa	species	27	0.017	0.000	0.503	
geographic range size	social learning frequency	Africa	species	27	0.024	0.000	0.436	
geographic range size	tool use frequency	Africa	species	27	0.006	0.000	0.699	
geographic range size	innovation frequency	Africa	CAIC	24	0.0712	9557196144	0.1974	1x1y
geographic range size	social learning frequency	Africa	CAIC	24	0.0993	1.2755E+12	0.1249	1x1y
geographic range size	tool use frequency	Africa	CAIC	24	0.0033	2354848563	0.7837	1x1y
ln(geographic range size)	innovation frequency	Africa	CAIC	24	0.1186	0.5031	0.0849	1x
ln(geographic range size)	social learning frequency	Africa	CAIC	24	0.1215	0.5775	0.081	1x
ln(geographic range size)	tool use frequency	Africa	CAIC	24	0.0234	0.2542	0.4552	1x
dependent variable	independent variable	location	method	n	r2	slope	p	assumption violations
geographic range size	innovation frequency	South America	species	27	0.321	2.40E+12	0.002	
geographic range size	social learning frequency	South America	species	27	0.103	1.60E+12	0.095	
geographic range size	tool use frequency	South America	species	27	0.242	1.70E+12	0.008	
geographic range size	innovation frequency	South America	CAIC	23	0.2904	2.0920E+12	0.0066	1x2y
geographic range size	social learning frequency	South America	CAIC	23	0.0872	1.3962E+12	0.1614	1y
geographic range size	tool use frequency	South America	CAIC	23	0.2134	1.6769E+12	0.0231	1x1y
ln(geographic range size)	innovation frequency	South America	CAIC	23	0.1199	0.3008	0.0974	2x
ln(geographic range size)	social learning frequency	South America	CAIC	23	0.0002	-0.016	0.9489	1x
ln(geographic range size)	tool use frequency	South America	CAIC	23	0.0018	0.0356	0.8422	1x
dependent variable	independent variable	location	method	n	r2	slope	p	assumption violations
geographic range size	innovation frequency	Asia	species	25	0.002	-8.40E+10	0.822	
geographic range size	social learning frequency	Asia	species	25	0.021	2.10E+11	0.478	
geographic range size	tool use frequency	Asia	species	25	0.029	-2.70E+11	0.406	
geographic range size	innovation frequency	Asia	CAIC	22	0.0284	-3566243020	0.4417	1x1y
geographic range size	social learning frequency	Asia	CAIC	22	0.0002	-2668921446	0.9504	1x1y
geographic range size	tool use frequency	Asia	CAIC	22	0.038	-3978511974	0.3725	1x1y
ln(geographic range size)	innovation frequency	Asia	CAIC	22	0.0032	-0.063	0.7928	2x
ln(geographic range size)	social learning frequency	Asia	CAIC	22	0.0006	0.0242	0.9133	1x1y
ln(geographic range size)	tool use frequency	Asia	CAIC	22	0.012	-0.117	0.6108	1x1y

Table 7
Results of regression of brain size measurements and geographic range size (whole dataset)

dependent variable	independent variable	method	n	r2	slope	p	x vs x@node	x vs SD	y vs y@node	y vs SD
ln(geographic range size)	ln(neocortex ratio)	species	30	0.022	-1.418	0.422				
spatial variation	ln(neocortex ratio)	species	24	0.004	-6.59E-02	0.769				
ln(seasonal variation)	ln(neocortex ratio)	species	24	0.026	-0.212	0.444				
ln(interannual variation)	ln(neocortex ratio)	species	24	0.033	-0.268	0.386				
ln(geographic range size)	ln(neocortex ratio)	CAIC	29	0	-6.06E-02	0.968			*	
ln(geographic range size)	ln(neocortex ratio)	CAIC	28	0.012	1.15E-00	0.577			*	
spatial variation	ln(neocortex ratio)	CAIC	23	0.0068	0.1915	0.701			*	
ln(seasonal variation)	ln(neocortex ratio)	CAIC	23	0.003	0.154	0.784			*	
ln(interannual variation)	ln(neocortex ratio)	CAIC	23	0.0018	0.1491	0.8429				
ln(geographic range size)	ln(absolute brain size)	species	88	0	-2.28E-02	0.875				
spatial variation	ln(absolute brain size)	species	63	0.012	1.31E-02	0.38				
ln(seasonal variation)	ln(absolute brain size)	species	63	0.069	4.77E-02	0.036				
ln(interannual variation)	ln(absolute brain size)	species	63	0.016	2.86E-02	0.317				
ln(geographic range size)	ln(absolute brain size)	CAIC	82	0.002	6.22E-02	0.659			*	**
ln(geographic range size)	ln(absolute brain size)	CAIC	80	0	-3.49E-02	0.845			*	**
spatial variation	ln(absolute brain size)	CAIC	58	0	-4.96E-03	0.877			***	*
spatial variation	ln(absolute brain size)	CAIC	57	0.009	2.35E-02	0.488			***	*
ln(seasonal variation)	ln(absolute brain size)	CAIC	58	0.015	1.93E-02	0.359			***	**
ln(seasonal variation)	ln(absolute brain size)	CAIC	57	0.016	2.25E-02	0.34			***	
ln(interannual variation)	ln(absolute brain size)	CAIC	58	0.004	1.41E-02	0.645			**	*
ln(interannual variation)	ln(absolute brain size)	CAIC	57	0.016	3.18E-02	0.35			**	*
ln(geographic range size)	Relative brain size	species	88	0.029	1.051	0.109				
spatial variation	Relative brain size	species	63	0	9.43E-03	0.885				
ln(seasonal variation)	Relative brain size	species	63	0.033	-0.145	0.149				
ln(interannual variation)	Relative brain size	species	63	0.024	-0.153	0.22				
ln(geographic range size)	Relative brain size	CAIC	82	0.019	1.189	0.21			**	
spatial variation	Relative brain size	CAIC	55	0.006	-5.33E-02	0.573			***	*
ln(seasonal variation)	Relative brain size	CAIC	58	0.022	-7.35E-02	0.257			***	**
ln(interannual variation)	Relative brain size	CAIC	58	0.027	-1.18E-01	0.21			**	*
							***p<0.001, **p<0.01, *p<0.05			

Table 8
Results of regression of brain size measurements and geographic range size (continental datasets)

dependent variable	independent variable	location	method	n	r2	slope	p	assumption violations
geographic range size	neocortex ratio	Africa	species	14	0.153	-1.20E+13	0.167	
geographic range size	absolute brain size	Africa	species	36	0.009	-3.00E+09	0.579	
geographic range size	relative brain size	Africa	species	36	0.003	5.10E+09	0.748	
geographic range size	neocortex ratio	Africa	CAIC	12	0.010	-5.03E+12	0.8128	1y
geographic range size	absolute brain size	Africa	CAIC	33	0.0008	-1.54E+09	0.8762	2x2y
ln(geographic range size)	neocortex ratio	Africa	CAIC	12	0.000	-0.3	0.9284	none
ln(geographic range size)	absolute brain size	Africa	CAIC	33	0.0001	-0.0001	0.9494	2x1y
geographic range size	relative brain size	Africa	CAIC	33	0.004	-8.10E+09	0.711	2y
ln(geographic range size)	relative brain size	Africa	CAIC	33	0.004	-1.21E-03	0.726	1y
geographic range size	neocortex ratio	South America	species	11	0.366	5.20E+13	0.049	
geographic range size	absolute brain size	South America	species	24	0.044	1.50E+10	0.327	
geographic range size	relative brain size	South America	species	24	0.029	2.20E+10	0.427	
geographic range size	neocortex ratio	South America	CAIC	10	0.2768	5.323E+13	0.0964	none
geographic range size	absolute brain size	South America	CAIC	23	0.0004	-2.93E+09	0.9221	1y
geographic range size	relative brain size	South America	CAIC	23	0.004	1.90E+10	0.762	1y
ln(geographic range size)	neocortex ratio	South America	CAIC	10	0.1344	5.6621	0.2674	none
ln(geographic range size)	absolute brain size	South America	CAIC	23	0.0284	-0.0047	0.431	none
ln(geographic range size)	relative brain size	South America	CAIC	23	0.003	-3.37E-03	0.788	none
geographic range size	absolute brain size	Asia	species	28	0.002	-1.00E+09	0.798	
geographic range size	neocortex ratio	Asia	species	6	0.017	3.60E+12	0.803	
geographic range size	relative brain size	Asia	species	28	0.000	-5.60E+08	0.959	
geographic range size	absolute brain size	Asia	CAIC	24	0.0009	806752832	0.8855	1x1y
geographic range size	neocortex ratio	Asia	CAIC	5	0.0249	-6.752E+12	0.7652	none
geographic range size	relative brain size	Asia	CAIC	24	0.001	3.20E+09	0.883	1y
ln(geographic range size)	absolute brain size	Asia	CAIC	24	0.0119	0.0014	0.6036	1x1y
ln(geographic range size)	neocortex ratio	Asia	CAIC	5	0.0506	-2.1096	0.6683	none
ln(geographic range size)	relative brain size	Asia	CAIC	24	0.028	8.55E-03	0.428	1y

Table 9. Results of regression of observation frequencies and home range

dependent variable	independent variable	method	n	r2	slope	p	x vs x@node	x vs SD	y vs y@node	y vs SD
ln(individual home range size)	innovation frequency	species	72	0.121	0.564	0.003				
ln(individual home range size)	social learning frequency	species	72	0.128	0.585	0.002				
ln(individual home range size)	tool use frequency	species	72	0.214	0.643	0				
ln(individual home range size)	innovation frequency	CAIC	67	0.0001	-0.0097	0.9541	***	**		
ln(individual home range size)	innovation frequency	CAIC	64	0	3.70E-02	0.867	***	**		
ln(individual home range size)	social learning frequency	CAIC	67	0.0015	0.0555	0.7572	***	**		
ln(individual home range size)	tool use frequency	CAIC	67	0.006	0.1138	0.5287	***			
ln(individual home range size)	ln(brain weight)	species	72	0.416	0.843	0				
ln(individual home range size)	relative brain weight	species	72	0.584	1	0				
ln(individual home range size)	neocortex ratio	species	25	0.368	13.582	0.001				
ln(individual home range size)	ln(brain weight)	CAIC	66	0.303	1.484	0	*		*	
ln(individual home range size)	relative brain weight	CAIC	66	0.001	-0.241	0.803			*	
ln(individual home range size)	neocortex ratio	CAIC	25	0.246	21.58	0.01			*	
							***p<0.001, **p<0.01, *p<0.05			

Table 10. Results of regression of threat status and observation frequencies

dependent variable	independent variable	method	n	r2	slope	p	assumption tests	x vs x@node	x vs SD	y vs y@node	y vs SD
innovation frequency	threat status	species	80	0.001	-1.02E-02	0.833					
social learning frequency	threat status	species	80	0	2.89E-03	0.952					
tool use frequency	threat status	species	80	0.005	3.48E-02	0.532					
innovation frequency	threat status	CAIC	70	0.0053	-0.0325	0.5461	2y			***	*
innovation frequency	threat status	CAIC	68	0.008	-3.25E-03	0.462	2y			***	*
social learning frequency	threat status	CAIC	70	0.0018	-0.0187	0.7264	1y			***	
tool use frequency	threat status	CAIC	70	0.0025	-0.0218	0.6792	1y			***	
								***p<0.001, **p<0.01, *p<0.05			

Table 11
Results of regression of observation frequencies and climatic variables (whole dataset)

dependent variable	independent variable	method	n	r2	slope	p	x vs x@node	x vs SD	y vs y@node	y vs SD
spatial variation in rainfall	innovation frequency	species	55	0.029	3.03E-02	0.208				
spatial variation in rainfall	social learning frequency	species	55	0.02	2.85E-02	0.298				
spatial variation in rainfall	SF controlling for IF	species	55	0.001	9.38E-03	0.803				
spatial variation in rainfall	tool use frequency	species	55	0.016	1.96E-02	0.345				
spatial variation in temperature	innovation frequency	species	55	0.011	5.48E-03	0.436				
spatial variation in temperature	social learning frequency	species	55	0.009	5.47E-03	0.494				
spatial variation in temperature	SF controlling for IF	species	55	0.001	2.29E-03	0.834				
spatial variation in temperature	tool use frequency	species	55	0.009	4.21E-03	0.488				
ln(seasonal variation)	innovation frequency	species	55	0.003	1.48E-02	0.7				
ln(seasonal variation)	social learning frequency	species	55	0.006	2.30E-02	0.58				
ln(seasonal variation)	SF controlling for IF	species	55	0.003	2.25E-02	0.691				
ln(seasonal variation)	tool use frequency	species	55	0	-5.87E-04	0.985				
ln(interannual variation)	innovation frequency	species	55	0.001	-1.02E-02	0.814				
ln(interannual variation)	SF controlling for IF	species	55	0.001	-1.42E-02	0.832				
ln(interannual variation)	social learning frequency	species	55	0.002	-1.55E-02	0.752				
ln(interannual variation)	tool use frequency	species	55	0.009	-2.64E-02	0.476				
spatial variation in rainfall	innovation frequency	CAIC	48	0.099	5.13E-09	0.027	***	**	*	
spatial variation in rainfall	innovation frequency	CAIC	47	0.142	5.62E-02	0.008	***	*	***	*
spatial variation in rainfall	innovation frequency	CAIC	45	0.069	5.10E-02	0.078	***		***	
spatial variation in rainfall	social learning frequency	CAIC	48	0.021	2.74E-02	0.323	***		**	
spatial variation in rainfall	social learning frequency	CAIC	47	0.005	1.66E-02	0.635	***		**	
spatial variation in rainfall	SF controlling for IF	CAIC	48	0.001	-7.97E-03	0.812				
spatial variation in rainfall	tool use frequency	CAIC	48	0.107	5.76E-02	0.022	***		**	
spatial variation in rainfall	tool use frequency	CAIC	47	0.145	6.16E-02	0.008	***		***	**
spatial variation in temperature	innovation frequency	CAIC	48	0.0444	0.0099	0.1463	***	*		
spatial variation in temperature	social learning frequency	CAIC	48	0.0043	0.0036	0.6545	***			
spatial variation in temperature	SF controlling for IF	CAIC	48	0	-1.19E-03	0.902				
spatial variation in temperature	tool use frequency	CAIC	48	0.0155	0.0062	0.3946	***			
ln(seasonal variation)	innovation frequency	CAIC	48	0.027	3.19E-02	0.26	***	*	***	
ln(seasonal variation)	innovation frequency	CAIC	47	0.048	3.89E-02	0.134	***	*	**	
ln(seasonal variation)	innovation frequency	CAIC	45	0.002	-1.07E-02	0.78	***		***	
ln(seasonal variation)	social learning frequency	CAIC	48	0.029	3.90E-02	0.246	***		***	
ln(seasonal variation)	social learning frequency	CAIC	46	0.018	3.45E-02	0.374	**		**	
ln(seasonal variation)	SF controlling for IF	CAIC	48	0.019	3.73E-02	0.35				
ln(seasonal variation)	tool use frequency	CAIC	48	0.023	3.29E-02	0.294	***		***	
ln(interannual variation)	innovation frequency	CAIC	48	0.061	6.90E-02	0.088	***	*	**	
ln(interannual variation)	innovation frequency	CAIC	46	0	4.20E-03	0.938	***		***	
ln(interannual variation)	social learning frequency	CAIC	48	0.0299	0.0561	0.2347	***		*	
ln(interannual variation)	social learning frequency	CAIC	47	0	-1.73E-03	0.998	**		***	
dependent variable	independent variable	method	n	r2	slope	p	x vs x@node	x vs SD	y vs y@node	y vs SD
ln(interannual variation)	SF controlling for IF	CAIC	48	0.008	3.47E-02	0.546				
ln(interannual variation)	tool use frequency	CAIC	48	0.055	7.02E-02	0.105	***		**	
							***p<0.001, **p<0.01, *p<0.05			

Table 12
Results of regression of observation frequencies and climatic variables (continental datasets)

dependent variable	independent variable	location	method	n	r2	slope	p	assumption violations
spatial variation in rainfall	innovation frequency	Africa	species	27	0.046	3.623	0.276	
spatial variation in rainfall	social learning frequency	Africa	species	27	0.006	1.555	0.689	
spatial variation in rainfall	tool use frequency	Africa	species	27	0.002	0.623	0.834	
spatial variation in temperature	innovation frequency	Africa	species	27	0.017	0.609	0.504	
spatial variation in temperature	social learning frequency	Africa	species	27	0.000	-0.002	0.978	
spatial variation in temperature	tool use frequency	Africa	species	27	0.000	-0.008	0.914	
seasonal variation	innovation frequency	Africa	species	27	0.000	-0.586	0.948	
seasonal variation	social learning frequency	Africa	species	27	0.001	-1.948	0.851	
seasonal variation	tool use frequency	Africa	species	27	0.015	-4.859	0.538	
interannual variation	innovation frequency	Africa	species	27	0.001	0.189	0.874	
interannual variation	social learning frequency	Africa	species	27	0.003	-0.355	0.797	
interannual variation	tool use frequency	Africa	species	27	0.003	-0.306	0.771	
spatial variation in rainfall	innovation frequency	Africa	CAIC	24	0.2119	7.2699	0.0206	1x1y
spatial variation in rainfall	social learning frequency	Africa	CAIC	24	0.0521	4.0719	0.2726	1x1y
spatial variation in rainfall	tool use frequency	Africa	CAIC	24	0.2179	8.3821	0.0187	1x1y
spatial variation in temperature	innovation frequency	Africa	CAIC	24	0.1239	1.6834	0.0844	1x
spatial variation in temperature	social learning frequency	Africa	CAIC	24	0.0239	0.8357	0.4604	1x
spatial variation in temperature	tool use frequency	Africa	CAIC	24	0.027	0.8934	0.4326	1x
seasonal variation	innovation frequency	Africa	CAIC	24	0.035	7.4141	0.3704	1x1y
seasonal variation	social learning frequency	Africa	CAIC	24	0.0289	7.6166	0.4162	1x1y
seasonal variation	tool use frequency	Africa	CAIC	24	0.0114	4.8167	0.611	1x1y
interannual variation	innovation frequency	Africa	CAIC	24	0.1628	2.2297	0.0455	1x1y
interannual variation	social learning frequency	Africa	CAIC	24	0.023	0.9463	0.4695	1x1y
interannual variation	tool use frequency	Africa	CAIC	24	0.1298	2.2635	0.0769	1x1y
ln(interannual variation)	innov_c	Africa	CAIC	24	0.185	0.122	0.025	1x
spatial variation in rainfall	innovation frequency	South America	species	27	0.000	0.008	0.983	
spatial variation in rainfall	social learning frequency	South America	species	27	0.028	3.792	0.395	
spatial variation in rainfall	tool use frequency	South America	species	27	0.051	3.678	0.247	
spatial variation in temperature	innovation frequency	South America	species	27	0.001	-0.170	0.891	
spatial variation in temperature	social learning frequency	South America	species	27	0.021	1.054	0.461	
spatial variation in temperature	tool use frequency	South America	species	27	0.041	1.056	0.300	
seasonal variation	innovation frequency	South America	species	27	0.011	-3.278	0.601	
seasonal variation	social learning frequency	South America	species	27	0.007	-2.993	0.681	
seasonal variation	tool use frequency	South America	species	27	0.001	-0.694	0.894	
interannual variation	innovation frequency	South America	species	27	0.056	-1.149	0.226	

Table 12 continued

dependent variable	independent variable	location	method	n	r2	slope	p	assumption violations
interannual variation	social learning frequency	South America	species	27	0.021	-0.814	0.463	
interannual variation	tool use frequency	South America	species	27	0.009	-0.387	0.627	
dependent variable	independent variable	location	method	n	r2	slope	p	assumption violations
spatial variation in rainfall	innovation frequency	South America	CAIC	23	0.0046	1.1657	0.7531	2x
spatial variation in rainfall	social learning frequency	South America	CAIC	23	0.0001	-0.2473	0.9576	none
spatial variation in rainfall	tool use frequency	South America	CAIC	23	0.0188	2.3822	0.5228	1x
spatial variation in temperature	innovation frequency	South America	CAIC	23	0.0048	-0.3158	0.7468	2x
spatial variation in temperature	social learning frequency	South America	CAIC	23	0.0153	-0.7374	0.5641	none
spatial variation in temperature	tool use frequency	South America	CAIC	23	0.0031	0.2455	0.7951	1x
seasonal variation	innovation frequency	South America	CAIC	23	0	0.0951	0.9832	2x
seasonal variation	social learning frequency	South America	CAIC	23	0	-0.1737	0.9774	none
seasonal variation	tool use frequency	South America	CAIC	23	0.0327	4.3192	0.3977	1x
interannual variation	innovation frequency	South America	CAIC	23	0.0259	-0.729	0.4527	2x
interannual variation	social learning frequency	South America	CAIC	23	0.0036	-0.3332	0.781	none
interannual variation	tool use frequency	South America	CAIC	23	0.0018	0.1855	0.8455	1x

Table 13
Life history results

dependent variable	independent variable	method	n	r2	slope	p	x vs x@node	x vs SD	y vs y@node	y vs SD
ln(geographic range size)	ln(gestation length)	species	54	0.021	-1.38E-00	0.296				
ln(geographic range size)	GL controlling for BM	species	54	0.037	-2.536	0.161				
ln(geographic range size)	ln(maximum lifespan)	species	38	0.01	0.32	0.541				
ln(geographic range size)	lifespan controlling for BM	species	38	0.029	0.814	0.304				
ln(geographic range size)	ln(emale body mass)	species	96	0.01	-1.19E-01	0.33				
ln(geographic range size)	ln(gestation length)	CAIC	51	0.0004	0.2629	0.8894			**	
ln(geographic range size)	GL controlling for BM	CAIC	50	0.002	0.659	0.77			**	
ln(geographic range size)	ln(maximum lifespan)	CAIC	37	0.0088	0.4095	0.576				
ln(geographic range size)	lifespan controlling for BM	CAIC	37	0.003	0.256	0.757				
ln(geographic range size)	ln(female body mass)	CAIC	86	0.0003	0.0385	0.8782			**	*
ln(geographic range size)	ln(female body mass)	CAIC	85	0.004	-0.168	0.553			**	*
							***p<0.001, **p<0.01, *p<0.05			

DIGITAL DATA

Primate species geographic range databases

Data sources

Primate geographic ranges
Source: *Primates of the world* by J. Wolfheim (1983)
Format: paper maps
Resolution: 1: 50,000,000 km
Projection information: none provided in source.
Method of acquisition: scanned and heads up digitized in AutoCAD Map (1997).

World Vector Shoreline
Source: USGS, http://crusty.er.usgs.gov/coast/getcoast.html
Format: Arc Ungenerate
Resolution: 1:250,000
Accuracy: 90% of all identifiable shoreline features are located within 500 metres (2.0mm at 1:250,000) circular error of their true geographic positions with respect to the preferred datum.
Projection information: Latitude, longitude co-ordinates. Spheroid WGS84.

Processing

The digitized maps were imported to GRASS version 4.1 (1993) as .dxf files and converted to raster images. The geographic range maps were rectified using the World Vector Shoreline map. The GRASS function of i.vpoints was used to identify corresponding points on the two maps. The co-ordinates derived from this were used to rectify all of the range maps using a third order transformation in i.rectify2. The rectified maps were converted to raster data using v.to.rast in GRASS.

For the African and South American databases, the range maps and the land cover map were exported as ASCII files and imported to ArcInfo (1999), where they were converted to GRID format. They were projected from latitude, longitude co-ordinates to Lambert Azimuthal Equal Area projection using the ArcInfo PROJECT function. The projected files were exported as ASCII data and imported back into GRASS.

For the Asian primate geographic range database, one stage of the rectification process was bypassed by rectifying the maps using a vector coastline map previously projected in Arcview (1999). The range maps were then rectified with reference to the projected vector coastline map in GRASS. The Asian raster maps have a higher resolution; this was necessary to represent distribution on some of the smaller islands.

African primates GRASS database
Projection: Lambert Azimuthal Equal Area
Radius of sphere of reference: 6370997.0 (default)
Longitude and Latitude of centre of projection: 0, 20.
Resolution: 9229.086m

Asian primates GRASS database
Projection: Equal Area Cylindrical
Spheroid: sphere
Longitude of central meridian: 0
Latitude of standard parallel: 0
Resolution: 5000m

South American primates GRASS database:
Projection: Lamberts Azimuthal Equal Area
Spheroid: sphere
Central Meridian: -65
Reference Latitude: -10
Resolution: 10000m

Database names

African primates
Primates2/africa/PERMANENT/
All_nig – ther_gal: reclassified species geographic range maps (range = 1, rest = 0).
Allnigx - thergalx: original projected geographic range maps.
B.allnig – b.thergal: geographic range maps reclassified by body mass.
P.allnig – p.thergal: geographic range maps reclassified by population density.
R.allnig – r.thergal: habitats within species geographic ranges.
Hab.allnig – hab.thergal: habitats within species geographic range.
Eco, ecodata: Olson (1983) ecosystem classes.
Rain: precipitation.
Density9: total density of primates (sum of GR maps).
Bmass: mean primate body mass.
Damuth: mean primate population density.

Species geographic range maps:

All_nig2: *Allenopithecus nigroviridis*
Arc_cal: *Arctocebus calabarensis*
Cerc_aeth: *Cercopithecus aethiops*
Cerc_alb: *Cercocebus albigena*
Cerc_asc: *Cercopithecus ascanius*
Cerc_at: *Cercocebus aterrimus*
Cerc_ceph: *Cercopithecus cephus*
Cerc_di: *Cercopithecus diana*
Cerc_eryg: *Cercopithecus erythrogaster*
Cerc_eryt: *Cercopithecus erythrotis*
Cerc_gal: *Cercocebus galeritus*
Cerc_ham2: *Cercopithecus hamlyni*
Cerc_lho: *Cercopithecus l'hoesti*
Cerc_mit: *Cercopithecus mitis*
Cerc_mon: *Cercopithecus mona*
Cerc_neg: *Cercopithecus neglectus*
Cerc_nict: *Cercopithecus nictitans*
Cerc_pet: *Cercopithecus petaurista*
Cerc_pog: *Cercopithecus pogonias*
Cerc_tor: *Cercocebus torquatus*
Col_ang: *Colobus angolensis*
Col_bad: *Colobus badius*
Col_guer: *Colobus guereza*
Col_pol: *Colobus polykomos*
Col_sat: *Colobus satanas*
Col_ver: *Colobus vellerosus*
Ery_pat: *Erythrocubus patas*
Euo_ele: *Euoticus elegantulus*
Euo_inu: *Euoticus inustus*
Gal_all: *Galago alleni*
Gal_crass: *Otolemur crassicaudatus*
Gal_dem: *Galagoides demidovii*
Gal_sen: *Galago senegalensis*
Gor_gor: *Gorilla gorilla*
Mac_syl: *Macaca sylvanus*
Mio_tal: *Miopithecus talapoin*
Pan_pan: *Pan paniscus*
Pan_trog: *Pan troglodytes*
Pap_anu: *Papio anubis*
Pap_cyn: *Papio cynocephalus*
Pap_ham2: *Papio hamadryas*
Pap_leu: *Mandrillus leucophaeus*
Pap_pap: *Papio papio*
Pap_sph: *Mandrillus sphinx*
Pap_urs: *Papio ursinus*
Per_pot: *Perodictus potto*
Ther_gal: *Theropithecus galago*

South American primates
Saprim2/samerica/PERMANENT/
Al_bel – saim_sciu: Raster maps or primate species geographic ranges.
Density: primate species richness.

Species geographic range maps:

Al_bel: *Alouatta belzebul*
Al_car: *Alouatta caraya*
Al_fusc: *Alouatta fusca*
Al_pal: *Alouatta palliata*
Al_pig: *Alouatta villosa*
Al_sen: *Alouatta seniculus*
Ao_triv: *Aotus trivirgatus*
At_pan: *Ateles paniscus*
Brach_arach: *Brachyteles arachnoides*
Cac_cal: *Cacajao calvus*
Cac_mel: *Cacajao melanocephalus*
Cal_arg: *Callithrix argentata*
Cal_go: *Callimico goeldii*
Cal_hum: *Callithrix humeralifer*
Cal_jac: *Callithrix jacchus*
Cal_mol: *Callicebus moloch*
Cal_pers: *Callicebus personatus*
Cal_tor: *Callicebus torquatus*
Ceb_alb: *Cebus albifrons*
Ceb_ap: *Cebus apella*
Ceb_cap: *Cebus capucinus*
Ceb_nig: *Cebus olivaceus*
Ceb_pyg: *Cebuella pygmaea*
Chir_alb: *Chiropotes albinasus*
Chir_sat: *Chiropotes satanas*
Lag_flav: *Lagothrix flavicauda*
Lag_lag: *Lagothrix lagothricha*
Leo_ros: *Leontopithecus rosalia*
Pith_mon: *Pithecia monachus*
Pith_pith: *Pithecia pithecia*
Sag_bi: *Saguinus bicolor*
Sag_fusc: *Saguinus fuscicollis*
Sag_imp: *Saguinus imperator*
Sag_inu: *Saguinus inustus*
Sag_lab: *Saguinus labiatas*
Sag_leu: *Saguinus leucopus*
Sag_mid: *Saguinus midas*
Sag_mys: *Saguinus mystax*
Sag_nig: *Saguinus nigricollis*
Sag_oe: *Saguinus oedipus*
Saim_sciu: *Saimiri sciureus*

Asian primates
Asiaprim/asia2/PERMANENT/
Coastmres2: coastline of Asia.
Density1: primate species density (sum of primate geographic range maps).
Mac1: macaque species density
Disturbedground: raster map of ecosystem classes related to disturbed ground (such as fields and towns).
Ecohres2: map of ecosystem classes from Olson (1983).
Hylcon – tarsyr: raster maps of primate species geographic ranges.

Species geographic range maps:

Hylcon: *Hylobates concolor*
Hylhoo: *Hylobates hoolock*
Hylklo: *Hylobates klossi*
Hyllar: *Hylobates lar*
Hylpil: *Hylobates pileatus*
Hylsyn: *Hylobates syndactylus*
Lortar: *Loris tardigradus*
Macarc: *Macaca arctoides*
Macass: *Macaca assamensis*
Maccyc: *Macaca cyclopis*
Macfasc: *Macaca fascicularis*
Macfusc: *Macaca fuscata*
Macmul: *Macaca mulatta*
Macnem: *Macaca nemestrina*
Macnig: *Macaca nigra*
Macrad: *Macaca radiata*
Macsil: *Macaca silenus*
Macsin: *Macaca sinica*
Naslarv: *Nasalis larvatus*
Nyccou: *Nycticebus coucang*
Nycpyg: *Nycticebus pygmaeus*
Pongpyg: *Pongo pygmaeus*
Presayg: *Presbytis comata*
Prescris: *Presbytis cristata*
Presen(t): *Presbytis entellus*
Presfranc: *Presbytis francoisi*
Presfron: *Presbytis frontata*
Presgeei: *Presbytis geei*
Presjo2: *Presbytis johnii*
Presobs: *Presbytis obscura*
Presphay: *Presbytis phayrei*
Prespil: *Presbytis pileatus*
Prespot: *Presbytis potenziani*
Presrub: *Presbytis rubicunda*
Pressen: *Presbytis vetulus*
Pygnem: *Pygathrix nemaeus*
Rhinavun: *Pygathrix avunculus*
Rhinrox: *Pygathrix roxellana*
Simcon: *Simias concolor*
Tarban: *Tarsius bancanus*
Tarspec: *Tarsius spectrum*
Tarsyr: *Tarsius syrichta*

Climatic variability maps

Data source

CRU05 0.5 degree 1961-1990 Mean Monthly Climatology dataset
CRU05 0.5 degree 1901-1995 Monthly Climate Time-Series
This data has been supplied by the Climate Impacts LINK Project (UK Department of the Environment Contract EPG 1/1/16) on behalf of the Climatic Research Unit, University of East Anglia.
Format: ascii data
Resolution: 0.5°
Projection: Geographic
Method of acquisition: Digital data, downloaded from the Internet.

References

New, M., M. Hulme and P. Jones, 1999: Representing twentieth-century space-time climate variability. Part I: Development of a 1961-90 mean monthly terrestrial climatology. J. Climate, 12, 829-856.

New, M. G., M. Hulme and P. D. Jones, 2000: Representing twentieth-century space-time climate variability. Part II: Development of 1901-1996 monthly grids of terrestrial surface climate. J. Climate, 13, 2217-2238.

Data processing

Mean yearly climatology: rainfall and temperature
Dataset: CRU05 0.5 degree 1961-1990 Mean Monthly Climatology dataset
Variables: precipitation and mean temperature. Precipitation data is in units of mm/day*10, temperature in °C*10.
Mean monthly data was imported as ascii to 16 bit rasters using Erdas Imagine. New yearly mean maps were created using a model that multiplied data for each month by the number of days in the year and then divided by the number of days in the year (taking the number of leap years during the period 1961-1990 into account). An equation in the

following form was used:
Meantemp = (31*(temp1 + temp3 + temp5 + temp7 + temp8 + temp10 + temp12) + 30*(temp4 + temp6 + temp9 + temp11) + 28.23*temp2) /365.23,
where numbers on files equal order of months.

Seasonality: annual temperature range.
Dataset: CRU05 0.5 degree 1961-1990 Mean Monthly Climatology dataset.
Variables: maximum and minimum temperatures in °C*10
The data was imported as 16 bit rasters to Erdas Imagine. A model based on a conditional statement was used to determine the lowest and highest values for temperature for year from monthly values for each cell. Minimum temperature maps were subtracted from maximum temperature maps to derive the yearly temperature range map.

Interannual variability

I chose to look at interannual variability within a period of 30 years (1961-1990). This unit of time is often used in studies of climate. There are a number of important patterns in climatic variability that fall in distinct cycles, e.g. El Niño. The thirty-year period includes several or one full cycle of a number of these patterns. Thus the variability identified in this map should not be due to my having used data from years in one stage of a cycle. However there are a number of longer cycles (in terms of centuries or more) which are not accounted for by this selection of a time period.

Total annual rainfall was calculated for each year from 1961-1990. To do this data was imported for each month of year as a raster. The source for this data was the CRU05 0.5° 1901-1995 Monthly Climate Time-Series dataset. Rasters were imported as 16 bit unsigned data to avoid losing higher values. A model was used to multiply the monthly means by the number of days in the month, allowing for leap years, and add the products. This produced a total for annual rainfall in mm/year *10. Due to the size of values these maps were then divided by ten giving data in values of mm/year.
Standard deviation (SD) in annual rainfall was calculated for each cell over the 30-year period. A model was used that carried out the following equation:

$$s = \sqrt{(3x_i^2 - ((3x_1)^2/n)))/n-1)} \qquad \text{(Shennan 1988)}$$

where x represents the annual rainfall maps and n is the number of years. This produced a map of standard deviation in rainfall over the thirty-year period worldwide. At this stage there is a methodological query: the signed 32-bit data type doesn't allow decimal point precision, and this may have simplified the data.
The measure of standard deviation may be related to the size of the mean, so I also produced a map of the coefficient of variation of total annual rainfall. This was produced using a model that divided the map of standard deviation obtained above by a map of mean annual rainfall. The map of mean annual rainfall was produced by summing the maps of total rainfall for each year from 1961-1990 and dividing by 30. The new map of coefficient of variation is in 32 bit floating point data type to include decimal points to a single precision. This map of the coefficient of variation of total annual rainfall shows a markedly different pattern from the map of standard deviation. The most noticeable difference is that the desert areas (eg Sahara) have low values of standard deviation but high coefficients of variation. This could be because the size of the mean does affect the standard deviation value, and the low SDs for desert areas are a result of the low mean total rainfall in these areas.

There is a gap in the data in the interannual variation map (Figure 4.15), but this occurs outside the range of all African primates and does not cause any problems with the analyses presented below.

Maps for analysis with species ranges

Climate maps were projected into Lambert Azimuthal equal area format covering South America and Africa, using Erdas Imagine.

Database names

All climate maps are in ERDAS Imagine Image format.

Climatemaps/
Clitemp_range/
Maxtemp6190.img: raster map of minimum yearly temperature (°C*10).
Mintemp6190.img: raster map of maximum yearly temperature (°C*10).
Temprange26190.img: raster map of annual temperature range (°C*10).
Maxtemp.gmd: model for calculation of maximum temperature.
Mintemp.gmd: model for calculation of minimum temperature.
Temprange.gmd: model for calculation of temperature range.

Clivar/
Mean2.img: raster map of mean annual rainfall (mm/year*10).

Mrstdevup.img: raster map of standard deviation of mean annual rainfall.
Mrcvarpccor1up.img: raster map of coefficient of interannual variation (%).
Af...img: projected climate maps for Africa.
As...img: projected climate maps for Asia.
Sa...img: projected climate maps for South America.
Cvar.gmd, cvarcor.gmd, cvarpercent.gmd: models for calculating coefficient of variation.
Mean2.gmd: model for calculating total mean annual rainfall.
2stdev.gmd: model for calculating standard deviation of rainfall.
Leapyearmean.gmd, /nonleapyearmean.gmd: models for calculating mean annual rainfall for leapyears and non-leapyears.

Cliyearmns/
Meanrain61902.img: raster map of annual mean of daily rainfall (mm/day*10).
Meantemp61902.img: raster map of annual mean of daily temperature (°C*10).
Precipyrmn.mdl: model for calculating mean for year.
Yeardaymean.mdl: model for calculating annual mean of daily temperature.

African mammals maps

Data source

Source: The African Mammals Databank
Istituto Ecologia Applicata
http://gorilla.bio.uniroma1.it/amd/index.htm
Format: Polygon coverage, ArcInfo export files (.e00).
Projection: Lambert Equal-Area Azimuthal
Central meridian: 20
Reference latitude: 5

African elephant database
Barnes et al (1998). African Elephant Specialist Group, Species Survival Commission.
http://iucn.org/themes/ssc/sgs/afesg/aed/aed98.html

Processing

The ArcInfo export files were imported to ERDAS Imagine (1997), and the polygon coverage files were extracted. These vector files were converted to raster format (ArcInfo GRIDS) in Imagine. The maps were reclassified in ArcView GIS 3.2 (1999) to have values of 1 for species' extent of occurrence and 0 elsewhere. This simplified the data available for the original maps, which included different values for areas based on the degree of certainty that populations were currently present.
New maps were produced showing spatial patterns in the distribution of ecological variables using the map calculator facilities of ArcView GIS 3.2 (1999) and ERDAS Imagine 8.4 (1997). The details of production of particular maps are given below.

Diversity: individual species geographic range maps were added together using the Map Calculator in ArcView, to produce maps showing total species richness.

Boundaries: ArcView was used to create a simple edge detecting neighbourhed filter, and ran the filter on each species' range map. This produced new maps reclassified with values of 1 for cells at the edge of the range and 0 elsewhere. The new maps were added together using the Map Calculator in ArcView.

Biomass: geographic range maps were reclassified according to the biomass estimate for each species, using ArcView GIS 3.2's 'Reclassify' option. I added together the resulting maps using the Map Calculator facility to obtain an estimate of the total carnivore biomass present in an area. I converted the raster map to contours for comparison with Thackeray (1995).

Body mass: geographic range maps were reclassified with body mass values as above. Equations were written in Imagine's Modeller facility to derive maximum and range of values based on different values for species present in each cell.

Home range maps: geographic range maps were reclassified with home range values as above. The sum of home range values as well as number of species was calculated using ArcView's Map Calculator. The mean value was obtained by dividing the sum of the home range values map by the number of species map in ArcView. Maps showing the distribution of species for which home range data was absent were also produced using Map Calculator.

Format: raster maps are in Arc-Info GRID format. Vector maps are in shapefile format. Both formats can be opened in ESRI ArcView or ArcInfo.
Projection: Lambert Equal-Area Azimuthal
Central meridian: 20
Reference latitude: 5
Cell size for raster maps: 25000m

References

AutoCAD Map 2 (1997) Autodesk, Inc.
Erdas Imagine 8.4 (1997) Erdas, Inc.

ArcInfo 8.0.1 (1999) Environmental Systems Research Institute, Inc.
ArcView GIS 3.2 (1999) Environmental Systems Research Institute, Inc.

Database names

Carnivores/
Aaedgedetect/carboundaries: raster map of carnivore boundaries.
Aagridbiomass/carbiomass3: raster map of carnivore biomass in kg/ha.
Aagridbiomass/carbiomass.shp: vector map of contours of carnivore biomass in kg/ha.
Aagridbmass/carbmassmaxkg: raster map of maximum carnivore body mass in kg.
Aagridbmass/carbmrangekg: raster map of range of carnivore body mass in kg.
Aagridbmass/carbmassmax: raster map of maximum carnivore body mass in g.
Aagridbmass/carbmassrange: raster map of range of carnivore body mass in g.
Aagridbmass/carbmassmin: raster map of minimum carnivore body mass in g.
Aagridbmass/carbmass3.gmd, carbodmass.gmd, carminbmass.gmd: models for calculating maximum minimum and range of carnivore body mass.
Aagridfiles/cardensity: raster map of carnivore diversity.
Aagridfiles3/acinonyx_g6 – xeno_nas6: raster maps of carnivore species' extent of occurrence.
Homerange/carhrmeankm: raster map of mean carnivore individual home range size in km.
Homerange/carhrmeanha1: raster map of mean carnivore individual home range size in ha.
Homerange/carhrsumha: raster map of sum of individual home range size in ha.
Homerange/specieswnohr: raster map of distribution of species for which there was no home range data.
Homerange/specieswithhr: raster map of species for which home range data is available.

Environment/
Olson_carbon: raster map of carbon in live vegetation, from Olson (1983).
Olsonecosyst: raster map of ecosystem classes, from Olson (1983).
Afmeanrain: raster map of average annual rainfall (mm/day*10), from New et al (1999).
Avanntemp4: raster map of average daily temperature (°C*10), from New et al (1999).

Primates/
Aagridbiomass/pbiomass: raster map of primate biomass in kg/ha.
Aagridbiomass/pbiomcontour1.shp: vector map of primate biomass in kg/ha.
Aagridbmass/pmaxbmass2kg: raster map of maximum primate body mass in kg.
Aagridbmass/prangebmasskg: raster map of range of primate body mass in kg.
Aagridbmass/pmaxbmass2: raster map of maximum primate body mass in g.
Aagridbmass/pminbmass2: raster map of minimum primate body mass in g.
Aagridbmass/prangebmass2: raster map of range of primate body mass in g.
Aagridfiles3/all-nig3 – ther_galthre: raster maps of primate species extent of occurrence.
Homerange/primhrmeanha: raster map of mean primate individual home range size in ha.
Homerange/specieswithhr: raster map of primate species for which home range data was available.
Homerange/specieswnohr: raster map of primate species for which no home range data was available.
Homerange/pmaxhr2: raster map of maximum primate individual home range size in metres.
Homerange/sumofhr: raster map of sum of primate individual home range size in metres.
Pboundarygrids/pboundaries: raster map of primate geographic range boundaries.
Primdensity2: raster map of primate species richness.

Ungulates/
Aagridboundaries/ungboundaries: raster map of ungulate geographic range boundaries.
Aagridfiles/addax_naso2 – trich_sene2: raster maps of ungulate species extent of occurrence.
Aagridfiles/ungdensitym3: raster map of ungulate species richness.
Ungbiomass/unbiomass: raster map of ungulate biomass in kg/ha.
Ungbiomass/ungbiomass10.shp: vector map of ungulate biomass with 10kg/ha contours.
Ungbiomass/ungbiomass20.shp: vector map of ungulate biomass with 20kg/ha contours.

LIST OF REFERENCES

AutoCAD Map 2 (1997) Autodesk, Inc.

Erdas Imagine 8.4 (1997) Erdas, Inc.

ArcInfo 8.0.1 (1999) Environmental Systems Research Institute, Inc.

ArcView GIS 3.2 (1999) Environmental Systems Research Institute, Inc.

Abbate, E., Albianelli, A., Azzaroli, A., Benvenuti, M., Tesfamariam, B., Bruni, P., Cipriani, N., Clarke, R. J., Ficcarelli, G., Macchiarelli, R., Napoleone, G., Papini, M., Rook, L., Sagri, M., Tecle, T. M., Torre, D. and Villa, I. (1998) A one-million-year-old *Homo* cranium from the Danakil (Afar) Depression of Eritrea. *Nature,* **393,** 458-460.

Aguirre, E. and Carbonell, E. (2001) Early human expansions into Eurasia: The Atapuerca evidence. *Quaternary International,* **75,** 11-18.

Aiello, L. C. and Dunbar, R. I. M. (1993) Neocortex size, group size, and the evolution of language. *Current Anthropology,* **34,** 184-193.

Aiello, L. C. and Wheeler, P. (1995) The expensive-tissue hypothesis: the brain and the digestive system in human and primate evolution. *Current Anthropology,* **36,** 199-221.

Alemseged, Z., Wynn, J. G., Kimbel, W. H., Reed, D., Geraads, D. and Bobe, R. (2005) A new hominin from the Basal Member of the Hadar Formation, Dikika, Ethiopia, and its geological context. *Journal of Human Evolution* **49**(4), 499-514

Andrews, P. (1984) On the characters that define *Homo erectus. Courier Forschungsintitut Senckenberg,* **69,** 167-178.

Anton, S. C., Aziz, F. and Zaim, Y. (2001) Plio-Pleistocene *Homo*: Patterns and determinants of dispersal. In *Humanity from African naissance to coming millenia* (Eds, Tobias, P. V., Rath, M. A., Moggi-Cecchi, J. and Doyle, G. A.) Florence University Press, Florence, pp. 91-102.

Anton, S. C., Leonard, W. R. and Robertson, M. L. (2002) An ecomorphological model of the initial hominid dispersal from Africa. *Journal of Human Evolution,* **43,** 773-785.

Aoki, K. and Feldman, M. V. (1987) Toward a theory for the evolution of cultural communication: Coevolution of signal transmission and reception. *Proceedings of the National Academy of Sciences of the United States of America,* **84,** 7164-7168.

Arambourg, C. (1955) A recent discovery in human paleontology: *Atlanthropus* of Ternifine (Algeria). *American Journal of Physical Anthropology,* **13,** 191-202.

Arambourg, C. and Biberson, P. (1956) The fossil human remains from the Palaeolithic site of Sidi Abderrahman. *American Journal of Physical Anthropology,* **14,** 267-290.

Arribas, A. and Palmqvist, P. (1999) On the ecological connection between sabre-tooths and hominids: Faunal dispersal events in the Lower Pleistocene and a review of the evidence for the first human arrival in Europe. *Journal of Archaeological Science,* **26,** 571-585.

Ascenzi, A., Biddutu, I., Cassoli, P. F., Segre, A. G. and Segre-Naldini, E. (1996) A calvarium of late *Homo erectus* from Ceprano, Italy. *Journal of Human Evolution,* **31,** 409-423.

Asfaw, B. (1987) The Belohdelie frontal: New evidence of early hominid cranial morphology from the Afar of Ethiopia. *Journal of Human Evolution,* **16,** 611-624.

Asfaw, B., Beyene, Y., Suwa, G., Waiter, R. C., White, T. D., WoldeGabriel, G. and Yemane, T. (1992) The earliest Acheulean from Konso-Gardula. *Nature,* **360,** 732-735.

Asfaw, B., Gilbert, W. H., Beyene, Y., Hart, W. K., Renne, P. R., Wolde-Gabriel, G., Vrba, E. S. and White, T. D. (2002) Remains of *Homo erectus* from Bouri, Middle Awash, Ethiopia. *Nature,* **416,** 317-320.

Asfaw, B., White, T., Lovejoy, O., Latimer, B., Simpson, S. and Suwa, G. (1999) *Australopithecus garhi*: A new species of early hominid from Ethiopia. *Science,* **284,** 629-635.

Avery, D. M. (1995) Southern savannas and Pleistocene hominid adaptations: The micromammalian perspective. In *Paleoclimate and evolution with emphasis on early human origins* (Eds, Vrba, E. S., Denton, D. H., Partridge, T. C. and Burckle, L. H.) Yale University Press, Yale, pp. 459-478.

Backwell, L. R. and d'Errico, F. (2001) Evidence of termite foraging by Swartkrans early hominids. *Proceedings of the National Academy of Sciences USA,* **98,** 1358-1363.

Baker, R. R. (1978) *The evolutionary ecology of animal migration.* Hodder & Stoughton, London.

Balout, L., Biberson, P. and Tixier, J. (1967) L'Acheuléen de Ternifine (Algérie): Gisement de l'Atlanthrope. *Anthropologie,* **71,** 217-238.

Barnes, R. F. W., Craig, G. C., Dublin, H. T., Overton, G., Simons, W. and Thouless, C. R. (1998) *African Elephant Database 1998.* The IUCN Species Survival Commission.

Barton, N. H. and Dunbar, R. I. M. (1997) Evolution of the social brain. In *Machiavellian intelligence.*, Vol. 2 (Eds, Whiten, A. and Byrne, R.) Cambridge University Press, Cambridge, pp. 240-263.

Barton, N. H. and Harvey, P. H. (2000) Mosaic evolution of brain structure in mammals. *Nature,* **405,** 1055-1058.

Bar-Yosef, O. and Belfer-Cohen, A. (2001) From Africa to Eurasia - early dispersals. *Quaternary International,* **75,** 19-28.

Bar-Yosef, O. and Goren-Inbar, N. (1993) The lithic assemblages of 'Ubeidiyah: A Lower Palaeolithic site in the Jordan Valley. *Oedem,* **45,** 1-266.

Beauchamp, G. and Cabana, G. (1990) Group-size variability in Primates. *Primates,* **31,** 171-182.

Belmaker, M., E. Tchernov, et al. (2002). New evidence for hominid presence in the Lower Pleistocene of the Southern Levant. *Journal of Human Evolution,* **43,** 43-56.

Benefit, B. R. (1999) Biogeography, dietary specialization, and the diversification of African Plio-Pleistocene monkeys. In *African biogeography, climate change and human evolution* (Eds, Bromage, T. G. and Schrenk, F.) Oxford University Press, Oxford, pp. 172-188.

Berger, L. R. and Tobias, P. V. (1994) New discoveries at the early hominid site of Gladysvale, South Africa. *South African Journal of Science,* **90,** 223-226.

Bergman, A. and Feldman, M. V. (1995) On the evolution of learning: Representation of a stochastic environment. *Theoretical population biology.,* **48,** 251-276.

Blackburn, T. and Gaston, K. (1996) Spatial patterns in the geographic range sizes of bird species in the New World. *Philosophical Transactions of the Royal Society of London, ser.B,* **351,** 897-912.

Bloch, M. (1991) Language, anthropology and cognitive science. *Man,* **26,** 183-198.

Blumenschine, R. J., Cavallo, J. A. and Capaldo, S. D. (1994) Competition for carcasses and early hominid behavioural ecology: a case study and conceptual framework. *Journal of Human Evolution,* **27,** 197-213.

Blurton-Jones, N. G., Smith, L. C., O'Connell, J. F., Hawkes, K. and Kamuzora, C. L. (1992) Demography of the Hadza, an increasing and high density population of savanna foragers. *American Journal of Physical Anthropology,* **89,** 159-181.

Boaz, N. T. (1979) Early hominid population densities: New estimates. *Science,* **206,** 592-595.

Boesch, C. and Boesch, H. (1989) Hunting behaviour of wild chimpanzees in the Tai National Park. *American Journal of Physical Anthropology,* **78,** 547-573.

Bonnefille, R. (1976) Végétations et climats des temps oldowayens et acheuléens à Melka-Kunture (Ethiopie). In *L'ethiopie avant l'histoire* CNRS cahier n.1, Paris, pp. 55-71.

Bonnefille, R. (1995) A reassessment of the Plio-Pleistocene pollen record of East Africa. In *Palaeoclimate and evolution with emphasis on human origins* (Eds, Vrba, E. S., Denton, G. H., Partridge, T. C. and Burckle, L. H.) Yale University Press, New Haven and London, pp. 299-310.

Box, H. O. (1984) *Primate behaviour and social ecology.* Chapman and Hall, London.

Box, H. O. and Gibson, K. R. (Eds.) (1999) *Mammalian social learning: Comparative and ecological perspectives,* Cambridge University Press, Cambridge.

Boyd, R. and Richerson, P. J. (1985) *Culture and the evolutionary process.* University of Chicago Press, Chicago.

Boyd, R. and Richerson, P. J. (1988) An evolutionary model of social learning: the effects of spatial and temporal variation. In *Social learning: Psychological and Biological Perspectives* (Eds, Zentall, T. R. and Galef, G. B.) Erlbaum, Hillsdale, N.J., pp. 29-48.

Boyd, R. and Richerson, P. J. (2000) Climate, culture and the evolution of cognition. In *The evolution of cognition.* (Eds, Heyes, C. and Huber, L.) Erlbaum, Hillsdale, N.J., pp. 329-346.

Brain, C. K. (1981) *The hunters or the hunted? An introduction to African cave taphonomy.* University of Chicago Press, Chicago.

Brauer, G. (1994) How different are Asian and African *Homo erectus*? *Courier Forschungsintitut Senckenberg,* **171,** 301-318.

Brauer, G. and Mbua, E. (1992) *Homo-Erectus* features used in cladistics and their variability in Asian and African hominids. *Journal of Human Evolution,* **22,** 79-108.

Broecker, W. S. and Denton, D. H. (1990) The role of ocean atmosphere reorganisations in glacial cycles. *Quaternary Science Review,* **9,** 305-343.

Bromage, T. G. and Schrenk, F. (1995) Biogeographic and climatic basis for a narrative of early hominid evolution. *Journal of Human Evolution,* **28,** 109-114.

Broom, R. (1938) The Pleistocene anthropoid apes of South Africa. *Nature,* **142,** 377-379.

Brown, B., Walker, A., Ward, C. V. and Leakey, R. E. F. (1993) New *Australopithecus boisei* calvaria from east Lake Turkana, Kenya. *American Journal of Physical Anthropology,* **91,** 137-159.

Brown, F., Harris, J., Leakey, R. and Walker, A. (1985) Early *Homo erectus* skeleton from west Lake Turkana, Kenya. *Nature,* **316,** 788-792.

Brown, F. H. (1994) Development of Pliocene and Pleistocene chronology of the Turkana Basin, East

Africa, and its relation to other sites. In *Integrative paths to the past: Palaeoanthropological advances in honor of F. Clark Howell* (Eds, Corrucini, R. S. and Ciochon, R. L.) Prentice Hall, Eaglewood Cliffs, N.J., pp. 285-312.

Brown, F. H. and Feibel, C. S. (1991) Stratigraphy, depositional environments and paleogeography of the Koobi Fora Formation. In *Koobi Fora Research Project Vol. 3: the fossil ungulates: geology, fossil artiodactyls and paleoenvironments* (Ed, Harris, J. M.) Clarendon Press, Oxford.

Brown, J. H. (1995) *Macroecology.* Chicago University Press, Chicago.

Brown, J. H. and Maurer, B. A. (1989) Macroecology: the division of food and space among species on continents. *Science,* **243,** 1145-50.

Brown, J. H., Stevens, G. C. and Kaufman, D. M. (1996) The geographic range: size, shape, boundaries, and internal structure. *Annual Review of Ecology and Systematics,* **27,** 597-623.

Brunet, M., Beauvilain, A., Coppens, Y., Heintz, E., Moutaye, A. H. E. and Pilbeam, D. (1996) *Australopithecus bahrelghazali*, a new species of early hominid from Koro Toro region, Chad. *Comptes Rendus De L Academie Des Sciences Serie Ii Fascicule a- Sciences De La Terre Et Des Planetes,* **322,** 907-913.

Brunet, M., Beauvilain, A., Geraads, D., Guy, F., Kasser, M., Mackaye, H. T., MacLatchy, L. M., Mouchelin, G., Sudre, J. and Vignaud, P. (1997) Chad: a new Pliocene hominid site. *Comptes Rendus De L Academie Des Sciences Serie IIA - Sciences De La Terre Et Des Planetes,* **324,** 341-345.

Brunet, M., Guy, F., Pilbeam, D., Mackaye, H. T., Likius, A., Ahounta, D., Beauvilain, A., Blondel, C., Bocherens, H., Boisserie, J.-R., de Bonis, L., Coppens, Y., Dejax, J., Denys, C., Duringer, P., Eisemann, V., Fanones, G., Fronty, P., Geraads, D., Lehmann, T., Lihoreau, F., Louchart, A., Mahamat, A., Merceron, G., Mouchelin, G., Otero, O., Campomanes, P. P., Ponce de Leons, M., Rage, J.-C., Sapanet, M., Schuster, M., Sudre, J., Tassy, P., Valentin, X., Vignaud, P., Viriot, L., Zazzo, A. and Zollikofer, C. (2002) A new hominid from the Upper Miocene of Chad, Central Africa. *Nature,* **418,** 145-151.

Bunn, H. T. (1981) Archaeological evidence for meat-eating by Plio-Pleistocene hominids from Koobi Fora, Kenya. *Nature,* **291,** 574-577.

Bunn, H. T. and Kroll, E. M. (1986) Systematic butchery by Plio-Pleistocene hominids at Olduvai Gorge, Tanzania. *Current Anthropology,* **5,** 431-452.

Byrne, R. (1995) *The thinking ape.* Oxford University Press, Oxford.

Byrne, R. W. and Whiten, A. (1988) *Machiavellian intelligence: Social expertise and the evolution of intellect in monkeys.* Oxford University Press, Oxford.

Cachel, S. and Harris, J. W. K. (1997) The lifeways of *Homo erectus* inferred from archaeology and evolutionary ecology: a perspective from East Africa. In *Early human behavior in global context: the rise and diversity of the lower Palaeolithic record* (Eds, Petraglia, M. D. and Korisettar, R.) Routledge, London.

Calvin, W. H. (2002) *A brain for all seasons: Human evolution and abrupt climate change.* University of Chicago Press, Chicago.

Carbone, C. and Gittleman, J. L. (2002) A common rule for the scaling of carnivore density. *Science,* **295,** 2273-2276.

Carbone, C., Mace, G. M., Roberts, S. C. and Macdonald, D. W. (1999) Energetic constraints on the diet of terrestrial carnivores. *Nature,* **402,** 286-288.

Carbonell, E., Bermudez de Castro, J. M., Arsuaga, J. L., Diez, J. C., Rosas, A., Cuenca-Bescos, G., Sala, R., Mosquera, M. and Rodriguez, X. P. (1995) Lower Pleistocene hominids and artefacts from Atapuerca-TD6 (Spain). *Science,* **269,** 826-830.

Carney, J., Hill, A., Miller, J. A. and Walker, A. (1971) Late Australopithecine from Baringo District, Kenya. *Nature,* **230,** 509-514.

Cavalli-Sforza, L. L. and Feldman, M. V. (1983) Cultural versus genetic adaptation. *Proceedings of the National Academy of Sciences of the United States of America,* **79,** 1331-1335.

Cerling, T. E. (1992) Development of grasslands and savannas in East Africa during the Neogene. *Paleogeography, Paleoclimatology, Paleoecology,* **97,** 241-247.

Cerling, T. E., Bowman, J. R. and O'Neil, J. R. (1988) An isotopic study of a fluviatile lacustrine sequence: The Plio-Pleistocene Koobi Fora sequence, East Africa. *Paleogeography, Paleoclimatology, Paleoecology,* **63,** 335-356.

Cerling, T. E. and Hay, R. L. (1986) An isotopic study of paleosol carbonates from Olduvai Gorge. *Quaternary Research,* **25,** 63-78.

Cerling, T. E., Quade, J., Ambrose, S. H. and Sikes, N. E. (1991) Fossil soils, grasses and carbon isotopes from Fort Ternan: Grassland or woodland? *Journal of Human Evolution,* **21,** 295-306.

Chamberlain, A. T. (1989) Variations within *Homo habilis*. In *Hominidae: Proceedings of the Second International Congress of Human Paleontology.* (Ed, Giacobini, G.) Jaca Book, Milan, pp. 175-181.

Chapman, C. A. and Chapman, L. J. (1990) Dietary variability in primate populations. *Primates,* **31,** 121-128.

Chapman, C. A., Gautier-Hion, A., Oates, J. F. and Onderdonk, D. A. (1999) African primate communities: determinants of structure and threats to survival. In *Primate communities* (Eds, Fleagle, J. G., Janson,

C. and Reed, K. E.) Cambridge University Press, Cambridge, pp. 1-37.

Chavaillon, J. (1979) Stratigraphie du site archaéologique de Melka-Kunturé (Éthiopie). *Bulletin de la Société Géologique de la France.*, **21,** 227-232.

Chavaillon, J. (1982) Position chronologique des hominidés fossiles d'Éthiopie. In *L'Homo erectus et la place de l'homme de Tautavel parmi les hominidés fossiles* (Ed, de Lumley, M. A.) Premier Congrès International de Paléontologie Humaine, Nice, pp. 766-797.

Chavaillon, J., Brahimi, C. and Coppens, Y. (1974) Première découverte d'un hominide dans l'un des sites acheuléens de Melka-Kunturé (Éthiopie). *Comptes Rendus De L Academie Des Sciences Ser. D,* **278,** 3299-3302.

Chivers, D. J. (1991) Species differences in tolerance to environmental change. In *Primate responses to environmental change.* (Ed, Lee, P. C.) Chapman & Hall, London, pp. 5-38.

Clark, A. B. (1991) Individual variation in responsiveness to environmental change. In *Primate responses to environmental change* (Ed, Box, H. O.) Chapman & Hall, London, pp. 91-114.

Clark, J. D. (1993) Stone artefact assemblages from Members 1-3, Swartkrans Cave. *Transvaal Museum Monograph,* **8,** 167-194.

Clark, J. D. and Kurashina, H. (1979) Hominid occupation of the east-central highlands of Ethiopia in the Plio-Peistocene. *Nature,* **282,** 33-39.

Clarke, R. J. (1977) A juvenile cranium and some adult teeth of early *Homo* from Swartkrans, Transvaal. *South African Journal of Science,* **73,** 46-49.

Clarke, R. J. (1988) Habiline handaxes and paranthropine pedigree. *World Archaeology,* **20,** 1-12.

Clarke, R. J. (1994a) On some new interpretations of Sterkfontein stratigraphy. *South African Journal of Science,* **90,** 211-214.

Clarke, R. J. (1994b) The significance of the Swartkrans *Homo* to the *H.erectus* problem. *Courier Forschungsintitut Senckenberg,* **171,** 185-193.

Clarke, R. J. (1995) Mr, Mrs and Miss: conceptions in studies of human ancestry. In *Hominids and their environment during the Lower and Middle Pleistocene of Eurasia: Proceedings of the International Conference of Human Palaeontology, Orce, Spain* (Ed, Gilbert, J.), pp. 327-329.

Clarke, R. J., Howell, F. C. and Brain, C. K. (1970) More evidence of an advanced hominid at Swartkrans. *Nature,* **225,** 1219-1222.

Clarke, R. J. and Tobias, P. V. (1995) Sterkfontein Member 2 foot bones of the oldest South African hominid. *Science,* **269,** 521-524.

Coffing, K., Feibel, C., Leakey, M. and Walker, A. (1994) Four-million-year-old hominids from east Lake Turkana, Kenya. *American Journal of Physical Anthropology,* **93,** 55-65.

Collard, I. F. and Foley, R. A. (2002) Latitudinal patterns and environmental determinants of recent human cultural diversity: do humans follow biogeographical rules? *Evolutionary Ecology Research,* **4,** 371-383.

Conroy, G. C. (1997) *Reconstructing human origins. A modern synthesis.* W.W.Norton and Company, London.

Corbet and Hill (1991) *A world list of mammalian species.*

Cowlishaw, G. (1999) Predicting the pattern of decline of African primate diversity: an extinction debt from historical deforestation. *Conservation biology,* **13,** 1183-1193.

Cowlishaw, G. and Dunbar, R. (2000) *Primate conservation biology* Chicago University Press, Chicago.

Cowlishaw, G. and Hacker, J. E. (1997) Distribution, diversity, and latitude in African primates. *American Naturalist,* **150,** 505-512.

Damuth, J. (1981) Population density and body mass in mammals. *Nature,* **290,** 699-700.

Damuth, J. (1987) Interspecific allometry of population density in mammals and other animals: the independence of body mass and population energy use. *Biological Journal of the Linnaean Society.,* **31,** 193-246.

Dawkins, R. (1976) *The selfish gene.* Oxford University Press, Oxford.

de Heinzelin, J., Clark, J. D., White, T., Hart, W., Renne, P., WoldeGabriel, G., Beyene, Y. and Vrba, E. (1999) Environment and behavior of 2.5-million-year-old Bouri hominids. *Science,* **284,** 625-629.

de Menocal, P. B. and Bloemendal, J. (1995) Plio-Pleistocene climatic variability in subtropical Africa and the palaeoenvironment of hominid evolution. In *Paleoclimate and evolution with emphasis on human origins.* (Eds, Vrba, E. S., Denton, D. H., Partridge, T. C. and Burckle, L. H.) Yale University Press, New Haven, pp. 262-288.

de Ruiter, D. (2000) *A methodological analysis of the relative abundance of hominids and other macromammals from the site of Swartkrans, South Africa.* Unpublished PhD thesis. Science Faculty, University of the Witwatersrand

Deino, A. and Potts, R. (1990) Single crystal 40Ar/39Ar dating of the Olorgesailie Formation, southern Kenya Rift. *Journal of Geophysical Research,* **95,** 8453-8470.

Delson, E, (1997) One skull does not a species make. *Nature,* **389,** 446-447.

Dennell, R. W. (1998) Grasslands, tool making and the hominid colonization of southern Asia: a reconsideration. In *Early human behaviour in global context: The rise and diversity of the Lower Palaeolithic record.* (Eds, Petraglia, M. D. and Korisettar, R.) Routledge, London.

Dennell, R. W., Rendell, H. and Hailwood, E. (1988) Late Pliocene sites from northern Pakistan. *Current Anthropology,* **29**.

Dennell, R. and Roebroeks, W. (2005) An Asian perspective on early human dispersal from Africa. *Nature,* **438,** 1099-1104.

Dennett, D. C. (1996) *Darwin's dangerous idea: evolution and the meanings of life.* Penguin, London.

Dominguez-Rodrigo, M. (2002) Hunting and scavenging by early humans: the state of the debate. *Journal of World Prehistory,* **16,** 1-54.

Dominguez-Rodrigo, M., de Luque, L., Alcala, L., Mora, R., Serrallonga, J. and Medina, V. (2002) The ST site complex at Peninj, West Lake Natron, Tanzania: Implications for early hominid behavioural models. *Journal of Archaeological Science,* **29,** 639-665.

Dominguez-Rodrigo, M., Lopez-Saez, J. A., Vincens, A., Alcala, L., Luque, L. and Serrallonga, J. (2001) Fossil pollen from the Upper Humbu Formation of Peninj (Tanzania): hominid adaptation to a dry open Plio-Pleistocene savanna environment. *Journal of Human Evolution,* **40,** 151-157.

Dunbar (1996) *Grooming, gossip and the evolution of language.* Faber, London.

Dunbar, R. I. M. (1992) Behavioural ecology of the extinct papionines. *Journal of Human Evolution,* **22,** 407-421.

Dunbar, R. I. M. (1993) Socioecology of the extinct theropiths: a modelling approach. In *Theropithecus: The rise and fall of a primate genus.* (Ed, Jablonski, N. G.) Cambridge University Press, Cambridge, pp. 465-486.

Dunbar, R. I. M. (1998) The social brain hypothesis. *Evolutionary Anthropology,* **6,** 178-190.

Eeley, H. A. C. (1994) *Ecological and evolutionary patterns of primate species area.* Unpublished PhD thesis. Department of Biological Anthropology, University of Cambridge

Eeley, H. A. C. and Foley, R. A. (1999) Species richness, species range size and ecological specialisation among African primates: geographical patterns and conservation implications. *Biodiversity and Conservation,* **8,** 1033-1056.

Eeley, H. A. C. and Lawes, M. J. (1999) Large-scale patterns of species richness and species range size in anthropoid primates. In *Primate communities* (Eds, Fleagle, J. G., Hanson, C. H. and Reed, K. E.) Cambridge University Press, Cambridge.

Eldredge, N. and Gould, S. J. (1972) Punctuated equilibria: an alternative to phyletic gradualism. In *Models in paleobiology* (Ed, Schopf, T. J. M.) Freeman, Cooper and Company, San Fransisco, pp. 82-115.

Fagen, R. (1982) Evolutionary issues in the development of behavioural flexibility. In *Perspectives in ethology.* (Eds, Bateson, P. P. G. and Klopfer, P. H.) Plenum Press, New York, pp. 365-383.

Feblot-Augustins, J. (1997a) *La circulation des matieres premieres au paleolithique. Synthese des donnees prespectives comportmentales.* Etudes et recherches archeologiques de l'Universite de Liege, Liege.

Feblot-Augustins, J. (1997b) Anciens hominides d'Afrique et premiers temoignages de circulation des matieres premieres. *Archeo-Nil,* **7,** 17-45.

Fedigan, L. M. (1990) Vertebrate predation in *Cebus capucinus*: Meat eating in a Neotropical monkey. *Folia Primatologica,* **54,** 196-205.

Feibel, C., Harris, J. M. and Brown, F. H. (1991) Palaeoenvironmental context for the late Neogene of the Turkana Basin. In *Koobi Fora Research Project,* Vol. 3 (Ed, Harris, J. M.) Clarendon Press, Oxford, pp. 321-370.

Feibel, C. S., Brown, F. H. and McDougall, I. (1989) Stratigraphic context of fossil hominids from the Omo Group deposits: Northern Turkana Basin, Kenya and Ethiopia. *American Journal of Physical Anthropology,* **78,** 595-622.

Feldman, M. V., Aoki, K. and Kumm, J. (1996) Individual versus social learning: Evolutionary analysis in a fluctuating environment. *Anthropological Science,* **104,** 209-232.

Feldman, M. W. and Cavalli-Sforza, L. L. (1976) Cultural and biological evolutionary processes: Selection for a trait under complex transmission. *Theoretical population biology,* **9,** 238-259.

Felzenstein, J. (1985) Phylogenies and the comparative method. *American Naturalist,* **125,** 1-15.

Fleagle (1999) *Primate adaptation and evolution.* Academic Press, San Diego.

Fleagle, J. G., Rasmusson, D. T., Yirga, S., Bown, T. M. and Grine, F. E. (1991) New hominid fossils from Fejej, southern Ethiopia. *Journal of Human Evolution,* **21,** 145-152.

Foley, R. A. (1987a) Hominid species and stone-tool assemblages: how are they related? *Antiquity,* **61,** 380-392.

Foley, R. A. (1987b) *Another unique species: patterns in human evolutionary ecology.* Longman Scientific & Technical, Harlow.

Foley, R. A. (1992) Studying human evolution by analogy. In *The Cambridge Encyclopedia of Human Evolution* (Eds, Jones, S., Martin, R. and Pilbeam, D.) Cambridge University Press, Cambridge, pp. 335-340.

Foley, R. A. (1993) African terrestrial primates: the comparative evolutionary biology of *Theropithecus* and the *Hominidae.* In *Theropithecus* (Ed, Jablonski, N. G.) Cambridge University Press, Cambridge, pp. 245-270.

Foley, R. A. (1999) Evolutionary geography of Pliocene hominids. In *African biogeography, climate change and human evolution* (Eds, Bromage, T. G. and Schrenk, F.) Oxford University Press, Oxford, pp. 328-348.

Foley, R. A. and Lahr, M. M. (1997) Mode 3 technologies and the evolution of modern humans. *Cambridge Archaeological Journal,* **7,** 3-36.

France, R. (1992) The North American latitudinal gradient in species richness and geographical range of freshawater crayfish and amphipods. *American Naturalist,* **139,** 342-354.

Gabunia, L., Anton, S. C., Lordkipanidze, D., Vekua, A., Justus, A. and Swisher, C. C. I. (2001) Dmanisi and dispersal. *Evolutionary Anthropology,* **10,** 158-170.

Gabunia, L., de Lumley, M.-A., Vekua, A., Lordkipanidze, D.D., and de Lumley, H. (2002) D'écouverte d'un nouvel hominidé à Dmanisi (transcaucasie, Georgie). C. R. Palevol. **1,** 243-253.

Gabunia, L., Vekua, A., Lordkipanidze, D., Swisher, C. C. I., Ferring, R., Justus, A., Nioradze, M., Tvalchrelidze, M., Anton, S. C., Bosinski, G., Joris, O., de Lumley, M.-A., Majsuradze, G. and Mouskhelishvili, A. (2000a) Earliest Pleistocene cranial remains from Dmanisi, Republic of Georgia: taxonomy, geological setting and age. *Science,* **288,** 1019-1025.

Gabunia, L., Vekua, A. and Lordkipanidze, D. (2000b) The environmental context of early human occupation of Georgia (Transcaucasia). *Journal of Human Evolution,* **38,** 785-802.

Galef, B. G. (1988) Imitation in animals: history, definition and interpretation of data from the Psychological laboratory. In *Social learning: Psychological and biological perspectives.* (Eds, Zentall, T. R. and Galef, B. G.) Ehrlbaum, Hillsdale, NJ.

Gamble, C. (1993) *Timewalkers: the prehistory of global colonization.* Alan Sutton, Stroud.

Gamble, C. and Steele, J. (1999) Hominid ranging patterns and dietary strategies. In *Hominid evolution: Lifestyle and survival strategies.* (Ed, Ullrich, H.) Edition Archaea, Gelsenkirchen, pp. 396-409.

Garland, T., Harvey, P. H. and Ives, A. R. (1992) Procedures for the analysis of comparative data using phylogenetically independent contrasts. *Syst. Biol.,* **41,** 18-32.

Gaston, K. J. (1994) *Rarity.* Chapman & Hall, London.

Gaston, K. J. (1996) Species range size distributions: patterns, mechanisms and implications. *Trends in Ecology & Evolution,* **11,** 197-201.

Gaston, K. J. (1998) Species-range size distributions: products of speciation, extinction and transformation. *Philosophical Transactions of the Royal Society of London Series B-Biological Sciences,* **353,** 219-230.

Gaston, K. J., Blackburn, T. M. and Spicer, J. I. (1998a) Rapoport's rule: time for an epitaph? *Trends in Ecology & Evolution,* **13,** 70-74.

Gaston, K., Blackburn, T. and Spicer, J. (1998b) Too much latitude for Rapoport's rule? Reply from K.J. Gaston, T.M. Blackburn and J.I. Spicer. *Trends in Ecology & Evolution,* **13,** 242.

Gaston, K. J. and Lawton, J. H. (1988) Patterns in the distribution and abundance of insect populations. *Nature,* **311,** 709-712.

Geraads, D. (1980) La faune des sites á '*Homo erectus*' des carrières Thomas (Casablanca, Maroc). *Quaternaria,* **22,** 65-94.

Geraads, D., Amani, F. and Hublin, J.-J. (1992) Le gisement Pléistocène moyen de l'Aïn Maarouf près de El Hajeb, Maroc: Présence d'un hominidé. *Comptes Rendus De L Académie Des Sciences, Paris, ser. 2,* **314,** 319-323.

Geraads, D., Amani, F., Raynal, J. P. and Sbihi-Alaoui, F.-Z. (1998) The Late Pliocene mammalian fauna of Ahl al Oughlam, Casablanca, Morocco. *Comptes Rendus De L Academie Des Sciences Serie II A,* **326,** 671-676.

Geraads, D., Berrio, P. and Roche, H. (1980) La faune et l'industrie des sites à *Homo erectus* des carrières Thomas (Maroc): Précisions sur l'âge de ces hominidés. *Comptes Rendus De L Académie Des Sciences Ser. D,* **291,** 195-198.

Geraads, D., Hublin, J. J., Jaeger, J. J., Tong, H., Sen, S. and Toubeau, P. (1986) The Pleistocene hominid site of Ternifine, Algeria - new results on the environment, age, and human industries. *Quaternary Research,* **25,** 380-386.

Gittleman, J. L. (1985) Carnivore body size: ecological and taxonomic correlates. *Oecologia,* **67,** 540-554.

Gittleman, J. L. (1989) *Carnivore behaviour, ecology and evolution.* Cornell University Press, Ithaca, New York.

Gittleman, J. L. and Harvey, P. H. (1982) Carnivore home range size, metabolic needs and ecology. *Behavioural Ecology and Sociobiology,* **10,** 57-63.

Gowlett, J. A. J. (1978) Kilombe - an Acheulean site complex in Kenya. In *Geological background to early man* (Ed, Bishop, W. W.) Scottish Academic Press, Edinburgh, pp. 337-360.

Gowlett, J. A. J. (1991) Kilombe - review of an African site complex. In *Approaches to understanding early hominid lifeways in the African savannah.* (Ed, Clark, J. D.) Römisch-Germanisches Zentralmuseum, Mainz.

Gowlett, J. A. J. and Crompton, R. H. (1994) Kariandusi: Acheulean morphology and the question of allometry. *African Archaeological Review,* **12,** 3-42.

Gowlett, J. A. J., Harris, J. W. K., Walton, D. A. and Wood, B. A. (1981) Early archaeological sites, further hominid remains and traces of fire from Chesowanja, Kenya. *Nature,* **294,** 125-129.

Grant, J. W. A., Chapman, C. A. and Richardson, K. S. (1992) Defended versus undefended home range size of carnivores, ungulates and primates. *Behavioral Ecology and Sociobiology,* **31,** 149-161.

GRASS 4.2 (1997) GRASS Research Group Baylor University

Grine, F. E. (1986) Dental evidence for dietary differences in *Australopithecus* and *Paranthropus*: A quantitative analysis of permanent molar microwear. *Journal of Human Evolution,* **15,** 783-822.

Grine, F. E. (1989) New hominid fossils from the Swartkrans Formation (1979-1986 excavations): Craniodental specimens. *American Journal of Physical Anthropology,* **79,** 409-449.

Grine, F. E. (1993) Description and preliminary analysis of new hominid craniodental fossils from the Swartkrans Formation. *Transvaal Museum Monograph,* **8.**

Grine, F. E., Demes, B., Jungers, W. L. and Cole, T. M. (1993) Taxonomic affinity of the early *Homo* cranium from Swartkrans, South Africa. *American Journal of Physical Anthropology,* **92,** 411-426.

Groves, R. H. (1989) *A theory of primate and human evolution.* Clarendon Press, Oxford.

Hacker, J. E., Cowlishaw, G. and Williams, P. H. (1998) Patterns of African primate diversity and their evaluation for the selection of conservation areas. *Biological Conservation,* **84,** 251-262.

Haffer, J. (1982) General aspects of the refuge theory. In *Biological diversification in the tropics* (Ed, Prance, G. T.) Columbia University Press, New York, pp. 6-24.

Haile-Selassie, Y. (2001) Late Miocene hominids from the Middle Awash, Ethiopia. *Nature,* **412,** 178-181.

Hamilton, A. C. (1976) The significance of patterns of distribution shown by forest plants and animals in tropical Africa for the reconstruction of upper Pleistocene paleoenvironments, a review. *Palaeoecology of Africa,* **9,** 63-97.

Hamilton, A. C. (1989) African forests. In *Rain forest ecosystems: Biogeographical and ecological studies,* Vol. 2 (Eds, Leith, H. and Werger, M. J. A.) Elsevier, Amsterdam, pp. 155-182.

Harcourt, A. H. (2000) Latitude and latitudinal extent: a global analysis of the Rapoport effect in a tropical mammalian taxon: primates. *Journal of Biogeography,* **27,** 1169-1182.

Harestad, A. S. and Bunnell, F. L. (1979) Home range and body weight - a reevaluation. *Ecology,* **60,** 389-402.

Harris, J. W. K. (1983) Cultural beginnings: Plio-Pleistocene archaeological occurrences from the Afar, Ethiopia. *African Archaeological Review,* **1,** 3-31.

Harris, J. W. K. and Capaldo, S. D. (1993) The earliest stone tools: their implications for an understanding of the activities and behaviour of late Pleistocene hominids. In *The use of tools by human and non-human primates.* (Eds, Berthelet, A. and Chavaillon, J.) Clarendon Press, Oxford, pp. 197-220.

Harris, J. W. K. and Harris, K. (1981) A note on the archaeology at Laetoli. *Nyame Akuma,* **18,** 18-21.

Harrison, G. A., Tanner, J. M., Pilbeam, D. R. and Baker, P. T. (1988) *Human biology: an introduction to human evolution, variation, growth and adaptability.* Oxford University Press, Oxford.

Harvey, P. H., Martin, R. D. and Clutton-Brock, T. H. (1987) Life histories in comparative perspective. In *Primate societies* (Eds, Smuts, B. B., Cheney, D. L., Seyfarth, R. M., Wrangham, R. W. and Struhsaker, T. T.) University of Chicago Press, Chicago, pp. 181-196.

Harvey, P. H. and Pagel, M. D. (1991) *The comparative method in evolutionary biology.* Oxford University Press, Oxford.

Harvey, P. H. and Rambaut, A. (2000) Comparative analyses for adaptive radiations. *Philosophical Transactions of the Royal Society of London Series B-Biological Sciences,* **355,** 1599-1605.

Hay, R. L. (1976) *Geology of the Olduvai Gorge.* University of Chicago Press, Berkeley.

Hendey, Q. B. (1982) *Langebaanweg: A record of past life.* The South African Museum, Cape Town.

Hengeveld, R. (1990) *Dynamic Biogeography.* Cambridge University Press, Cambridge.

Hill, A. (1985) Early hominid from the Baringo district, Kenya. *Nature,* **315,** 222-224.

Hill, A., Ward, C. V. and Brown, B. (1992) Anatomy and age of the Lothagam mandible. *Journal of Human Evolution,* **22,** 439-451.

Holmes, K. (2002) *Geographical Information System (GIS) simulation of the earliest hominid colonisation of Eurasia.* Unpublished PhD thesis. Archaeology, University of Sheffield

Howell, F. C. (1970) European and northwest African Middle Pleistocene hominids. *Current Anthropology,* **1,** 195-232.

Howell, F. C. (1978) *Hominidae* In *Evolution of African Mammals* (Eds, Maglio, V. J. and Cooke, H. B. S.) Harvard University Press, Cambridge, pp. 154-248.

Howell, F. C., Heasaerts, P. and de Heinzelin, P. (1987) Depositional environment, archaeological occurrences and hominids from members E and F of Shungura Formation (Omo Basin, Ethiopia). *Journal of Human Evolution,* **16,** 643-664.

Howell, N. (1979) *Demography of the Dobe !Kung.* Academic Press, New York.

Huang, W., Ciochon, R. L., Yumin, G., Larick, R., Qiren, F., Schwartcz, H., Yonge, C., de Vos, J. and Rink, W. (1995) Early *Homo* and associated artefacts from Asia. *Nature,* **378,** 275-278.

Hublin, J.-J. (1986) Human fossils from the north African middle Pleistocene and the origins of *Homo sapiens.* In *Ancestors: the hard evidence.* (Ed, Delson, E.) Alan R. Liss, New York, pp. 282-288.

Huffman, O. F. (2001) Geologic context and age of the Perning/Mojokerto *Homo erectus,* East Java. *Journal of Human Evolution,* **40,** 353-362.

Huffman, F., Shipman, P., Hertler, C., de Vos, J. and Aziz, F. (2005) Historical evidence of the 1936 Mojokerto skull discovery. *Journal of Human Evolution*, **48**, 321-363.

Hughes, A. R. and Tobias, P. V. (1977) A fossil skull, probably of the genus *Homo* from Sterkfontein, Transvaal. *Nature*, **265**, 310.

Humphrey, N. K. (1976) The social function of intellect. In *Growing points in ethology* (Eds, Bateson, P. P. G. and Hinde, R. A.) Cambridge University Press, Cambridge.

IEA (1998) *AMD African Mammals Databank - A databank for the conservation and management of African mammals. Vol 1 and 2.* Directorate-General for Development, European Commission.

Isaac, G. L. (1978) The food sharing behavior of proto-human hominids. *Scientific American*, **238**, 90-108.

Isaac, G. L. (1981) Archaeological tests of alternative models of early human behavior: Excavation and experiments. *Philosophical Transactions of the Royal Society of London, ser.B*, **292**, 177-188.

Isaac, G. L. and Curtis, G. H. (1974) The age of early Acheulean industries in East Africa - new evidence from the Peninj Group. *Nature*, **249**, 624-627.

Isaac, G. L. and Harris, J. W. K. (1978) Archaeology. In *Koobi Fora Research Project* (Eds, Leakey, M. D. and Leakey, R. E. F.) Clarendon Press, Oxford.

Jablonski, N. G., Whitfort, M. J., Roberts-Smith, N. and Xu, Q. Q. (2000) The influence of life history and diet on the distribution of catarrhine primates during the Pleistocene in eastern Asia. *Journal of Human Evolution*, **39**, 131-157.

Janson, C. H. and Chapman, C. A. (1999) Resources and primate community structure. In *Primate communities* (Eds, Fleagle, J. G., Janson, C. H. and Reed, K. E.) Cambridge University Press, Cambridge, pp. 237-267.

Johanson, D. C., Taieb, M. and Coppens, Y. (1982) Pliocene hominids from the Hadar Formation, Ethiopia (1973-1977): Stratigraphic, chronologic and paleoenvironmental contexts, with notes on hominid morphology and systematics. *American Journal of Physical Anthropology*, **57**, 373-402.

Johanson, D. C., White, T. D. and Coppens, Y. (1978) A new species of the genus *Australopithecus* (Primates: Hominidae) from the Pliocene of eastern Africa. *Kirtlandia*, **28**, 2-14.

Jolly, A. (1966) Lemur social behaviour and primate intelligence. *Science*, **153**, 501-506.

Jones, K. E. and Purvis, A. (1997) An optimum body size for mammals? Comparative evidence from bats. *Functional ecology*, **11**, 751-756.

Jouzel, J., Barkov, N. I., Barnola, J. M., Bender, M., Chappellaz, J., Genthon, C., Kotlyakov, V. M., Lipenkov, V., Lorius, C., Petit, J. R., Raynaud, D., Raisbeck, G., Ritz, C., Sowers, T., Stievenard, M., Yiou, F. and Yiou, P. (1993) Extending the Vostok ice-core record for paleoclimate to the penultimate glacial period. *Nature*, **365**, 203-207.

Kay, R. F. and Grine, F. E. (1988) Early hominid diets from quantitative image analysis of dental microwear. *Nature*, **333**, 765-768.

Keeley, L. H. (1977) The function of Palaeolithic stone tools. *Nature*, **237**, 108-126.

Key, C. (2000) The evolution of human life history. *World Archaeology*, **31**, 329-350.

Keyser, A. W., Menter, C. G., Moggi-Cecchi, J., Pickering, T. R. and Berger, L. R. (2000) Drimolen: a new hominid-bearing site in Gauteng, South Africa. *South African Journal of Science*, **96**, 193-197.

Kibunjia, M. (1994) Pliocene archaeological occurrences in the Lake Turkana basin. *Journal of Human Evolution*, **27**, 159-171.

Kimbel, W. H., Johanson, D. C. and Rak, Y. (1994) The first skull and other new discoveries of *Australopithecus afarensis* at Hadar, Ethiopia. *Nature*, **368**, 449-451.

Kimbel, W. H., Walter, R. C., Johanson, D. C., Reed, K. E., Aronson, J. L., Assefa, Z., Marean, C. W., Eck, G. G., Bobe, R., Hovers, E., Rak, Y., Vondra, C., Yemane, T., Chen, Y., Evensen, N. M. and Smith, P. E. (1996) Late Pliocene *Homo* and Oldowan tools from the Hadar Formation (Kada Hadar Member), Ethiopia. *Journal of Human Evolution*, **31**, 549-561.

Kingdon, J. (1997) *The Kingdon field guide to African mammals.* Academic Press, London.

Kingston, J. D., Marino, B. D. and Hill, A. (1994) Isotopic evidence for Neogene hominid paleoenvironments in the Kenya Rift-Valley. *Science*, **264**, 955-959.

Kirkpatrick, M. and Barton, N. H. (1997) Evolution of a species range. *The American Naturalist*, **150**, 1-23.

Klein, R. G. (1994) Southern Africa before the Iron Age. In *Integrative paths to the past* (Eds, Corrucini, R. S. and Ciochon, R. L.) Prentice Hall, Englewood Cliffs, N.J., pp. 471-519.

Klein, R. G. (1999) *The human career: human biological and cultural origins.* University of Chicago Press, Chicago.

Kullmer, O., Sandrock, O., Abel, R., Schrenk, F., Bromage, T. G. and Juwayeyi, Y. M. (1999) The first *Paranthropus* from the Malawi Rift. *Journal of Human Evolution*, **37**, 121-127.

Kuman, K. (1998) The earliest South African industries. In *Early human behaviour in global context. The rise and diversity of the Lower Palaeolithic Record* (Eds, Petraglia, M. D. and Korisettar, R.) Routledge, London.

Kuman, K. and Clarke, R. J. (2000) Stratigraphy, artefact industries and hominid associations for Sterkfontein, Member 5. *Journal of Human Evolution*, **38**, 827-847.

Laland, K. N., Odling-Smee, J. and Feldman, M. W. (2000) Niche construction, biological evolution, and cultural change. *Behavioral and Brain Sciences,* **23,** 131-175.

Laland, K. N., Richerson, P. J. and Boyd, R. (1996) Developing a theory of animal social learning. In *Social learning in animals: The roots of culture* (Eds, Heyes, C. and Galef, B. G.) Jr. Academic Press.

Larick, R., Ciochon, R.L., Zaim, Y., Suminto, S., Rizal, Y., Aziz, F., Reagan, M., Heizler, M. (2001) Early Pleistocene $^{40}Ar/^{39}Ar$ ages for Bapang Formation hominins, Central Java, Indonesia. *Proceedings of the National Academy of Science USA*, **98**, 4866-4871

Leakey, M. and Hay, R. L. (1982) The chronological position of the fossil hominids of Tanzania. In *L'Homo erectus et la place de l'homme de Tautavel parmi les hominidés fossiles* (Ed, de Lumley, M. A.) Premier Congrès International de Paléontologie Humaine, Nice, pp. 753-765.

Leakey, M. D. (1971) *Olduvai Gorge. Vol. 3 : Excavations in Beds 1 and 2, 1960-1963.* Cambridge University Press, London.

Leakey, M. D. (1975) Cultural patterns in the Olduvai sequence. In *After the Australopithecines.* (Eds, Butzer, K. W. and Isaac, G. L.) Mouton, The Hague, pp. 477-494.

Leakey, M. D. (1977) The archaeology of the early hominids. In *A survey of the prehistory of eastern Africa* (Ed, Wilson, T. H.) Eighth Panafrican Congress of Prehistory and Quaternary Studies, Nairobi, pp. 61-79.

Leakey, M. D. and Roe, D. A. (1994) *Olduvai Gorge: Excavations in Beds III, IV and the Masek Beds, 1968-1971.* Cambridge University Press, Cambridge.

Leakey, M. G., Feibel, C. S., McDougall, I. and Walker, A. (1995) New four-million-year-old hominid species from Kanapoi and Allia Bay, Kenya. *Nature,* **376,** 565-571.

Leakey, M. G., Spoor, F., Brown, F. H., Gathogo, P. N., Kiarie, C., Leakey, L. N. and McDougall, I. (2001) New hominin genus from eastern Africa shows diverse middle Pliocene lineages. *Nature* **410**, 433-440

Leakey, R. E. F. and Walker, A. (1988) New *Australopithecus boisei* specimens from East and West Lake Turkana, Kenya. *American Journal of Physical Anthropology,* **76,** 1-24.

Lee, P. C. (1991) Adaptations to environmental change: an evolutionary perspective. In *Primate responses to environmental change* (Ed, Box, H. O.) Chapman & Hall, London, pp. 39-56.

Lee-Thorp, J., Thackeray, J. F. and van der Merwe, N. (2000) The hunters and the hunted revisited. *Journal of Human Evolution,* **39,** 565-576.

Lee-Thorp, J. A., Vandermerwe, N. J. and Brain, C. K. (1994) Diet of *Australopithecus-Robustus* at Swartkrans from Stable Carbon Isotopic analysis. *Journal of Human Evolution,* **27,** 361-372.

Lefebvre, L., Gaxiola, A., Dawson, S., Timmermans, S., Rosza, L. and Kabai, P. (1998) Feeding innovations and forebrain size in Australasian birds. *Behaviour,* **135,** 1077-1097.

Leonard, W. R. and Robertson, M. L. (1996) On diet, energy metabolism, and brain size in human evolution. *Current Anthropology,* **37,** 125-128.

Leonard, W. R. and Robertson, M. L. (2000) Ecological correlates of home range variation in primates: implications for hominid evolution. In *On the move: how and why animals travel in groups* (Eds, Boinski, S. and Garber, P.) University of Chicago Press, Chicago, pp. 628-648.

Letcher, A. J. and Harvey, P. H. (1994) Variation in geographical range size among mammals of the Palearctic. *American Naturalist,* **144,** 30-42.

Lieberman, D. E. (2001) Another face in our family tree. *Nature* **410**, 419-420

Lieberman, D. E., Pilbeam, D. and Wood, B. A. (1988) A probabilistic approach to the problem of sexual dimorphism in *Homo habilis*: A comparison of KNM-ER 1470 and KNM-ER 1813. *Journal of Human Evolution,* **17,** 503-511.

Lordkipanidze, D., Vekua, A., Ferring, R., Rightmire, G. P., Agusti, J., Kiladze, G., Mouskhelishvili, A., Nioradze, M., de Leon, M. S. P., Tappen, M. and Zollikofer, C. P. E. (2005) Anthropology: The earliest toothless hominin skull. *Nature* **434**(7034), 717-718.

Lott, D. F. (1991) *Intraspecific variation in the social systems of wild vertebrates.* Cambridge University Press, Cambridge.

MacArthur, R. H. and Wilson, E. O. (1967) *Island biogeography.* Princeton University Press, Princeton, New Jersey.

Martinez Navarro, B., Turq, A., Agusti Ballester, J. and Oms, O. (1997) Fuente Nueva-3 (Orce, Granada, Spain) and the first human occupation of Europe. *Journal of Human Evolution,* **33,** 611-620.

McHenry, H. M. (1994) Behavioral ecological implications of early hominid body-size. *Journal of Human Evolution,* **27,** 77-87.

McKee, J. K., Tobias, P. V. and Clarke, R. J. (1996) Faunal evidence and Sterkfontein Member 2 foot bones of Early Hominid. *Science,* **271,** 1301-1302.

Milton, K. and May, M. L. (1976) Body weight, diet and home range area in primates. *Nature,* **259,** 459-462.

Mithen, S. and Reed, M. (2002) Stepping out: a computer simulation of hominid dispersal from Africa. *Journal of Human Evolution,* **43,** 433-462.

Moore, J. L., Manne, P., Brooks, T., Burgess, N. D., Davies, R., Rahbek, C., Williams, P. and Balmford, A. (2002) The distribution of cultural and biological diversity in Africa. *Proceedings of the Royal Society of London Series B-Biological Sciences,* **269,** 1645-1653.

New, M. G., Hulme, M. and Jones, P. D. (1999) Representing 20th century space-time climate variability. I: Development of 1961-1990 mean monthly terrestrial climatology. *J. Climate,* **12,** 829-856.

New, M. G., Hulme, M. and Jones, P. D. (2000) Representing 20th century space-time climate variability. II: Development of 1901-1996 monthly terrestrial climate fields. *J. Climate,* **13,** 2217-2238.

Nicolakakis, N., Sol, D. and Lefebvre, L. (2003) Behavioural flexibility predicts species richness in birds, but not extinction risk. *Animal behaviour,* **65,** 445-452.

Nishida, T. (1987) Local traditions and cultural transmission. In *Primate societies* (Eds, Smuts, B. B., Cheney, D. L., Seyfarth, R. M., Wrangham, R. W. and Struhsaker, T. T.) University of Chicago Press, Chicago, pp. 462-474.

Nunn, C. L. and Barton, N. H. (2001) Comparative methods for studying primate adaptation and allometry. *Evolutionary Anthropology,* **10,** 81-98.

Nunn, C. L. and Barton, R. A. (2000) Allometric slopes and independent contrasts: A comparative test of Kleiber's law in primate ranging patterns. *American Naturalist,* **156,** 519-533.

Oates, J. F., Whitesides, G. H., Davies, A. G., Waterman, P. G., Green, S. M., Dasilva, G. L. and Mole, S. (1990) Determinants of variation in tropical forest primate biomass: new evidence from West Africa. *Ecology,* **71,** 328-343.

O'Connell, J. F., Hawkes, K. and Jones, N. G. B. (1999) Grandmothering and the evolution of *Homo erectus. Journal of Human Evolution,* **36,** 461-485.

Odling-Smee, F. J. (1988) Niche-constructing phenotypes. In *The role of behaviour in evolution* (Ed, Plotkin, H. C.) The MIT Press, London.

Olff, H., Ritchie, M. E. and Prins, H. H. T. (2002) Global environmental controls of diversity in large herbivores. *Nature,* **415,** 901-904.

Oliver, J. S., Sikes, N. E. and Stewart, K. M. (1994) Introduction to 'Early Hominid Behavioural Ecology': new looks at old questions. *Journal of Human Evolution,* **27,** 1-6.

Olson, D. M. and Dinerstein, E. (1998) *The Global 200: a representation approach to conserving the earths distinctive ecoregions.* World Wildlife Fund

Olson, J. S., Watts, J. A. and Allison, L. J. (1983) *Carbon in live vegetation of major world ecosystems.* Oak Ridge National Laboratory.

Oms, O., Parés, J., Martínez-Navarro, B., Agustí, J., Toro, I., Martínez-Fernández, G. and Turq, A. (2000) Early human occupation of western Europe: paleomagnetic dates for two paleolithic sites in Spain. *Proceedings of the National Academy of Sciences of the United States of America* **97**, 10666-10670.

Pagel, M. D., May, R. M. and Collie, A. R. (1991) Ecological aspects of the geographical-distribution and diversity of mammalian-species *American Naturalist,* **137,** 791-815.

Parfitt, S. A., R. W. Barendregt, M. Breda, I. Candy, M. J. Collins, G. R. Coope, P. Durbidge, M. H. Field, J. R. Lee, A. M. Lister, R. Mutch, K. E. H. Penkman, R. C. Preece, J. Rose, C. B. Stringer, R. Symmons, J. E. Whittaker, J. J. Wymer and A. J. Stuart (2005). "The earliest record of human activity in northern Europe." *Nature,* **438**: 1008-1012.

Patterson, B. and Howells, W. W. (1967) Hominid humeral fragment from early Pleistocene of north-western Kenya. *Science,* **156,** 64-66.

Pelegrin, J. (1991) Les savoir-faire: une très longue histoire. *Terrain,* **16,** 106-113.

Pelegrin, J. (1993) A framework for analysing prehistoric stone tool manufacture and a tentative application to some early stone tool industries. In *The use of tools by human and non-human primates* (Eds, Berthelet, A. and Chavaillon, J.) Clarendon Press, Oxford.

Peres, C. A. (1999a) Effects of subsistence hunting and forest types on the structure of Amazonian primate communities. In *Primate Communities* (Eds, Fleagle, J. G., Janson, C. and Reed, K. E.) Cambridge University Press, Cambridge, pp. 268-283.

Peres, C. A. (1999b) Primate community structure at twenty western Amazonian flooded and unflooded forests. *Journal of Tropical Ecology,* **13,** 381-405.

Perry, S. and Rose, L. (1994) Begging and transfer of Coati meat by white-faced capuchin monkeys, *Cebus-Capucinus. Primates,* **35,** 409-415.

Pianka, E. R. (1970) On *r*- and K-selection. *American Naturalist,* **104,** 592-597.

Pickering, T. R. (1999) *Taphonomic interpretations of the Sterkfontein Early Hominid sites (Guateng, South Africa) reconsidered in light of recent evidence.* Unpublished PhD thesis. University of Wisconsin

Pinker, S. (1995) *The language instinct.* Penguin, London.

Potts, R. (1989) Olorgesailie: New excavations and findings in Early and Middle Pleistocene contexts, southern Kenya Rift Valley. *Journal of Human Evolution,* **18,** 477-484.

Potts, R. (1994) Variables versus models of early Pleistocene hominid land use. *Journal of Human Evolution,* **27,** 7-24.

Potts, R. (1998a) Variability selection in hominid evolution. *Evolutionary Anthropology,* **7,** 81-96.

Potts, R. (1998b) Environmental hypotheses of hominin evolution. *Yearbook of Physical Anthropology,* **41,** 93-136.

Potts, R. and Shipman, P. (1981) Cutmarks made by stone tools from Olduvai Gorge. *Nature,* **291,** 577-580.

Prat, S., Brugal, J. P., Tiercelin, J. J., Barrat, J. A., Bohn, M., Delangnes, A., Harmand, S., Kimeu, K., Kibunjia, M., Texier, P. J. and Roche, H. (2005) First occurence of early Homo in the Nachukui Formation (West Turkana, Kenya) at 2.3-2.4 Myr. *Journal of Human Evolution* **49**(2), 230-240.

PRISM_Project_Members (1995) Middle Pliocene paleoenvironments of the northern hemisphere. In *Paleoclimate and evolution, with emphasis on human origins* (Eds, Vrba, E. S., Denton, D. H., Partridge, T. C. and Burckle, L. H.) Yale University Press, New Haven.

Purvis, A. (1995) A composite estimate of primate phylogeny. *Philosophical Transactions of the Royal Society of London Series B-Biological Sciences,* **348,** 405-421.

Purvis, A., Gittleman, J. L., Cowlishaw, G. and Mace, G. M. (2000) Predicting extinction risk in declining species. *Proceedings of the Royal Society of London Series B-Biological Sciences,* **267,** 1947-1952.

Purvis, A., Gittleman, J. L. and Luh, H.-K. (1994) Truth or consequences: effects of phylogenetic accuracy on two comparative methods. *Journal of Theoretical Biology,* **167,** 293-200.

Purvis, A. and Rambaut, A. (1995) Comparative-analysis by independent contrasts (CAIC) - an Apple-Macintosh application for analysing comparative data. *Computer Applications in the Biosciences,* **11,** 247-251.

Purvis, A. and Webster, A. J. (1999) Phylogenetically independent comparisons and primate phylogeny. In *Comparative primate socioecology* (Ed, Lee, P. C.) Cambridge University Press, Cambridge.

Pusey, A. E. (1991) The primate perspective on dispersal. In *Animal dispersal: small animals as a model* (Eds, W.Z., L. and Stenseth, N. C.) Chapman Hall, New York, pp. 243-259.

Rapoport, E. H. (1982) Areography: Geographical strategies of species, Pergamon Press.

Raynal, J.-P., Sbihi-Alaoui, F.-Z., Geraads, D., Magoga, L. and Mohi, A. (2001) The earliest occupation of North-Africa: the Moroccan perspective. *Quaternary International,* **75,** 65-75.

Reader, S. M. (2000) *Social learning and innovation: individual differences, diffusion dynamics and evolutionary issues.* Unpublished PhD thesis. Department of Zoology, University of Cambridge

Reader, S. M. and Laland, K. N. (1999) Forebrain size, opportunism and the evolution of social learning in non-human primates. *Ethology,* **S34,** 50.

Reader, S. M. and Laland, K. N. (2002) Social intelligence, innovation and enhanced brain size in primates. *Proceedings of the National Academy of Sciences of the United States of America,* **99,** 4436-4441.

Reader, S. M. and MacDonald, K. (2003) Environmental variability and primate behavioural flexibility. In *Animal Innovation* (Eds, Laland, K. N. and Reader, S. R.) Oxford University Press, Oxford.

Reed, K. E. (1997) Early hominid evolution and ecological change through the African Plio-Pleistocene. *Journal of Human Evolution,* **32,** 289-322.

Rhine, R. J., Norton, G. W., Wynn, G. M., Wynn, R. D. and Rhine, H. B. (1986) Insect and meat eating among infant and adult baboons (*Papio-Cynocephalus*) of Mikumi-National-Park, Tanzania. *American Journal of Physical Anthropology,* **70,** 105-118.

Rhode, K., Heap, M. and Heap, D. (1993) Rapoport's rule does not apply to marine teleosts and cannot explain latitudinal gradients in species diversity. *American Naturalist,* **142,** 1-16.

Richard, A. F. (1985) *Primates in nature.* W.H. Freeman, New York.

Richard, A. F., Goldstein, S. J. and Dewar, R. E. (1989) Weed Macaques - the evolutionary implications of Macaque feeding ecology. *International Journal of Primatology,* **10,** 569-597.

Rightmire, G. P. (1990) *The evolution of Homo erectus: Comparative anatomical studies of an extinct human species.* Cambridge University Press, Cambridge.

Rightmire, G. P. (1993) Variation among Early *Homo* Crania from Olduvai-Gorge and the Koobi Fora Region. *American Journal of Physical Anthropology,* **90,** 1-33.

Roche, H. (1995) Les industries de la limite Plio-Pleistocene et du Pleistocene ancien en Afrique. In *Congreso Internacional de Paleontologia Humana (Orce, September 1995), 3a Circular* Orce, Spain.

Roche, H., Delagnes, A., Brugal, J.-P., Feibel, C., Kibunjia, M., Mourre, V. and Texier, P. J. (1999) Early hominid stone tool production and technical skill 2.34 Myr ago in West Turkana, Kenya. *Nature,* **399,** 57-60.

Roebroeks, W. (2001) Hominid behaviour and the earliest occupation of Europe: an exploration. *Journal of Human Evolution,* **41,** 437-461.

Roebroeks, W. and van Kolfschoten, T. (1994) The earliest occupation of Europe: A short chronology. *Antiquity,* **68,** 489-503.

Rogers, A. (1988) Does biology constrain culture? *American Anthropologist,* **90,** 819-831.

Rogers, M. J., Harris, J. W. K. and Feibel, C. S. (1994) Changing patterns of land-use by Plio-pleistocene hominids in the Lake Turkana Basin. *Journal of Human Evolution,* **27,** 139-158.

Rolland, N. (1998) The Lower Palaeolithic settlement of Eurasia, with special reference to Europe. In *Early human behaviour in global context. The rise and diversity of the Lower Palaeolithic record* (Eds, Petraglia, M. D. and Korisettar, R.) Routledge, London, pp. 187-220.

Rose, L. M. (1997) Vertebrate predation and food-sharing in *Cebus* and *Pan*. *International Journal of Primatology*, **18**, 727-765.

Rowe, N. (1996) *The pictorial guide to the living primates*. Pogonias Press, Charlestown, Rhode Island.

Ruggiero, A. (1994) Latitudinal correlates of the sizes of mammalian geographical ranges in South-America. *Journal of Biogeography*, **21**, 545-559.

Ruggiero, A., Lawton, J. H. and Blackburn, T. M. (1998) The geographic ranges of mammalian species in South America: spatial patterns in environmental resistance and anisotropy. *Journal of Biogeography*, **25**, 1093-1103.

Sahnouni, M. and de Heinzelin, J. (1998) The site of Ain Hanech Revisited: New investigations at this Lower Pleistocene site in Northern Algeria. *Journal of Archaeological Science*, **25**, 1083-1101.

Saragusti, I. and Goren-Inbar, N. (2001) The biface assemblage from Gesher Benot Ya'aqov, Israel: illuminating patterns in 'Out of Africa' dispersal. *Quaternary International*, **75**, 85-89.

Schaik, C. P., Ancrenaz, M., Borgen, G., Galdikas, B., Knott, C. D., Singleton, I., Suzuki, A., Utami, S. S. and Merrill, M. Y. (2003) Orangutan cultures and the evolution of material culture. *Science*, **299**, 102-105.

Schaik, C. P., Deaner, R. O. and Merrill, M. Y. (1998) The conditions for tool use in primates: implications for the evolution of material culture. *Journal of Human Evolution*, **36**, 719-741.

Schick, K. D. (1987) Modelling the formation of Early Stone Age artefact concentrations. *Journal of Human Evolution*, **16**, 789-807.

Schick, K. D. and Dong, Z. (1993) Early Palaeolithic of China and eastern Asia. *Evolutionary Anthropology*, **2**, 22-35.

Schick, K. D. and Toth, N. (1993) *Making silent stones speak. Human evolution and the dawn of technology*. Phoenix, London.

Schoeninger, M. J., Moore, J. and Sept, J. M. (1999) Subsistence strategies of two "savanna" chimpanzee populations: the stable isotope evidence. *American journal of Primatology*, **49**, 297-314.

Schwartz, J. H. and Tattersall, I. (1996) Whose teeth? *Nature*, **381**, 201-202.

Semaw, S. (2000) The world's oldest stone artefacts from Gona, Ethiopia: Their implications for understanding stone technology and patterns of human evolution between 2.6-1.5 million years ago. *Journal of Archaeological Science*, **27**, 1197-1214.

Semaw, S., Renne, P., Harris, J. W. K., Feibel, C. S., Bernor, R. L., Fesseha, N. and Mowbray, K. (1997) 2.5-million-year-old stone tools from Gona, Ethiopia. *Nature*, **385**, 333-336.

Semaw, S., Simpson, S. W., Quade, J., Renne, J. P., Butler, R. F., McIntosh W. C., Levin, N., Dominguez-Rodrigo, M. and Rogers, M. J. (2005) Early Pliocene hominids from Gona, Ethiopia. *Nature* **433** (7023), 301-305

Senut, B., Pickford, M., Gommery, D., Mein, P., Cheboi, K. and Coppens, Y. (2001) First hominid from the Miocene (Lukeino Formation, Kenya). *Comptes Rendus De L Academie Des Sciences Serie IIA*, **332**, 137-144.

Senut, B., Pickford, M., Ssemmanda, I., Elepu, D. and Obwana, P. (1987) First discovery of fossil hominінae (*Homo* sp.) from the Pleistocene of Nyabusosi (Western Uganda). *Comptes Rendus De L Academie Des Sciences Ser. 2*, **305**, 819-822.

Shea, J. J. (2003). Neandertals, competition, and the origin of modern human behavior in the Levant. *Evolutionary Anthropology*, **12**, 173-187.

Shennan, S. (1997) *Quantifying archaeology*. Edinburgh University Press, Edinburgh.

Shennan, S. J. and Steele, J. (1999) Cultural learning in hominids: a behavioural ecological approach. In *Mammalian Social Learning: Comparative and ecological perspectives* (Ed, Box, H. O.) Cambridge University Press, Cambridge, pp. 367-388.

Shipman, P. and Harris, J. M. (1988) Habitat preference and paleoecology of *Australopithecus boisei* in East Africa. In *Evolutionary history of the Robust Australopithecines* (Ed, Grine, F. E.) Aldine de Gruyter, New York, pp. 343-382.

Shipman, P. and Walker, A. (1989) The costs of becoming a predator. *Journal of Human Evolution*, **18**, 373-92.

Sibly, R. M. (1999) Evolutionary biology of skill and information transfer. In *Mammalian social learning: comparative and ecological perspectives* (Eds, Box, H. O. and Gibson, K. R.) Cambridge University Press, Cambridge.

Sikes, N. E. (1994) Early hominid habitat preferences in East-Africa - paleosol carbon isotopic evidence. *Journal of Human Evolution*, **27**, 25-45.

Sikes, N. E. (1999) Plio-Pleistocene floral context and habitat preference of sympatric hominid species in East Africa. In *African biogeography, climate change and human evolution* (Eds, Bromage, T. G. and Schrenk, F.) Oxford University Press, Oxford, pp. 301-315.

Sikes, N. E., Potts, R. and Behrensmeyer, A. K. (1999) Early Pleistocene habitat in Member 1 Olorgesailie based on paleosol stable isotopes. *Journal of Human Evolution*, **37**, 721-746.

Sillen, A. (1992) Strontium-calcium ratios (Sr/Ca) of *Australopithecus robustus* and associated fauna from Swartkrans. *Journal of Human Evolution*, **23**, 495-516.

Sillen, A., Hall, G. and Armstrong, R. (1995) Strontium calcium ratios (Sr/Ca) and strontium isotopic ratios (87Sr/86Sr) of *Australopithecus robustus* and *Homo* sp. from Swartkrans. *Journal of Human Evolution*, **28**, 277-285.

Silva, M. and Dowling, J. A. (1995) *CRC Handbook of mammalian body masses.* CRC Press, Boca Raton, Florida.

Singhvi, A. A., Wagner, G. A. and Korisettar, R. (1998) Techniques for the chronometry of the Palaeolithic: evidence for global colonization. In *Early human behaviour in global context. The rise and diversity of the Lower Palaeolithic Record* (Eds, Petraglia, M. D. and Korisettar, R.) Routledge, London, pp. 23-83.

Smith, B. H. and Tompkins, R. L. (1995) Toward a life history of the hominidae. *Annual Review of Anthropology,* **24,** 257-279.

Smith, R. J. and Jungers, W. L. (1997) Body mass in comparative primatology. *Journal of Human Evolution,* **32,** 523-559.

Sol, D. and Lefebvre, L. (2000) Behavioural flexibility predicts invasion success in birds introduced to New Zealand. *Oikos,* **90,** 599-605.

Sponheimer, M. and Lee-Thorp, J. A. (1999) Isotopic evidence for the diet of an early hominid, *Australopithecus Africanus. Science,* **283,** 368-370.

Stanford, C. B., Wallis, J., Matama, H. and Goodall, J. (1994) Patterns of predation by chimpanzees on red colobus monkeys in Gombe National Park, 1982-1991. *American Journal of Physical Anthropology,* **94,** 213-228.

Steele, J. (1996) On predicting hominid group sizes. In *The archaeology of human ancestry. Power, sex and tradition* (Eds, Steele, J. and Shennan, S. J.) Routledge, London.

Steele, J. and Shennan, S. (1996) Introduction. In *The evolution of human ancestry: Power, sex and tradition* (Eds, Steele, J. and Shennan, S.) Routledge, London, pp. 1-42.

Stephan, H., Frahm, H. D. and Baron, G. (1981) New and revised data on volumes of brain structures in insectivores and primates. *Folia Primatologica,* **35,** 1-29.

Stephens, D. W. (1991) Change, regularity, and value in the evolution of animal learning. *Behavioral Ecology,* **2,** 77-89.

Stevens, G. C. (1989) The latitudinal gradient in geographical range - how so many species coexist in the tropics. *American Naturalist,* **133,** 240-256.

Stevens, G. C. (1992) The elevational gradient in altitudinal range - an extension of Rapoport's latitudinal rule to altitude. *American Naturalist,* **140,** 893-911.

Stiner, M. C. (2002) Carnivory, coevolution and the geographic spread of the Genus *Homo. Journal of Archaeological Research,* **10,** 1-64.

Strait, D. S. and Wood, B. A. (1999) Early hominid biogeography. *Proceedings of the National Academy of Sciences of the United States of America,* **96,** 9196-9200.

Stringer, C. B. (1984) The definition of *Homo erectus* and the existence of the species in Africa and Europe. *Courier Forschungsintitut Senckenberg,* **69,** 131-144.

Stringer, C. B. (1986) The credibility of *Homo habilis.* In *Major topics in human evolution.* (Eds, Wood, B., Martin, L. and Andrews, P.) Cambridge University Press, Cambridge, pp. 266-294.

Susman, R. L. and Stern, J. T. (1982) Functional morphology of *Homo habilis. Science,* **297,** 931-934.

Suwa, G., Asfaw, B., Beyene, Y., White, T. D., Katoh, S., Nagaoka, S., Nakaya, H., Uzawa, K., Renne, P. and WoldeGabriel, G. (1997) The first skull of *Australopithecus boisei. Nature,* **389,** 489-492.

Suwa, G., White, T. and Howell, F. C. (1996) Mandibular postcanine dentition from the Shungura Formation, Ethiopia: Crown morphology, taxonomic allocations, and Plio-Pleistocene hominid evolution. *American Journal of Physical Anthropology,* **101,** 247-282.

Swisher, C. C., Curtis, G. H., Jacob, T., Getty, A. G., Suprijo, A. and Widiasmoro (1994) Age of the earliest known hominids in Java, Indonesia. *Science,* **263,** 1118-1121.

Szalay, F. S. (1999) Paleontology and macroevolution. On the theoretical conflict between an expanded synthesis and hierarchic punctuationism. In *African biogeography, climate change and evolution* (Eds, Bromage, T. G. and Schrenk, F.) Oxford University Press, Oxford, pp. 35-56.

Tattersall, I. (1986) Species recognition in human paleontology. *Journal of Human Evolution,* **15,** 165-175.

Teaford, M. F. and Ungar, P. S. (2000) Diet and the evolution of the earliest human ancestors. *Proceedings of the National Academy of Sciences of the United States of America,* **97,** 13506-13511.

Thackeray, J. F. (1995) Exploring ungulate diversity, biomass and climate in modern and past environments. In *Paleoclimate and evolution with emphasis on human origins* (Eds, Vrba, E. S., Denton, G. H., Partridge, T. C. and Burckle, L. H.) Yale University Press, New Haven and London, pp. 479-482.

Thackeray, J. F. and Reynolds, S. (1997) Variability in Plio-Pleistocene climates, habitats and ungulate biomass in southern Africa. *South African Journal of Science,* **93,** 171-2.

Toth, N. (1987) Behavioural inferences from early stone artefact assemblages. An experimental model. *Journal of Human Evolution,* **16,** 763-787.

Turner, A. (1984) Hominids and fellow travellers: human migration into high latitudes as part of a large mammal community. In *Hominid evolution and community ecology* (Ed, Foley, R. A.) Academic Press, London.

Turner, A. (1992) Large carnivores and earliest European hominids: changing determinants of resource availability during the Lower and Middle Pleistocene. *Journal of Human Evolution,* **22,** 109-126.

Turner, A., Bishop, L. C., Denys, C. and McKee, J. K. (1999) Appendix: A locality based listing of African Plio-Pleistocene mammals. In *African biogeography, climate change and human evolution* (Eds, Bromage, T. G. and Schrenk, F.) Oxford University Press, Oxford, pp. 369-399.

Turner, A. and Chamberlain, A. (1989) Speciation, morphological change and the status of African *Homo-Erectus. Journal of Human Evolution,* **18,** 115-130.

Turner, A. and Wood, B. (1993) Taxonomic and geographic diversity in Robust Australopithecines and other African Pliopleistocene larger mammals. *Journal of Human Evolution,* **24,** 147-168.

GRASS 4.1 (1993) U.S. Army Corps of Engineers

Uehara, S. (1997) Predation on mammals by the chimpanzee (*Pan troglodytes*). *Primates,* **38,** 193-214.

Vekua, A., Lordkipanidze, D., Rightmire, G. P., Agusti, J., Ferring, R., Maisuradze, G., Mouskhelishvili, A., Nioradze, M., Ponce de Leon, M., Tappen, M. and Tvalchrelidze, M. (2002) A new skull of early *Homo* from Dmanisi, Georgia. *Science,* **297,** 85-89.

Vrba, E. S. (1985a) Early hominids in southern Africa: Updated observations on chronological and ecological background. In *Hominid evolution, past, present and future* (Ed, Tobias, P. V.) Alan R. Liss, New York, pp. 195-200.

Vrba, E. S. (1985b) Ecological and adaptive changes associated with early hominid evolution. In *Ancestors-the hard evidence* (Ed, Delson, E.) Alan R Liss, New York, pp. 63-71.

Vrba, E. S. (1995) On the connections between paleoclimate and evolution. In *Paleoclimate and evolution with emphasis on human origins.* (Eds, Vrba, E. S., Denton, D. H., Partridge, T. C. and Burckle, L. H.) Yale University Press, New Haven, CT, pp. 24-45.

Vrba, E. S. (1999) Habitat theory in relation to the evolution in African neogene biota and hominids. In *African biogeography, climate change and evolution* (Eds, Bromage, T. G. and Schrenk, F.) Oxford University Press, Oxford, pp. 19-34.

Walker, A. (1981) The Koobi Fora Hominids and their bearing on the origins of the genus *Homo.* In *Homo erectus - Papers in honour of Davidson Black* (Eds, Sigmon, B. A. and Cybulski, J. S.) University of Toronto Press, Toronto, pp. 193-215.

Walker, A., Leakey, M. D., Harris, J. M. and Brown, F. H. (1986) 2.5-myr *Australopithecus boisei* from West of Lake Turkana, Kenya. *Nature,* **322,** 517-522.

Walker, A. and Leakey, R. E. F. (1978) The hominids of East Turkana. *Scientific American,* **239,** 54-66.

Walker, A. and Leakey, R. E. F. (1993) Perspectives on the Nariokotome discovery. In *The Nariokotome Homo erectus skeleton* (Eds, Walker, A. and Leakey, R. E. F.) Harvard University Press, Cambridge.

Ward, C. V., Walker, A. and Leakey, M. G. (1997) New fossils of *Australopithecus anamensis* from Kanapoi and Allia Bay, Kenya. *American Journal of Physical Anthropology, suppl.,* **24,** 235.

Watson, V. (1993) Composition of the Swartkrans bone accumulations, in terms of skeletal parts and animals represented. In *Swartkrans: A cave's chronicle of early man* (Ed, Brain, C. K.) Transvaal Museum Monographs, Transvaal.

White, T., Suwa, G. and Asfaw, B. (1994) *Australopithecus ramidus*, a new species of early hominind from Aramis, Ethiopia. *Nature,* **371,** 306-312.

White, T. D. (1984) Pliocene hominids from the Middle Awash, Ethiopia. *Courier Forschungsintitut Senckenberg,* **69,** 57-68.

White, T. D., Suwa, G., Hart, W. K., Walter, R. C., Wolde-Gabriel, G., de Heinzelin, J., Clark, J. D., Asfaw, B. and Vrba, E. S. (1993) New discoveries of *Australopithecus* at Maka in Ethiopia. *Nature,* **366,** 261-265.

White, T. D., WoldeGabriel, G., Asfaw, B., Ambrose, S., Beyene, Y., Bernor, R.L., Boisserie, J.-R., Currie, B., Gilbert, H., Haile-Selassie, Y., Hart, W. K., Hlusko, L. J., Howell, F. C., Kono, R. T., Lehmann, T., Louchart, A., Lovejoy, C. O., Renne, P. R., Saegusa, H., Vrba, E. S., Wesselman, H. and Suwa G. (2006) Asa Issie, Aramis and the origin of Australopithecus. *Nature* **440**(7086), 883-889

Whiten, A. and Byrne, R. W. (1997) *Machiavellian intelligence II: Extensions and evaluations.* Cambridge University Press, Cambridge.

Whiten, A., Custance, D. M., Gomez, J.-C., Teixidor, P. and Bard, K. A. (1996) Imitative learning of artificial fruit-processing in children (*Homo sapiens*) and chimpanzees (*Pan troglodytes*). *Journal of Comparative Psychology,* **110,** 3-14.

Whiten, A., Goodall, J., McGrew, W. C., Nishida, T., Reynolds, V., Sugiyama, Y., Tutin, C. E. G., Wrangham, R. W. and Boesch, C. (1999) Cultures in Chimpanzees. *Nature,* **399,** 682-685.

Wilson, A. C. (1985) The molecular basis of evolution. *Scientific American,* **253,** 148-157.

Wolde-Gabriel, G., Haile-Selassie, Y., Renne, P. R., Hart, W. K., Ambrose, S. H., Asfaw, B., Heiken, G. and White, T. (2001) Geology and paleontology of the Late Miocene Middle Awash valley, Afar rift, Ethiopia. *Nature,* **412,** 175-178.

Wolde-Gabriel, G., White, T. D., Suwa, G., Renne, P., de Heinzelin, J., Hart, W. and Heiken, G. (1994) Ecological and temporal placement of early Pliocene hominids at Aramis, Ethiopia. *Nature,* **371,** 330-333.

Wolfheim, J. H. (1983) *Primates of the world: distribution, abundance and conservation.* University of London Press, London.

Wolpoff, M. A. (1996) *Human evolution.* McGraw Hill, New York.

Wolpoff, M. A. (1999) *Palaeoanthropology.* McGraw Hill, New York.

Wood, B. (1984) The origins of *Homo erectus*. *Courier Forschungsintitut Senckenberg,* **69,** 99-112.

Wood, B. (1991) *Koobi Fora Research Project Volume 4: Hominid cranial remains.* Clarendon Press, Oxford.

Wood, B. (1992) Origin and evolution of the genus *Homo*. *Nature,* **355,** 783-790.

Wood, B. (1997) The oldest whodunnit in the world. *Nature,* **385,** 292-3.

Wood, B. and Brooks, A. S. (1999) Human evolution: we are what we ate. *Nature,* **400,** 219-220.

Wood, B. and Collard, M. (1999a) The Human Genus. *Science,* **284,** 65-71.

Wood, B. and Collard, M. (1999b) The changing face of genus *Homo. Evolutionary Anthropology,* **8,** 195-207.

Wood, J. W. (1994) *Dynamics of human reproduction: biology, biometry and reproduction.* Aldine de Gruyter, New York.

Wrangham, R. W. and Bergmann-Riss, E. (1990) Rates of predation on mammals by Gombe chimpanzees 1972-1975. *Primates,* **31,** 157-170.

Zhu, R. X., R. Potts, et al. (2004). New evidence on the earliest human presence at high northern latitudes in northeast Asia. *Nature*, **431**, 559-562.

www.ingramcontent.com/pod-product-compliance
Lightning Source LLC
LaVergne TN
LVHW071211210926
848482LV00015B/779